The
BELIEVING
BRAIN

The Mind of the Market

Why Darwin Matters

Science Friction

The Science of Good and Evil

In Darwin's Shadow: The Life and Science of Alfred Russel Wallace

The Skeptic Encyclopedia of Pseudoscience (general editor)

The Borderlands of Science

Denying History

How We Believe

Why People Believe Weird Things

The
BELIEVING
BRAIN

From Ghosts and Gods
to Politics and Conspiracies—
How We Construct Beliefs
and Reinforce Them as Truths

Michael Shermer

 St. Martin's Griffin 🙣 New York

www.stmartins.com

Design by Meryl Sussman Levavi

For further information on the Skeptics Society and Skeptic magazine, contact P.O. Box 338, Altadena, CA 91001; www.skeptic.com

The Library of Congress has cataloged the Henry Holt edition as follows:

Shermer, Michael.
    The believing brain : from ghosts and gods to politics and conspiracies—how we construct beliefs and reinforce them as truths / by Michael Shermer.
        p. cm.
    Includes bibliographical references and index.
    ISBN 978-0-8050-9125-0
    1. Belief and doubt.   2. Knowledge, Theory of.   3. Cognitive neuroscience.
    I. Title.
    BF773.S54 2011
    153.4—dc22   2010030706

ISBN 978-1-250-00880-0 (trade paperback)

First published in hardcover format by Times Books, an imprint of Henry Holt and Company

D   20   19   18   17   16   15   14

To Devin Ziel Shermer

For our small contribution—6,895 days or 18.9 years from birth to independence—to the metaphorically miraculous 3.5-billion-year continuity of life on Earth from one generation to the next, unbroken over the eons, glorious in its contiguity, spiritual in its contemplation. The mantle is now yours.

For the mind of man is far from the nature of a clear and equal glass, wherein the beams of things should reflect according to their true incidence; nay, it is rather like an enchanted glass, full of superstition and imposture, if it be not delivered and reduced.

—FRANCIS BACON, *Novum Organum*, 1620

Contents

Part IV: Belief in Things Seen

The
BELIEVING
BRAIN

Prologue

━━━━◆━━━━

I Want to Believe

THE 1990S' ÜBER CONSPIRACY-THEORY TELEVISION SERIES *THE X-FILES* was a decade-defining and culture-reflecting mosh pit of UFOs, extraterrestrials, psychics, demons, monsters, mutants, shape-shifters, serial killers, paranormal phenomena, urban legends turned real, corporate cabals and government cover-ups, and leakages that included a Deep Throat–like "cigarette smoking man" character played, ironically, by real-life skeptic William B. Davis. Gillian Anderson's skeptical FBI agent Dana Scully played off David Duchovny's believing character Fox Mulder, whose slogans became posterized pop-culture catchphrases: "I want to believe" and "The truth is out there."

As the show's creator-producer Chris Carter developed the series narrative arc, Scully and Mulder came to symbolize skeptics and believers in a psychological tug-of-war between reality and fantasy, fact and fiction, story and legend. So popular was *The X-Files* that it was parodied in a 1997 episode of *The Simpsons* entitled "The Springfield Files," in which Homer has an alien encounter in the woods after imbibing ten bottles of Red Tick Beer. The producers ingeniously employed Leonard Nimoy to voice the intro as he once did for his post-Spock run on the television mystery series *In Search of . . .* , a 1970s nonfiction version of *The X-Files*. Nimoy: "The following tale of alien encounters is true. And by true, I mean false. It's all lies. But they're entertaining lies, and in the end isn't that the real truth? The answer is no."

No squared. The postmodernist belief in the relativism of truth, coupled with the clicker culture of mass media, in which attention spans are measured in New York minutes, leaves us with a bewildering array of truth claims packaged in infotainment units. It must be true—I saw it on television, the movies, the Internet. *The Twilight Zone, The Outer Limits, That's Incredible!, The Sixth Sense, Poltergeist, Loose Change, Zeitgeist: The Movie.* Mysteries, magic, myths, and monsters. The occult and the supernatural. Conspiracies and cabals. The face on Mars and aliens on Earth. Bigfoot and Loch Ness. ESP and psi. UFOs and ETIs. OBEs and NDEs. JFK, RFK, and MLK Jr.—alphabet conspiracies. Altered states and hypnotic regression. Remote viewing and astroprojection. Ouija boards and tarot cards. Astrology and palm reading. Acupuncture and chiropractic. Repressed memories and false memories. Talking to the dead and listening to your inner child. It's all an obfuscating amalgam of theory and conjecture, reality and fantasy, nonfiction and science fiction. Cue dramatic music. Darken the backdrop. Cast a shaft of light across the host's face. *Trust no one. The truth is out there. I want to believe.*

I believe that the truth is out there but that it is rarely obvious and almost never foolproof. What I want to believe based on emotions and what I should believe based on evidence do not always coincide. I'm a skeptic not because I do not want to believe, but because I want to *know.* How can we tell the difference between what we would like to be true and what is actually true?

The answer is science. We live in the Age of Science, in which beliefs are supposed to be grounded in rock-solid evidence and empirical data. Why, then, do so many people believe in what most scientists would consider to be the unbelievable?

The Demographics of Belief

In a 2009 Harris Poll of 2,303 adult Americans, people were asked to "indicate for each [category below] if you believe in it, or not." The results were revealing.[1]

God	82 %
Miracles	76 %
Heaven	75 %
Jesus is God or the Son of God	73 %
Angels	72 %
Survival of the soul after death	71 %

The resurrection of Jesus Christ	70 %
Hell	61 %
The virgin birth (of Jesus)	61 %
The devil	60 %
Darwin's theory of evolution	45 %
Ghosts	42 %
Creationism	40 %
UFOs	32 %
Astrology	26 %
Witches	23 %
Reincarnation	20 %

More people believe in angels and the devil than believe in the theory of evolution. Disturbing. And yet, such results do not surprise me, as they match findings in similar surveys conducted over the past several decades,[2] including internationally.[3] For example, in a 2006 *Reader's Digest* survey of 1,006 adult Britons, 43 percent of respondents said that they can read other people's thoughts or have their thoughts read, more than half said that they have had a dream or premonition of an event that then occurred, more than two-thirds said they could feel when someone was looking at them, 26 percent said they had sensed when a loved one was ill or in trouble, and 62 percent said that they could tell who was calling before they picked up the phone. A fifth said they had seen a ghost, and nearly a third said they believe that near-death experiences (NDEs) are evidence for an afterlife.[4]

Although the specific percentages of belief in the supernatural and the paranormal across countries and decades vary slightly, the numbers remain fairly consistent: a majority of people hold some form of paranormal or supernatural belief.[5] Alarmed by such figures and concerned about the dismal state of science education and its role in fostering belief in the paranormal, the National Science Foundation (NSF) conducted its own extensive survey of beliefs in both the paranormal and pseudoscience, concluding "Such beliefs may sometimes be fueled by the media's miscommunication of science and the scientific process."[6]

I, too, would like to lay the blame at the feet of the media, because the fix then seems straightforward—just improve how we communicate science. But that's too easy, and it isn't even supported by the NSF's own data. Although belief in extrasensory perception (ESP) decreased from

65 percent among high school graduates to 60 percent among college graduates, and belief in magnetic therapy dropped from 71 percent among high school graduates to 55 percent among college graduates, that still leaves more than half of educated people fully endorsing such claims! And for embracing alternative medicine (a form of pseudoscience), the percentages actually *increased*, from 89 percent for high school grads to 92 percent for college grads.

Part of the problem may be that 70 percent of Americans still do not understand the scientific process, defined in the NSF study as grasping probability, the experimental method, and hypothesis testing. So one solution here is teaching *how science works* in addition to *what science knows*. A 2002 article in *Skeptic* magazine entitled "Science Education Is No Guarantee of Skepticism" presented the results of a study that found no correlation between science knowledge (facts about the world) and paranormal beliefs. "Students that scored well on these [science knowledge] tests were no more or less skeptical of pseudoscientific claims than students that scored very poorly," the authors concluded. "Apparently, the students were not able to apply their scientific knowledge to evaluate these pseudoscientific claims. We suggest that this inability stems in part from the way that science is traditionally presented to students: Students are taught *what* to think but not *how* to think."[7] The scientific method is a teachable concept, as evidenced in the previously referenced NSF study, which found that 53 percent of Americans with a high level of science education (nine or more high school and college science/math courses) understand the scientific process, compared to 38 percent with a middle level (six to eight such courses) of science education, and 17 percent with a low level (less than five such courses) of science education. So maybe the key to attenuating superstition and belief in the supernatural is in teaching *how* science works, not just what science has discovered.

The problem is deeper still and related to the fact that the majority of our most deeply held beliefs are immune to attack by direct educational tools, especially for those who are not ready to hear contradictory evidence. Belief change comes from a combination of personal psychological readiness and a deeper social and cultural shift in the underlying zeitgeist, which is affected in part by education but is more the product of larger and harder-to-define political, economic, religious, and social changes.

Belief-Dependent Realism: Why People Believe

Belief systems are powerful, pervasive, and enduring. I have devoted my career to understanding how beliefs are born, formed, nourished, reinforced, challenged, changed, and extinguished. This book synthesizes thirty years of research to answer the question of how and why we believe what we do in all aspects of our lives. Here I am interested in more than just why people believe weird things, or why people believe this or that claim, but why people believe anything at all. Why do people believe? My answer is straightforward:

> We form our beliefs for a variety of subjective, personal, emotional, and psychological reasons in the context of environments created by family, friends, colleagues, culture, and society at large; after forming our beliefs we then defend, justify, and rationalize them with a host of intellectual reasons, cogent arguments, and rational explanations. Beliefs come first, explanations for beliefs follow. I call this process belief-dependent realism, where our perceptions about reality are dependent on the beliefs that we hold about it. Reality exists independent of human minds, but our understanding of it depends upon the beliefs we hold at any given time.

The brain is a belief engine. From sensory data flowing in through the senses the brain naturally begins to look for and find patterns, and then infuses those patterns with meaning. The first process I call *patternicity: the tendency to find meaningful patterns in both meaningful and meaningless data.* The second process I call *agenticity: the tendency to infuse patterns with meaning, intention, and agency.* We can't help it. Our brains evolved to connect the dots of our world into meaningful patterns that explain why things happen. These meaningful patterns become beliefs, and these beliefs shape our understanding of reality.

Once beliefs are formed, the brain begins to look for and find confirmatory evidence in support of those beliefs, which adds an emotional boost of further confidence in the beliefs and thereby accelerates the process of reinforcing them, and round and round the process goes in a positive feedback loop of belief confirmation. As well, occasionally people form beliefs from a single revelatory experience largely unencumbered by their personal background or the culture at large. Rarer still, there are those who, upon carefully weighing the evidence for and against a position they already hold, or one they have yet to form a belief about, compute the odds and make a steely-eyed emotionless decision and never look back. Such belief reversals are so rare in religion and politics as to generate headlines if it is someone prominent, such as a cleric who changes religions or renounces his or her faith, or a politician who changes parties or goes independent. It happens, but it is as rare as a black swan. Belief reversals happen more often in science, but not at all as frequently as we might expect from the idealized visage of the exalted "scientific method" where only the facts count. The reason for this is that scientists are people too, no less subject to the whims of emotion and the pull of cognitive biases to shape and reinforce beliefs.

This process of *belief-dependent realism* is patterned after the philosophy of science called "model-dependent realism" presented by the University of Cambridge cosmologist Stephen Hawking and mathematician and science writer Leonard Mlodinow in their book, *The Grand Design*, in which they explain that because no one model is adequate to explain reality, we are free to use different models for different aspects of the world. Model-dependent realism "is based on the idea that our brains interpret the input from our sensory organs by making a model of the world. When such a model is successful at explaining events, we tend to attribute to it, and to the elements and concepts that constitute it, the quality of reality or absolute truth. But there may be different ways in which one could model the same physical situation, with each employing different fundamental elements and concepts. If two such physical theories or models accurately predict the same events, one cannot be said to be more real than the other; rather, we are free to use whichever model is most convenient."[8]

I take this one step further to argue that even these different models of physics and cosmology used by scientists to explain, say, light as a particle or light as a wave, are themselves beliefs, and when coupled to higher-order theories about physics, mathematics, and cosmology, form entire worldviews about nature, and therefore belief-dependent realism is a higher-order form of model-dependent realism. On top of this, our brains place a judgment value upon beliefs. There are good evolutionary reasons for why we form beliefs and judge them as good or bad that I will discuss in the chapter on political beliefs, but suffice it to say here that our evolved tribal tendencies lead us to form coalitions with fellow like-minded members of our group and to demonize others who hold differing beliefs. Thus, when we hear about the beliefs of others that differ from our own, we are naturally inclined to dismiss or dismantle their beliefs as nonsense, evil, or both. This propensity makes it even more difficult to change our minds in the teeth of new evidence.

In fact, *all* models of the world, not just scientific models, are foundational to our beliefs, and belief-dependent realism means that we cannot escape this epistemological trap. We can, however, employ the tools of science, which are designed to test whether or not a particular model or belief about reality matches observations made not just by ourselves but by others as well. Although there is no Archimedean point outside of ourselves from which we can view the Truth about Reality, science is the best tool ever devised for fashioning provisional truths about conditional realities. Thus, belief-dependent realism is not epistemological relativism where all truths are equal and everyone's reality deserves respect. The universe really did begin with a big bang, the earth really is billions of years old, and evolution really happened, and someone's belief to the contrary really is wrong. Even though the Ptolemaic earth-centered system can render observations equally well as the Copernican sun-centered system (at least in the time of Copernicus anyway), no one today holds that these models are equal because we know from additional lines of evidence that heliocentrism more closely matches reality than geocentrism, even if we cannot declare this to be an Absolute Truth about Reality.

That said, the evidence I present in this book demonstrates how dependent our beliefs are on a multitude of subjective, personal, emotional, and psychological factors that make our understanding of reality such an "enchanted glass" and so "full of superstition and imposture," in Francis Bacon's epigrammatic description. We begin with anecdotal evidence from three personal belief stories. The first story is about a man whom you will have never heard of but who had a profound and life-changing experience in the wee hours of the morning many decades ago that drove him to search for ultimate meaning in the cosmos. The second story is about a man whom you will most definitely have heard of, as he is one of the greatest scientists of our age, and he too had a life-changing early-morning experience that confirmed his decision to make a religious leap of faith. The third story is my own passage from believer to skeptic, and what I have learned along the way that drove me into a professional career of the scientific study of belief systems.

From narrative evidence we shall turn to the architecture of belief systems, how they are formed, nourished, reinforced, changed, and extinguished, first conceptually through the two theoretical constructs of *patternicity* and *agenticity*, and then delve deeper into how these cognitive processes evolved and what purpose they served in the lives of our ancestors as well as in our lives today. We shall then bore deeper into the brain, right down to the neurophysiology of belief system construction at the single neuron level, and then reconstruct from the bottom up how brains form beliefs. Then we shall examine how belief systems operate with regard to belief in religion, the afterlife, God, extraterrestrials, conspiracies, politics, economics, and ideologies, and then consider how a host of cognitive processes convince us that our beliefs are truths. In the final chapters we will consider how we know any of our beliefs are believable, which patterns are true and which false, which agents are real and which are not, and how science works as the ultimate pattern detection device that allows us a few degrees of freedom within belief-dependent realism, and some measurable progress away from its psychological trappings.

PART I

JOURNEYS OF BELIEF

Every man is the creature of the age in which he lives; very few are able to raise themselves above the ideas of the times.

—Voltaire

Mr. D'Arpino's Dilemma

THE VOICE WAS AS DISTINCT AS THE MESSAGE IT DELIVERED WAS
unmistakable. Emilio "Chick" D'Arpino bolted upright from his bed,
startled that the words he heard so clearly were not spoken by anyone in
the room. It was 4 a.m. on February 11, 1966, and Mr. D'Arpino was
alone in his bedroom, seemingly unperturbed by what he was hearing. It
wasn't a masculine voice, yet neither was it feminine. And even though
he had no reference guide built by experience from which to compare,
Mr. D'Arpino somehow knew that the source was not of this world.

⚫

I met Chick D'Arpino on my forty-seventh birthday, September 8, 2001,
just three days before the calamitous event that would henceforth cleave
history into pre- and post-9/11. Chick wanted to know if I would be willing to write an essay to answer this question: *Is it possible to know if there
is a source out there that knows we are here?*

"Uh? You mean God?" I queried.

"Not necessarily," Chick replied.

"ET?"

"Maybe," Chick continued, "but I don't want to specify the nature of
the source, just that it is out there and not here."

Who would ask such a question, I wondered, and more important,
why? Chick explained that he was a retired bricklayer interested in
pursuing answers to deep questions through essay contests and one-day

conferences he was sponsoring at San Jose State University and at Stanford University, near his home in Silicon Valley. I had never heard of a retired bricklayer sponsoring conferences before, so this got my attention, as I have long admired autodidacts.

Over the years, as Chick and I became close friends, I grew more and more curious to know why a bricklayer would spend what little money he had on funding essay contests and conferences to answer life's big questions. I had a sense that Chick already knew the answers to the questions he was posing, but for a decade he took the Fifth with me until one day, when I probed one more time, he gave me a hint:

I had an experience.

An experience. Okay! Now we're talking my language—the language of belief systems grounded in experiences. What type of experience?

Chick clammed up again, but I pushed and prodded for details. When was this experience?

Back in 1966.

What time of day did it happen?

Four in the morning.

Did you see or hear something?

I don't want to talk about that aspect of it.

But if it was a profound enough experience to be driving you to this day to explore such big questions, it is surely worth sharing with someone.

Nope, it's private.

Come on, Chick, I've known you practically a decade. We're the best of friends. I'm genuinely curious.

Okay, it was a voice.

A voice. Um.

I know what you're thinking, Michael—I've read all your stuff about auditory hallucinations, lucid dreams, and sleep paralysis. But that's not what happened to me. This was clearly, distinctly, unmistakably not from my mind. It was from an outside source.

Now we were getting somewhere. Here is a man I've come to know and love as a dear friend, a man who otherwise is as sane as the next guy and as smart as a whip. I needed to know more. Where did this happen?

At my sister's house.

What were you doing sleeping at your sister's house?

I was separated from my wife and going through a divorce.

Aha, right, the stress of divorce.

I know, I know, my psychiatrist thought the same thing you're thinking now—stress caused the experience.

A psychiatrist? How does a bricklayer end up in the office of a psychiatrist?

Well, see, the authorities sent me to see this psychiatrist up at Agnews State Hospital.

What?! Why?

I wanted to see the president.

Okay, let's see . . . 1966 . . . President Lyndon Johnson . . . Vietnam War protests . . . construction worker wants to see the president . . . mental hospital. There's a compelling story here for someone who studies the power of belief for a living, so I pressed for more.

Why did you want to see the president?

To deliver to him the message from the source of the voice.

What was the message?

That I will never tell you, Michael—I have never told anyone and I'm taking it to my grave. I haven't even told my children.

Wow, this must be some message, like Moses on the mountaintop taking dictation from Yahweh. Must have gone on for quite some time. How long?

Less than a minute.

Less than a minute?

It was thirteen words.

Do you remember the thirteen words?

Of course!

Come on, Chick, tell me what they were.

Nope.

Did you write them down somewhere?

Nope.

Can I guess what the theme of the message was?

Sure, go ahead, take a guess.

Love.

Michael! Yes! That's exactly right. Love. The source not only knows we're here, but it loves us and we can have a relationship with it.

The Source

I would like to understand what happened to my friend Chick D'Arpino on that early morning in February 1966 and how that experience changed his life in profound ways ever since. I want to comprehend what happened to Chick because I want to know what happens to all of us when we form beliefs.

In Chick's case the experience happened while separated from his wife and children. The details of the separation are not important (and he wishes to protect the privacy of his family), but its effects are. "I was a broken man," Chick told me.[1] "I was broke in every way you can think of: financially, physically, emotionally, and psychologically."

To this day Chick maintains that what he experienced was unquestionably outside of his mind. I strongly suspect otherwise, so what follows is my interpretation. Lying alone in bed, Chick was awake and perhaps anxious about the new dawn that would soon break over his day and life. Away from his beloved wife and children, Chick was troubled by the uncertainty of where his life would go from there, restless about which path before him to take, and especially apprehensive about whether he was loved. Those of us who have felt the sting of unrequited love, the anguish of relationship uncertainty, the torturous suffering of a troubled marriage, or the soul-shattering desolation of divorce, well know the painful inner turmoil that stirs the emotional lees—stomach-churning, heart-pounding, stress-hormone-pumping fight-or-flight emotional overdrive—especially in the wee hours of the morning before the sun signals the possibility of redemption.

I have experienced such emotions myself, so perhaps I am projecting. My parents divorced when I was four, and although detailed memories of the separation and disruption are foggy, one memory is as clear to me now as it was those late nights and early mornings while lying awake: I had an almost vertigo sense of spiraling down and shrinking into my bed, as the room I was in expanded outward in all directions, leaving me feeling ever smaller and insignificant, frightened and anxious about . . . well . . . everything, including and especially being loved. And although the ever-shrinking-room experience has mercifully receded, today there are still too many late nights and early mornings when lost-love anxieties return to haunt me, emotions that I usually wash away with productive work or physical exercise, sometimes (but not always) successfully.

What happened to Chick next can best be described as surreal, ethereal, and otherworldly. On that early morning in February 1966, a soothing, tranquil voice calmly delivered a message of what I imagine a mind racked in turmoil longed to hear:

You are loved by a higher source that wants your love in return.

I do not know if these are the exact thirteen words heard by Chick D'Arpino that morning, and he's still not talking, other than to exposit:

The meaning was love between the source and me. The source identified its relationship to me and my relationship to it. And it dealt with L-O-V-E. If I had to say what it was about, it was about the mutual love we have for one another, me and the source, the source and me.

⌁

How does one make sense of a supernatural occurrence with natural explanations? This is Mr. D'Arpino's dilemma.

I am burdened by no such dilemma because I do not believe in otherworldly forces. Chick's experience follows from the plausible causal scenario I am constructing here for what I believe to be an inner source of that outer voice. Since the brain does not perceive itself or its inner operations, and our normal experience is of stimuli entering the brain through the senses from the outside, when a neural network misfires or otherwise sends a signal to some other part of the brain that resembles an outside stimulus, the brain naturally interprets these internal events as external phenomena. This happens both naturally and artificially— lots of people experience auditory and visual hallucinations under varying conditions, including stress, and copious research that I will review in detail later demonstrates how easy it is to artificially trigger such illusory ephemera.

Regardless of the actual source of the voice, what does one do after such an experience? Chick picked up the story and recounted for me one of the most transfixing tales I've ever heard.

⌁

It happened on a Friday. The next Monday—I remember it was Valentine's Day—I went down to the Santa Clara Post Office because that's where the FBI office was located at the time. I wanted to see the president in order to deliver my message to him, but I didn't know how one is supposed to go about seeing the president. I figured that the FBI was a good place to start. So I walk in there and tell them what I want to do, and they asked me, "So Mr. D'Arpino, why do you want to see the president? You protesting something?" I said, "No sir, I've got good news!"

Had you thought through what you would tell the president?

Nope. I didn't know what I was going to say. I just figured it would come to me. Basically, I wanted to tell the president "There's a source out there that knows we're here, and that source really cares for us."

How did the FBI agent respond?

He says, "Well, I'll tell ya, if that's the case you need to go to the Secret Service office since they deal directly with the president." So I asked him, how do I go about that? He looked at his watch and said, "Well, Mr. D'Arpino, drive up to San Francisco and go to the federal building there, and on the sixth floor you'll find the Secret Service office. If you leave now, barring any traffic, you should be able to make it before they close." So that's exactly what I did! I got in my car and drove up to San Francisco, went to the federal building, got in the elevator and went up to the sixth floor, and sure enough, it was the Secret Service office!

They let you in?

Oh, sure. I met an agent, about six feet tall, and I told him my story about wanting to see the president. He immediately asked me, "Mr. D'Arpino, is the president in any danger?" I said, "Not that I know of." So he hands me a piece of paper with a phone number on it and says, "Well, then, here, call the Washington, D.C., White House switchboard operator and talk to the appointment secretary and see if you can make an appointment to see the president. That's how it's done."

Well, I couldn't believe it! It was going to be that simple. So I called. And I called. And I called again. And again. I never got through. So now I was stuck. I didn't know what else to do. Since I was a navy veteran, I went over to the Veterans Administration hospital and told them everything that I had done so far. As you can imagine, they tried to talk me out of it. "Now Mr. D'Arpino, why would you want to see the president?" Then they asked me to leave, but I was at the end of my options and I didn't know what else to do, so I took inspiration from those protestors the FBI guy was asking me about. I just sat down there at the VA hospital and refused to leave!

It was a sit-in!

Yeah. Then the clerk there says, "Come on, Mr. D'Arpino, if you don't leave I'm going to have to call the police and I don't want to do that. You seem like a nice guy." So I go back and forth with this guy. I remember his name was Marcy because that's my daughter's name. Five hours later he comes back and says, "You're still here, Mr. D'Arpino?" I said "Yup, and I'm staying here." He says, "Now doggone it, Mr. D'Arpino, if you don't leave I really am going to call the police." I said, "Marcy, you gotta do what you think is right, but I'm staying here."

So he called the police. Two officers showed up and they ask, "What's the problem?" Marcy replies, "This man wants to see the president." So the one cop says, "Mr. D'Arpino, you can't stay here. This is government property. This is for veterans." I say, "I'm a veteran." He says, "Oh, wow, okay, well . . ." Then he asks Marcy, "Is he causing any problems? Is he doing anything wrong?" And Marcy says, "No, sir, he's just sitting here." So the cop tells him, "I have no jurisdiction here." So they all kibitzed for a while and then decided that they would take me up to meet some people who could help me at Agnews State Hospital.

Now, as you can imagine, I had no idea what was going to happen once I entered a state mental institution. At first they talked to me for a while and they could see I wasn't crazy or anything like that, so one of the cops escorted me to my car and said, "Here you go, Mr. D'Arpino, here's your keys. If you promise that you will never try to see the president, you can just go home now." But I was still insistent on seeing the president, so they said they were going to hold me for seventy-two hours for observation. That was my biggest mistake. I thought I could just leave after that if I wanted, but no.

You spent three days in a mental hospital? What did you do?

They sent in several psychiatrists to talk to me, deciding that I needed additional observation and that I would need to appear before a superior court judge along with two court-appointed psychiatrists to determine if I would be committed to the mental institution for longer than three days. On February twenty-fourth, I appeared before the judge and two psychiatrists, who asked me some questions and recommended that I be committed. Diagnosis: psychosis. Time: to be decided.

At this point in the story I'm picturing Jack Nicholson's Randle McMurphy and Louise Fletcher's Nurse Ratched wrangling over patient privileges in Ken Kesey's famous novel cum Academy Award–winning film, a fancy I suggest to Chick.

Nah! One Flew Over the Cuckoo's Nest *was a piece of cake compared to this place. It was rough. For a year and a half I sat in my room and did all the little tasks they gave me to do and attended the group sessions and talked to the psychiatrists.*

❧

What should we make of all this? Is Chick D'Arpino just some crazy man out of touch with reality—a lunatic in a tinfoil hat? No. One thirty-second experience does not a psychotic make, let alone a lifetime spent pursuing science, theology, and philosophy in books, conferences, and university courses to better understand both himself and the human condition. Chick may be exceedingly ambitious, but he is not crazy. Perhaps he had a momentary break with reality triggered by an environmental stressor. Perhaps. And that is what I suspect happened . . . or something like it. Yet millions of people have gone through the emotional stressor of divorce without ever having such preternatural encounters.

Maybe it is a combination of an environmental stressor plus an anomalous brain hiccup—random neuronal firings, for example, or perhaps a minor temporal lobe seizure, the latter of which are well documented as causing both auditory and visual hallucinations along with hyperreligious behavior. Or maybe it was some sort of auditory hallucination triggered by who knows what. We might even chalk it up more broadly to the law of large numbers, where million-to-one odds happen three hundred times a day in America—given enough brains interacting with the environment over enough time, it is inevitable that even extraordinary incidents become ordinary. And thanks to our selective memory, we remember the anomalies and forget the mundane.

Most of us don't hear voices or see visions, yet all of our brains are wired in the same neural-chemical way as the visionaries who do, from Moses, Jesus, and Muhammad to Joan of Arc, Joseph Smith, and David Koresh. The model of how brains form beliefs and then act on them is what is of interest here, because this is something we all do—inevitably, inexorably, indisputably. Beliefs are what brains make. Whatever happened to Chick D'Arpino, I am even more interested in the power that belief systems lord over us once we form them and especially once we commit to follow through on them, whatever type of beliefs they are: personal, religious, political, economic, ideological, social, or cultural. Or psychiatric.

Sane in an Insane Land

When I was an undergraduate psychology student at Pepperdine University in the mid-1970s, for a course on abnormal psychology we were required to volunteer at a clinic or hospital in order to give us hands-on experience with mental illness. For one semester I drove up the Pacific Coast Highway every Saturday to spend the day at Camarillo State Mental Hospital. It was a grim experience. It was so depressing that even the transcendent beauty of the Pacific Ocean on the drive back did little to hoist my sagging spirits. Schizophrenics and other psychotic patients shuffled up and down the corridors, shuttling between bare and featureless bedrooms and barely equipped game rooms. Although Camarillo was a pioneer in the transition in mental health treatment from lobotomies to psychotropic drugs, stuporous brains seemed barely distinguishable from somnambulistic bodies.

In preparation for our hospital stint, our professor had us read (and listen to an interview with the author of) a paper published in the prestigious journal *Science* entitled "On Being Sane in Insane Places," by Stanford University psychologist David Rosenhan.[2] The article, now one of the most famous ever published in the annals of psychology, recounted an experiment by Rosenhan and his associates in which they entered a dozen mental hospitals in five different states on the East and West coasts, reporting having had a brief auditory hallucination. They stated that the voices were often unclear, but as far as they could tell said something like "empty," "hollow," and "thud." If pressed, they were to interpret the meaning of the voice's message as "My life is empty and hollow."

All eight were admitted, seven of them diagnosed as schizophrenic and one as manic-depressive. They were, in fact, a psychology grad student, three psychologists, a psychiatrist, a pediatrician, a housewife, and a painter (three women, five men), none of whom had any history of mental illness. Outside of the faux auditory hallucination and false names, they were instructed to tell the truth after admission, act normally, and claim that the hallucinations had stopped and that they now felt perfectly fine. Despite the fact that the nurses reported the patients as "friendly" and "cooperative" and said they "exhibited no abnormal indications," none of the hospital psychiatrists or staff caught on to the experiment, consistently treating these normals as abnormals. After an average stay of nineteen days (ranging from seven to fifty-two days—they had to get out by their own devices), all of Rosenhan's shills were discharged with a diagnosis of schizophrenia "in remission."

The power of the diagnostic belief engine was striking. In the recorded radio conversation,[3] Rosenhan recounted that in his admission interview the psychiatrist asked about his relationship with his parents and wife, and inquired if he ever spanked his children. Rosenhan answered that before adolescence he got on well with his parents but during his teen years he experienced some tension with them, that he and his wife got along fairly well but had occasional fights, and that he "almost never" spanked his kids, the exception being when he spanked his daughter for getting into a medicine cabinet and his son once for running across a busy street, adding that the psychiatrist never inquired into the context of either the spousal fights or the spankings. Instead, Rosenhan explained, this was all "interpreted as reflecting my enormous ambivalence in interpersonal relationships and a great deal of difficulty in impulse control, because in the main I don't spank my kids, but boy I get angry and I then spank them." The psychiatrist, Rosenhan concluded, "having decided that I was crazy, looked into my case history to find things that would support that view, and so ambivalence in interpersonal relationships was a damn good example."

The diagnostic belief bias was pervasive. Because Rosenhan's charges were bored out of their skulls in these institutions, to pass the time they kept detailed notes of their experiences. In one poignant descriptor, the staff reported that "patient engages in writing behavior" on a list of signs of pathology. The painter pseudopatient began churning out canvas after canvas, many of which were of such good quality that they were hung on the mostly barren walls of the institution she entered—which happened to be a hospital for which Rosenhan was a consultant.

> I come in one day for a case presentation conference to hear the staff going across her paintings over time saying, in effect, "look, here you can see real disturbances in her sensorium, you can see how things are erupting into consciousness, libidinous pushes, and here you can already see it sealed over," and so on. It's clear that in this matter, with regard to projectives of all kind, you read in what you want to read in to it. The statements that mental health professionals make about patients often tell us more about the professionals than they do about the patients.

Tellingly, the real patients—not privy to the psychiatrists' diagnoses of the pseudopatients—suspected something was up right away. Of the 118 patients whose remarks were recorded, 35 of them indicated that they knew what was really going on. As one exclaimed: "You're not crazy. You're a journalist, or a professor. You're checking up on the hospital." Of course. Who else would be sitting around a mental hospital taking copious notes?

How does the pathology belief system transmogrify normal into abnormal behavior? "Given that the patient is in the hospital, he must be psychologically disturbed," Rosenhan explained. "And given that he is disturbed, continuous writing must be behavioral manifestation of that disturbance, perhaps a subset of the compulsive behaviors that are sometimes correlated with schizophrenia." And not just writing. One pseudopatient, while pacing up and down the hall, was asked by a nurse, "Nervous, Mr. X?" He replied, "No, bored." Another ersatz patient overheard a psychiatrist explaining to some interns that the line of patients queued up for lunch thirty minutes early "was characteristic of the oral-acquisitive nature of the syndrome." Nah, it couldn't be that they were just bored and hungry.

If the diagnostic label is so powerful as to cause someone to judge sane people insane, could it work in the reverse? Might insane behavior be labeled sane under different circumstances? In a subsequent experiment to test the reverse power of diagnostic belief, Rosenhan contacted a mental institution after they wrote him to explain that *they* would never fall for such a ploy. Rosenhan told them that over the course of the next three months he would send in one or more pseudopatients, with the staff instructed to record which patients they thought were fake. Once again demonstrating the power of belief to interpret the data in light of the diagnostic tool, out of 193 patients admitted to this hospital, 41 were classified as impostors by at least one staff member, with an additional 42 classified as suspected fakes. In point of fact, *no* pseudopatients were sent to that institution! "It is clear that we cannot distinguish the sane from the insane in psychiatric hospitals," Rosenhan concluded. "The hospital itself imposes a special environment in which the meaning of behavior can easily be misunderstood."

What you believe is what you see. The label is the behavior. Theory molds data. Concepts determine percepts. Belief-dependent realism.

Know the Mind Itself and You Know Humanity

Now free on his own recognizance, Chick D'Arpino returned to work and began his journey of understanding. To what end?

Before I die I want to understand the human capacity to correctly answer such questions as "What am I?" "Who am I?" "Is there a source out there who knows we are here?" I think I have answers to these big questions that I want to share before I die.

Where did you get those answers?

I got these answers from the source.

What is the source?

The mind itself.

I am not the first to ask Chick D'Arpino such questions. When he initially approached Stanford University to sponsor essay contests on his big questions, some professors there had questions similar to mine. In a letter dated September 19, 2002, Chick explained himself to the Stanford professors thusly, and in the process offers us an epistemological golden nugget:

> *Basically, I was motivated to introduce the topic of this contest because I am profoundly aware that there is a correct answer to the question, "Who am I?" I want to do what I can to "bring out" affirmatively our human ability to understand correctly the whole extent of every person's individual self-identity. In regard to the original source that provided both the mental ability and the information that is necessary to achieve said understanding, I hereby also affirm that our built-in relationship to that source was epistemologically expressed as follows: "Know the mind itself and you know humanity."*

Herein lies what is arguably the greatest challenge science has ever faced, and it is the problem I am tackling in this book: *know the mind itself and you know humanity.*

For a materialist such as myself, there is no such thing as "mind." It ultimately reduces down to neurons firing and neurochemical transmitter substances flowing across synaptic gaps between neurons, combining in complex patterns to produce something we call *mind* but is actually just brain. Chick begged to differ.

That's a supposition, Michael. Your starting point is that there can be nothing more than brain, so of course you arrive at that conclusion.

Well, yes, I suppose that's true. But you have to start somewhere, so I start at the bottom, at neurons and their actions.

But the very choice to begin there is itself an article of faith, Michael. That's not a scientific induction, that's just a conscious choice on your part.

Sure, but why not start at the bottom? That's the principle of reductionism that is such an integral part of science.

But if you go that route you close yourself off to other possibilities: top-down instead of bottom-up possibilities. You could just as easily start at the top with mind and work your way down to neurons, which opens up other possibilities.

Isn't this just a roundabout way of explaining what happened to you as being something more than just a product of your brain—that there really is a source out there that knows we are here?

It is a different starting point of epistemology. Your conclusions are only as sound as your premises.

⌐

By now I'm beginning to feel like a character in *My Dinner with Andre*, the 1981 Louis Malle film in which Wallace Shawn and Andre Gregory converse for hours on profound philosophical issues in life, in which so much turns on how words are defined.

⌐

Like what?

You say that the brain can't perceive itself.

Yes.

Do you know who you are?

Sure, of course.

Then demonstrate it. Who is doing the asking? In terms of identity, someone is doing the perceiving in there. Who is the "I" doing the perceiving? For you, the mind is nothing more than the brain, but for me the mind is more than that. It is our identity. The fact that you know who you are means that the brain can perceive itself.

Okay, I see what you mean, but that can be explained by a neural feedback loop between a neural network that monitors the body, which is in the parietal lobe, and a neural network that monitors other parts of the brain, which is in the prefrontal cortex. So that's still a bottom-up neural explanation for mind. You seem to be talking about something more.

I am. The mind is universal—it extends beyond human beings, which also includes any form of ET or God or the source or whatever.

How do you know that? With what premises did you start to get to that conclusion?

I begin with our capacity to understand. Where did that come from? From the mind itself.

I don't understand. What do you mean by "understand"?

The mind perceives the mind. You perceive yourself in the act of perception. You are the subject and the object at the same time. We have the ability to perceive ourselves and to understand reality as it really is.

I think that this must be why I went into science instead of philosophy. You're losing me here. Isn't this just epistemology and the issue of how we know anything?

Yes, that's what I love about logic and epistemology. Where does logic come from? Aristotle? Where did he get it? Ultimately it is the mind itself, which is universal. Logic, like mathematics, is a priori. We don't create logic or mathematics. The syntax of logic and mathematics is invented, but the logical and mathematical principles were already there.

Einstein believed in logic and mathematics and the laws of nature, but he did not believe in a personal God or a supreme being of any kind. You seem to believe that in addition to logic and mathematics and the laws of nature, this universal mind also represents an intentional agent, a personal being who knows we're here and cares about us. How do you know *that*?

Because it talked to me.

So it does come down to personal experience.

Yes, and that's why I want to get past all this dialogue and debate about whether or not God or a higher power exists and bring it down to just three words: "Do an experiment."

What experiment?

The SETI experiment—the Search for Extraterrestrial Intelligence.

That's already being done.

Yes, and I think we need to do more, such as the METI program, or

Messaging to Extraterrestrial Intelligence, where we send signals out in hopes of them being detected. Or even the IETI program, or Invitation to Extraterrestrial Intelligence, which has an impressive collection of scientists and scholars who have already extended an invitation to ET online.

I've seen the IETI invitation. This presumes that ETs will be able to read English and navigate a web page on their computers, when only twenty years go—or twenty years from now—none of what we're using today worked or will work.

That's why I think we need to just extend the invitation to the source verbally through a global organization such as the United Nations.

What would you say?

I would say something like this: "We, the citizens of Earth, with peaceful intention, invite any and all extraterrestrial intelligences to make contact with us."

❧

Whether Chick D'Arpino ever realizes his dream of a UN-sponsored ET invitation event remains to be seen (if you want to read Chick's own statement on the ET invite, go to: http://www.chickdarpino.blog.com/). There is no harm in trying, and maybe it would even serve to bring humanity together for a brief respite between tribal disputes. There is, after all, no law of nature that says there cannot be an extraterrestrial intelligence out there, even one that knows we are here. I'm skeptical that we would get a response, as I am that what happened to Chick on that early morning those long gone decades ago represents a mind outside of the brain, but as a scientist I must always consider the possibility that I could be wrong. Either way, Chick D'Arpino's journey is a testament to the power of belief.

2

Dr. Collins's Conversion

By now you may be thinking to yourself, "oh come on! how does any of this apply to me? This D'Arpino guy is an uneducated brick-layer. My beliefs are based on reasoned analysis and educated consideration. I've never heard voices or tried to see the president. My brain and beliefs are just fine, thank you."

This is why I shall bookend Mr. D'Arpino's story with that of Dr. Francis Collins, an M.D. and Ph.D., former head of the Human Genome Project, current director of the National Institutes of Health, winner of the Presidential Medal of Freedom, and member of the prestigious National Academy of Sciences and the Pontifical Academy of Sciences, to name just a few of his accomplishments. Dr. Collins also had a life-changing epiphany, also in the early morning, propelling him to become an outspoken born-again evangelical Christian and write a bestselling book about both his experience and his journey from hard-core atheist to impassioned believer. You may reasonably think yourself immune to the power of belief as witnessed in the narrative arc of a brick mason, but few readers of this book can say that they have the intellectual horsepower or scientific credentials of Francis Collins, one of the greatest minds of our generation. If it can happen to him, it can happen to anyone. In fact, as I argue in this book, the power of belief happens to all of us, albeit at different levels of intensity and in varying parts and times of our lives. The particulars of Dr. Collins's belief path are radically different from that of Mr. D'Arpino's, but the process of how beliefs are formed and reinforced is what I wish to examine in the main.

In his bestselling 2006 book, *The Language of God: A Scientist Presents Evidence for Belief,*[1] Francis Collins recounts his journey from atheist to theist, which at first was a halting intellectual process filled with the internal debates scientists typically have with themselves while working on new ideas ("I hesitated, afraid of the consequences, and afflicted by doubts"). He read books on the existence of God and the divinity of Christ, most notably the works of the celebrated Oxford scholar and novelist C. S. Lewis, whose popular nonfiction works have become a staple of Christian apologetics, and whose children's book series *The Chronicles of Narnia*—filled with thinly disguised biblical allegories—are in steady production as Hollywood films. When I was studying at Pepperdine University, I took an entire course on the writings of C. S. Lewis and can attest firsthand to the power of his writings (although his science-fiction space trilogy lags behind the Narnia series in quality and is unlikely to see the light of film). Collins recalled his initial reaction to the argument that Jesus was God incarnate who had to come to Earth as a man in order to pay our debt of sin so that we may all be born again (famously posterized at sporting events in John 3:16: "For God so loved the world, that he gave his only begotten Son, that whosoever believeth in him should not perish, but have everlasting life."): "Before I became a believer in God, this kind of logic seemed like utter nonsense. Now the crucifixion and resurrection emerged as the compelling solution to the gap that yawned between God and myself, a gap that could now be bridged by the person of Jesus Christ." Again, as the principle of belief-dependent realism dictates, once the belief is formed, reasons can be found to support it.

Before Collins made the leap, however, his training in science and rationality kept religious belief at bay. "The scientist in me refused to go any further along this path toward Christian belief, no matter how appealing, if the biblical writings about Christ turned out to be a myth or, worse yet, a hoax." As long as belief was secondary to explanation, skepticism reigned supreme. But once you open your mind to the possibility of belief, explanations fall naturally into place. As he told a *Time* magazine reporter in a print debate with the celebrated atheist Richard Dawkins—who challenged Collins's claim that God is outside of the universe and called it "the mother and father of all cop outs"—Collins replied:

> I do object to the assumption that anything that might be outside
> of nature is ruled out of the conversation. That's an impoverished

view of the kinds of questions we humans can ask, such as "Why am I here?" "What happens after we die?" If you refuse to acknowledge their appropriateness, you end up with a zero probability of God after examining the natural world because it doesn't convince you on a proof basis. But if your mind is open about whether God might exist, you can point to aspects of the universe that are consistent with that conclusion.

The explanation-belief order was about to be reversed. Collins was poised on the precipice of the leap of faith that the Danish theologian Søren Kierkegaard claimed was necessary to circumvent the paradox of believing that a being could be both fully human and fully God. C. S. Lewis provided the catapult that Collins needed to hurl across that theological canyon. In *Mere Christianity*, Lewis famously presented what has come to be known as the "Liar, Lunatic, or Lord" argument:

> A man who was merely a man and said the sort of things Jesus said would not be a great moral teacher. He would either be a lunatic—on a level with a man who says He is a poached egg—or else He would be the Devil of Hell. You must make your choice. Either this man was, and is, the Son of God: or else a madman or something worse. You can shut Him up for a fool, you can spit at Him and kill Him as a demon; or you can fall at His feet and call Him Lord and God.

The intellectual arguments pro and con for the divinity of Christ that had so bedeviled Collins during his spiritual quest collapsed in one afternoon while communing with nature:

> Lewis was right. I had to make a choice. A full year had passed since I decided to believe in some sort of God, and now I was being called to account. On a beautiful fall day, as I was hiking in the Cascade Mountains during my first trip west of the Mississippi, the majesty and beauty of God's creation overwhelmed my resistance. As I rounded a corner and saw a beautiful and unexpected frozen waterfall, hundreds of feet high, I knew the search was over. The next morning, I knelt in the dewy grass as the sun rose and surrendered to Jesus Christ.

I wanted to know more about this experience and managed to catch Collins during a long drive to visit family, isolated in his car from the distractions that being head of the National Institutes of Health brings.[2] He was refreshingly (and revealingly) open about his beliefs and how he arrived at them, starting with what led to the frozen waterfall epiphany. Collins was a medical resident working hundred-hour weeks. "I was overworked and underslept, trying also to be a good husband and father, and I really had little time for deep reflection. So if there was anything to that moment in the mountains it was being set aside from those distractions and allowing myself to contemplate these profound questions." In this state of readiness, Collins explained, "I turned the corner of the trail and saw this frozen waterfall glistening in the sun. It wasn't so much a miraculous sign from God as it was a feeling that I was being called to a decision. I even remember thinking that if a bald eagle flew overhead at that moment that would be really cool, but that didn't happen. But I did experience a feeling of peace and of being ready and in the right place to make that decision. I just had a peaceful sense of 'I'm here. I made it.' "

After a "honeymoon period of about a year" in which Collins "felt great joy and relief and talked to lots of people about my conversion," doubts began to creep into his mind, making him wonder if "this had all been an illusion." One Sunday of particularly intense doubt, Collins "went up to the altar, knelt for a while in great distress, crying out in some voiceless prayer for help." Just then he felt a hand on his shoulder. "I turned and there was a man who had just joined the church that day. He asked me what I was going through. I told him, he took me to lunch, we talked, and we became good friends. It turns out that he was a physicist who had traveled a similar path to mine, and he helped me see that doubt is part of the faith journey." Reassured by a fellow scientist, Collins "was able to go back and reconstruct how I came to faith in the first place, and I concluded that my religious belief was real and not counterfeit."

Did it help that he was also a scientist?

It sure did! In talking to lots of people of faith I've discovered that I have intellectualized my belief far more than most people, so it was especially helpful to share my doubt with a fellow scientist.

Having doubts didn't set you back in your faith?

No, doubt is an opportunity to continue growing.

How can you tell the difference between the position that God exists and doubt is the normal part of faith, and the position that God does not exist and doubt is reasonable and appropriate?

There is a spectrum of belief, between absolute confidence in God's existence on one end and absolute confidence that there is no God on the other end. We are all living somewhere on this spectrum. I am over toward the belief end, but by no means all the way over there. And I know what it's like to live on the other end of the spectrum since that's where I was in my twenties. If you look at that spectrum from a purely rational perspective, neither extreme is defensible, although for all the reasons I describe in my book I conclude that the belief side is more rational than the disbelief side.

The Language of God is an honest and genuinely conciliatory effort at bridging the divide between science and religion. I quote it often in my debates with creationists because Collins—someone with considerable scientific status in his religious camp—nevertheless explains clearly why intelligent design creationism is bunk. And his chapter on the genetic evidence for human evolution is one of the most eloquent summaries ever penned on the subject. It is worth briefly summarizing here because it well captures Collins's integrity before the facts and sets up a conundrum that he (and all of us) must navigate around when it comes to ultimate questions about nature.

Collins begins by describing "ancient repetitive elements" (AREs) in DNA. AREs arise from "jumping genes," which are genes capable of copying and inserting themselves in other locations in the genome, usually without any function. "Within the genome, Darwin's theory predicts that mutations that do not affect function (namely, those located in 'junk DNA') will accumulate steadily over time," Collins explains. "Mutations in the coding region of genes, however, are expected to be observed less frequently, since most of these will be deleterious, and only a rare such event will provide a selective advantage and be retained during the evolutionary process. That is exactly what is observed." In fact, mammalian genomes are littered with AREs, with roughly 45 percent of the human genome made up of them. If you align sections of, say, human and mouse genomes, identical genes and many AREs are in the same location. Collins concludes his summation with this biting editorial: "Unless one is willing to take the position that God has placed these decapitated AREs in these precise positions to confuse and mislead us, the conclusion of a common ancestor for humans and mice is virtually inescapable."

If science is so good at explaining nature that we do not need to invoke the deity for such remarkable productions as DNA, why does Francis Collins believe in God? Indeed, why would any scientist or reasoning person believe in God? That question has two answers: intellectual and emotional. Intellectually, Collins is aligned tightly with his fellow scientists when it comes to explaining everything in the world by natural law, with two exceptions (in Immanuel Kant's poetic description): *the starry heavens above and the moral law within.*[3] Here—in the realm of the cosmic origin of the laws of nature and the evolutionary origins of morality—Collins stands on the craggy edge of the abyss. Instead of pushing the science even further, he makes a leap of faith. Why?

The number one predictor of anyone's religious beliefs is that of their parents and the religious environment of the family. Not so for Francis Collins, whose parents were Yale graduate secular freethinkers who home-schooled their four boys (Collins was the youngest) through sixth grade and neither encouraged nor discouraged religious thought. After parents, siblings, and family dynamics, peer groups and teachers play a powerful role in shaping one's beliefs, and in his middle school years—now enrolled in public schools—Collins encountered a compelling chemistry teacher and decided then and there that science was his calling. Assuming that religious skepticism was part and parcel of the scientific mind, Collins defaulted into agnosticism, not after careful analysis of the arguments and evidence, but "more along the lines of 'I don't want to know.'" Reading a biography of Einstein and the great scientist's rejection of the personal God of Abraham, "only reinforced my conclusion that no thinking scientist could seriously entertain the possibility of God without committing some sort of intellectual suicide. And so I gradually shifted from agnosticism to atheism. I felt quite comfortable challenging the spiritual beliefs of anyone who mentioned them in my presence, and discounted such perspectives as sentimentality and outmoded superstition."[4]

The intellectual edifice he had built on the skeptical side of the spectrum was gradually chipped away by emotional experiences as a medical student and resident, overwhelmed by the pain and suffering of his patients and impressed by how well their faith served them in their time of need. "What struck me profoundly about my bedside conversations with these good North Carolina people was the spiritual aspect of what many of them were going through. I witnessed numerous cases of individuals whose faith provided them with a strong reassurance of ultimate peace, be it in this world or the next, despite terrible suffering that in most instances they had done nothing to bring on themselves. If faith was a psychological crutch, I concluded, it must be a very powerful one. If it was nothing more than a veneer of cultural tradition, why were these people not shaking their fists at God and demanding that their friends and family stop all this talk about a loving and benevolent supernatural power?"

It's a fair question, as was the one asked of him by a woman suffering from severe and untreatable angina: what did he believe about God? Collins's skeptical convictions gave way to thoughtful sensitivity of the moment: "I felt my face flush as I stammered out the words 'I'm not really sure.' Her obvious surprise brought into sharp relief a predicament that I had been running away from for nearly all of my twenty-six years: I had never really seriously considered the evidence for and against belief."

Collins's family background, upbringing, and education led him to be a religious skeptic, a position reinforced through his scientific training and exposure to other skeptical scientists. Now an emotional trigger caused him to bolt upright and reexamine the evidence and arguments for religious belief from a different perspective. "Suddenly all my arguments seemed very thin, and I had the sensation that the ice under my feet was cracking," he recalled. "This realization was a thoroughly terrifying experience. After all, if I could no longer rely on the robustness of my atheistic position, would I have to take responsibility for actions that I would prefer to keep unscrutinized? Was I answerable to someone other than myself? The question was now too pressing to avoid."

It was at this crucial moment—an intellectual tipping point that an emotional trigger can send cascading down a different path—that Collins turned to the influential writings of C. S. Lewis, who himself was once lost but then found. The belief door now ajar, Lewis resonated with Collins, leading him inexorably to an emotional readiness where a frozen waterfall would close the door of skepticism. "For a long time I stood trembling on the edge of this yawning gap. Finally, seeing no escape, I leapt."

⊸

What was that leap like?

Obviously it was frightening, or I wouldn't have taken so long to get there. But when I finally made the leap there was a sense of peace and relief. I had been living with the tension of having already arrived at a confidence in the plausibility of belief but realizing that that could not be a stable position for the rest of my life. I was either going to have to deny that or go forward. Going forward seemed frightening and going back seemed intellectually irresponsible. That uneasy middle ground clearly wasn't going to be a place I could live for too long.

This does make me wonder that if you had been born at a different time or in a different place you might have had a different leap of faith with a different religion, and thus there is always going to be some cultural-historical component to belief.

There is, although I'm grateful that the journey that brought me to my faith didn't rest upon a heavy dose of childhood exposure to a particular religion. That has eased some of my doubts about whether this was my own decision or something culturally imposed.

As a believer who was once a nonbeliever, why do you suppose that God makes his existence so uncertain? If he wants us to believe in him, why not just make it obvious?

Because it apparently suited God to give us free will and ask us to choose. If God made his existence completely clear to everyone, we'd all be robots practicing a single universal faith. What would be the point of that?

Why do you suppose that there are lots of thoughtful people who look at the same evidence as you and come to a different conclusion? Maybe they're making emotional decisions the other way.

We all bring baggage to every decision we make, and there are aspects of what the evidence says and aspects of what we want the evidence to say. Certainly, there are lots of people who are unhappy with the idea of a God who has authority over them, or a God who expects something of them— that certainly rankled me when I was twenty-two, and I'm sure it rankles some people their whole lives. I had to become a believer to experience the freedom it brings.

You have debunked the intelligent design creationists for their "God of the Gaps" argument, and yet in a way you are saying that the ultimate origins of the universe and the moral law within are gaps that cannot be explained by science. Is it inevitable that there will always be gaps if we go back far enough?

I think that's right. There are gaps and there are Gaps. Gaps that science can fill with natural explanations don't need a God. But gaps that could never be filled with a natural explanation lend themselves to a supernatural explanation. They cry out for one. And that is where God comes in.

In *The Science of Good and Evil* I argue that the moral sense evolved within us because we are a social primate species and we need to get along with one another and therefore we are pro-social, cooperative, and even altruistic at times. And not altruistic in a game theory tit-for-tat calculating way, in which I help you and you owe me one, but in a deeper genuine sense of feeling good about helping others. That "small inner voice" of our moral conscience is something that evolution created. From a believer's perspective, why couldn't God have used evolution to create the moral sense within us, just like he used evolution to create the bacteria flagellum or DNA, which you argue did evolve?

I'm totally with you on that. My thinking has evolved on this question since writing The Language of God, *where I was more dismissive of the idea that radical altruism could have evolved. I now think that is a possibility. But that wouldn't rule out that God planned it, since for a theistic evolutionist like myself, evolution was God's awesome plan for all creation. If God's plan could give rise to toenails and temporal lobes, why not also a moral sense? And if one tries to dismiss altruism as purely naturalistic, there is still the question of why there are principles of right and wrong at all. If our moral sense is purely an artifact of evolutionary pressure, hoodwinking us into believing that morality matters, then ultimately right and wrong are an illusion. To say that good and evil have no meaning—that's a very hard place to go, even for a strict atheist. Does that bother you, Michael?*

Sometimes, yes, it does. If I were faced with that question from that dying woman you encountered in the hospital, I'm not sure what I would say. But I'm not an ethical relativist—that is a dangerous road to go down. I think that there really are moral principles that are nearly absolute—what I call *provisional moral truths*, where something is provisionally right or provisionally wrong. By this I mean that for most people in most places most of the time behavior X is right or wrong. I think this is as good as it can get without an outside source like God. But even if there is a God who objectifies right and wrong, how are we to learn what that is? Through holy books? Through prayer? How?

Through that still small voice within.

Yes, I hear that voice as well. The question is this: what is its source?

Right. For me, the source of that inner moral voice is God.

I understand. For me, the voice is part of our moral nature that evolved.

Sure, and maybe God gave us that moral nature through evolution.

So it really does come down to some ultimate unknown?

Yes, it does.

❦

I like and respect Francis Collins. He is a man who has bravely faced life's deepest questions, edged himself up to the cliff, looked over, and did what he thought was right. His path is not mine, but to thine own self be true. This is where belief is ultimately personal—belief-dependent realism. There are no ultimate answers to these eternal questions.

Where is the meaning of life under such elemental uncertainty? Whether you are a believer or a skeptic, the meaning of life is here. It is now. It is within us and without us. It is in our thoughts and in our actions. It is in our lives and in our loves. It is in our families and in our friends. It is in our communities and in our world. It is in the courage of our convictions and in the character of our commitments. Hope springs eternal, whether life is eternal or not.

Reason's Bit and Belief's Horse

A common myth most of us intuitively accept is that there is a negative correlation between intelligence and belief: as intelligence goes up belief in superstition or magic goes down. This, in fact, turns out not to be the case, especially as you move up the IQ spectrum. In professions in which everyone is above average in IQ (doctors, lawyers, engineers, and so forth), there is no relationship between intelligence and success because at that level other variables come into play that determine career outcomes (ambition, time allocation, social skills, networking, luck, and so on). Similarly, when people encounter claims that they know little about (which is most claims for most of us), intelligence is usually not a factor in belief, with one exception: once people commit to a belief, the smarter they are the better they are at rationalizing those beliefs. Thus: *smart people believe weird things because they are skilled at defending beliefs they arrived at for nonsmart reasons.*

Most people, most of the time, arrive at their beliefs for a host of reasons involving personality and temperament, family dynamics and cultural background, parents and siblings, peer groups and teachers, education and books, mentors and heroes, and various life experiences, very few of which have anything at all to do with intelligence. The Enlightenment ideal of *Homo rationalis* has us sitting down before a table of facts, weighing them in the balance pro and con, and then employing logic and reason to determine which set of facts best supports this or that theory. This is not at all how we form beliefs. What happens is that the facts of the world are filtered by our brains through the colored lenses of worldviews, paradigms, theories, hypotheses, conjectures, hunches, biases, and prejudices we have accumulated through living. We then sort through the facts and select those that confirm what we already believe and ignore or rationalize away those that contradict our beliefs.

Mr. D'Arpino's dilemma was to understand what happened to him—not to explain it away as an artifact of life trauma or neuro-misfiring, but to restructure it as giving an outer voice to inner meaning. Dr. Collins's conversion consisted of reconstructing his experiences into a meaningful case for belief, and his intellectual journey is an eloquent expression of the power of belief to drive reason and rationality to its ends, and vice versa. Reason's bit is in the mouth of belief's horse. The reins pull and direct, cajole and coax, wheedle and inveigle, but ultimately the horse will take its natural path.

3

A Skeptic's Journey

IN THE CORTEX OF OUR BRAINS THERE IS A NEURAL NETWORK THAT neuroscientists call the *left-hemisphere interpreter*. It is, in a manner of speaking, the brain's storytelling apparatus that reconstructs events into a logical sequence and weaves them together into a meaningful story that makes sense. The process is especially potent when it comes to biography and autobiography: once you know how a life turns out it is easy to go back and reconstruct how one arrived at that particular destination and not some other, and how this journey becomes almost inevitable once the initial conditions and final outcomes are established.

Although I have recounted in my various writings bits and pieces of autobiographical material in order to illustrate a particular point, I will narrate here how I arrived at my own religious, political, economic, and social beliefs, and along the way disclose some facts of my personal life that I've not written about before. With hindsight and the understanding that my own left-hemisphere interpreter is no less biased than anyone else's in reconstructing my own remembered past, here is one skeptic's journey.

Born Again

Over the years much has been made of the fact that I was once a born-again Christian who either lapsed (if you're a believer) or advanced (if you're a skeptic) into religious disbelief. Creationists have tried to pin my belief in evolution to my demise as a believer, thereby chalking up another lost soul to the evils of liberal secular education. Atheists have trumpeted my deconversion as evidence that education, especially in the sciences, demolishes ancient mythologies and antiquated faith-based beliefs. The truth is far more complex; rarely are important religious, political, or ideological beliefs attributable to single causal factors. Human thought and behavior are almost always multivariate in cause, and beliefs are no exception.

I was not born into a born-again family. None of my four parents (bio and step) were religious in the least; yet neither were they nonreligious. I think that they just didn't think about God and religion all that much. Like most children of the Great Depression who came of age during and fought in the Second World War, my parents just wanted to get on with life. None attended college, and all worked hard to support their children. My parents divorced when I was four and both remarried: my mother to a man with three kids who became my stepsiblings, my father to a woman with whom he had two daughters—my half sisters. Mine were the quintessential American blended families. Although I was periodically dropped off for the obligatory Sunday school classes (I still have my Bible from the Church of the Lighted Window in La Canada, California), religious services, prayer, Bible reading, and the usual style of God talk that one might find in religious families were absent in both of my homes. To this day, as far as I know, none of my siblings are very religious and neither are my two remaining stepparents. My father died of a heart attack in 1986, and my mother died of brain cancer in 2000; neither one of them ever embraced religion, not even my mom during her decadelong struggle through half a dozen brain surgeries and radiation treatments.

Imagine their surprise, then, when in 1971—at the start of my senior year in high school—I announced that I had become "born again," accepting Jesus as my savior. At the behest of my best friend George, reinforced the next day in church with him and his deeply religious parents, I repeated those words from John 3:16 as if they were gospel, which they are. I became profoundly religious, fully embracing the belief that Jesus suffered wretchedly and died, not just for humanity, but for me personally. Just for me! It felt good. It seemed real. And for the next seven years I walked the talk. Literally. I went door-to-door and person-to-person, witnessing for God and evangelizing for Christianity. I became a "Bible thumper," as one of my friends called me, a "Jesus freak" in the words of a sibling. A little religion is one thing, but when it is all one talks about it can become awkward and uncomfortable for family and friends who don't share your faith passion.

One solution to the problem of social appropriateness is to narrow the scope of one's peer group to like-minded believers, which I did. I hung around other Christians at my high school, attended Bible-study classes, and participated in singing and socializing at a Christian house of worship called *The Barn* (literally a red house with barnlike features). I matriculated at Pepperdine University, a Church of Christ institution that mandated chapel attendance twice a week, along with a curriculum that included courses in the Old and New Testaments, the life of Jesus, and the writings of C. S. Lewis. Although all this theological training would come in handy years later in my public debates on God, religion, and science, at the time I studied it because I believed it, and I believed it because I unquestioningly accepted God's existence as real, along with the resurrection of Jesus and all the other tenets of the faith. My years at Pepperdine—living in Malibu, sharing a dorm room with a professional tennis player (Paul Newman called once to arrange lessons, causing my mom to nearly faint when I told her that I actually spoke to her minor deity), playing Ping-Pong and Monopoly with a bunch of jocks in Dorm 10 (women were not allowed in the men's dorms, and vice versa), hearing speeches by President Gerald Ford and H-bomb father Edward Teller, and studying religion and psychology under exceptional professors—are among the most memorable of my life.

What happened next has become a matter of some curiosity among creationists and intelligent design proponents looking to bolster their belief that learning about the theory of evolution threatens religious faith.[1] There were a number of factors involved in my deconversion—in my becoming unborn, again—going back to my conversion experience. Shortly after I accepted Christ into my heart, I eagerly announced to another deeply religious high school friend of mine named Frank that I had become a Christian. Expecting an enthusiastic embrace of acceptance into the club he had long cajoled me to join, Frank instead was disappointed that I had gone to a Presbyterian church—and joined no less!—which he explained was a big mistake because that was the "wrong" religion. Frank was a Jehovah's Witness. After high school (but before Pepperdine) I attended Glendale College where my faith was tested by a number of secular professors, most notably Richard Hardison, whose philosophy course forced me to check my premises, along with my facts, which were not always sound or correct. But the Christian mantra was that when your belief is tested it is an opportunity for your faith in the Lord to grow. And grow it did, since there were some fairly serious challenges to my faith.

After Pepperdine, I began my graduate studies in experimental psychology at California State University–Fullerton. I was still a Christian, although the foundations of my faith were already cracking under the weight of other factors. Out of curiosity, I registered for an undergraduate course in evolutionary biology, which was taught by an irrepressible professor named Bayard Brattstrom, a herpetologist (one who studies reptiles) and showman extraordinaire. The class met on Tuesday nights from 7 to 10 p.m. I discovered that the evidence for evolution is undeniable and rich, and the arguments for creationism that I had been reading were duplicitous and hollow. After Brattstrom exhausted himself with a three-hour display of erudition and entertainment, the class adjourned to the 301 Club in downtown Fullerton, a nightclub where students hung out to discuss the Big Questions, aided by adult beverages. Although I had already been exposed to all sides in the great debates in my various courses and readings at Pepperdine, what was strikingly different in this context was the heterogeneity of my fellow students' beliefs. Since I was no longer exclusively surrounded by Christians, there were no social penalties for being skeptical—about anything. Except for the 301 Club discussions that went on into the wee hours of the morning, however, religion almost never came up in the classroom or lab. We were there to do science, and that is almost all we did. Religion was simply not part of the environment. So it was not the fact that I learned about evolutionary theory that rent asunder my Christian faith; it was that it was okay to challenge any and all beliefs without fear of psychological loss or social reprisal. There were other factors as well.

The Difference in Worldviews (and the Difference It Makes)

Over in the psychology department, where I was officially studying for a master's degree in experimental psychology, my adviser and mentor was Douglas Navarick, an old-school Skinnerian who preached the gospel of rigorous scientific methodology and who brooked no superstition or sloppy thinking in his students. As he reminded me in a recent letter in response to my query about his beliefs back then (memories do fade after three decades), "Within a scientific framework, I take a conventional, empiricist, cause-and-effect approach (i.e., independent and dependent variables). But outside that framework I try to keep an 'open mind' so I won't miss anything, such as the possibility that a coincidence could mean something more than a chance event, so I'll be alert for additional indications of some meaning, i.e., patterns of events, but recognizing that it's sheer speculation."

Indeed, I vividly recall inculcating this philosophy of science from Navarick because at the same time that we were conducting rigorous controlled-learning experiments in his lab, there was much hoopla about Thelma Moss's parapsychology lab at UCLA, where she studied "Kirlian photography" (photographing "energy fields" surrounding living organisms), along with hypnosis, ghosts, levitation, and the like. Since these were trained scientists smarter and more educated than myself, I figured that there might actually be something to the paranormal. But once I discovered the skeptical movement and its reasoned analysis of such claims, my skepticism overrode my belief.

As well, my current belief that there is no such thing as "mind," and that all mental processes can be explained only by understanding the underlying neural correlates of behavior, was primarily shaped by Navarick's Skinnerian philosophy: "I reject 'mentalistic' explanations of behavior," he reminded me, "i.e., attributing behavior to theoretical constructs that refer to internal states, like 'understands,' 'feels that,' 'knows,' 'gets it,' 'figures out,' 'wants,' 'needs,' 'believes,' 'thinks,' 'expects,' 'pleasure,' 'desire,' etc., the reified concepts that students routinely use in their papers despite instructions that they could lose points for doing so."[2] It isn't just students who reify mind out of behavior. Virtually everyone does, because "mind" is a form of dualism that I shall argue in a later chapter appears to be innate to our cognition. We are natural-born dualists, which is why behaviorists and neuroscientists struggle so mightily—and frustratingly—to rein in mind-talk.

Because of my newfound interest in evolutionary theory after Brattstrom's class, I studied ethology (the study of the evolutionary origins of animal behavior) under the deeply thoughtful and warmly advising Margaret White, who grounded me in the biology of human behavior and the evolution of social dynamics in primate groups. (She once sent me off to the San Diego Zoo to observe a silverback gorilla for an entire weekend, which both the gorilla and I—staring at each other for endless hours—found equally fruitless.) This was nearly two decades before the birth of evolutionary psychology as a full-fledged science, but the groundwork was laid for my later work on the evolutionary origins of religion and morality. I also took a course in cultural anthropology from the well-traveled and worldly Marlene Dobkin de Rios. Her lectures and books on her experiences in South America with hallucinogenic-imbibing shamans and the numerous animisms, spirits, ghosts, and gods made me realize just how insular my worldview was and how naive I was in assuming that my Christian beliefs were grounded in the One True Religion while all the others were so obviously culturally determined.

Together, these inputs led me to a personal exploration of comparative world religions and to the eventual realization that these often mutually incompatible beliefs were held by people who believed as firmly as I did that they were right and everyone else was wrong. Midway through my graduate training, I quietly gave up my religious belief and removed my silver *ichthus* (Greek for "fish," sometimes rendered as "Jesus Christ Son of God Savior") from around my neck. I didn't announce it to anyone because no one really cared one way or the other—with the possible exception of my siblings, who were probably relieved that I would now finally quit trying to save them.

One of the first things I noticed upon losing my religion was just how grating I must have been around people of different faiths (or no faith at all) with my incessant evangelizing—the logical product of believing that you have the One True Religion to which others must convert or forever lose a chance at eternal bliss. To nonbelievers, such a forced choice between belief, with its ultimate reward in heaven, and disbelief, with its ultimate punishment in hell, sounds so harsh and, well, Old Testament. But it wasn't meant to be that way. Earnest evangelicals—of which I was certainly one—evangelize not just on Sundays, but every day, in every way, never hiding their lantern under a bushel, as proclaimed in Matthew 5:16: "Let your light shine before others, so that they may see your good works and give glory to your Father who is in heaven." The primary point of being an evangelical Christian, in fact, is to love the Lord openly and try to bring to Christ as many people as possible; otherwise you wouldn't be an evangelical. I was doing God's work, and what could be more important than that? In the evangelical worldview there really is no separation of church and state. Yes, Jesus told us (in Matthew 22:21) to "Render unto Caesar the things which are Caesar's, and unto God the things that are God's," but we believed that this applies to specific things, such as taxes and tithings, not the general goal of bringing all people to the Lord.

Even more important, as a nonbeliever I realized the power that the believing paradigm has in filtering everything that happens through a religious lens. Chance, randomness, and contingencies dissolve into insignificance in the Christian worldview. Everything happens for a reason, and God has a plan for each and every one of us. When something good happens, God is rewarding us for our faith, our good works, or our love of Christ. When something bad happens, well, God works in mysterious ways, don't you know? Who am I to doubt, question, or challenge the Almighty? This belief filter operates on every level, from the sublime to the ridiculous, from career opportunities to sports scores. I thanked God for everything, from getting me into Pepperdine (I hardly had the grades or SAT scores for admission, that's for sure) to finding a parking place at the YMCA where I worked. In the Christian worldview there is a place for everything and everything is in its place, "a time to be born, and a time to die" (Ecclesiastes 3:2), a message rendered even into a 1960s pop tune that, when I was a believer, did not sound nearly as saccharine as it does today.

In this belief-dependent realism, even political, economic, and social events unfold by the logic of biblical end times—I had the *Los Angeles Times* open in my left hand and the books of Daniel, Ezekiel, or Revelation open in my right. Was the Ayatollah Khomeini the Antichrist, or was it Henry Kissinger? The four horsemen of the apocalypse were surely going to be nuclear war, overpopulation, pollution, and disease. The modern state of Israel was founded in 1948, so if we crunch the numbers correctly the second coming should be coming . . . very soon. When I became a nonbeliever, such political and economic events made more sense as machinations grounded in human nature and cultural history. A secular worldview led me to see that the laws of nature and the contingencies of chance unfold by their own logic along the carved channels of history largely independent of our actions and irrespective of our wishes.

In the end, though, what finally tipped my belief into skepticism was the problem of evil—if God is all knowing, all powerful, and all good, then why do bad things happen to good people? First, there was the intellectual consideration, where the more I thought about things such as cancer, birth defects, and accidents, the more I came to believe that God is either impotent or evil, or simply nonexistent. Second, there was an emotional consideration that I was forced to confront on the most primal of levels. I've never told anyone this before, but the last time I ever prayed to God was in early 1980, shortly after I decided that I no longer believed in God. What happened to bring me back one last time?

My college sweetheart, Maureen, a brilliant and beautiful Alaskan whom I met at Pepperdine and whom I was still dating, was in a horrific automobile accident in the middle of the night in the middle of nowhere. Maureen worked for an inventory company that vanned their employees around the state during off hours; they slept supine on bench seats between jobs. The van veered off the highway and rolled several times, snapping Maureen's back and rendering her paralyzed from the waist down. When she called me in the wee hours of the morning from a Podunk hospital hours from Los Angeles, I figured it couldn't be too bad since she sounded as lucid and sanguine as ever. It wasn't until days later, after we had her transported to the Long Beach Medical Center so she could be put into a hyperbaric chamber to try to coax some life into her severely bruised spinal cord, did the full implications of what this meant for her begin to dawn on me. The cognizance of Maureen's prospects generated a sickening feeling in the pit of my stomach, an indescribable sense of dread—what's the point if it can all be taken away in the flash of a moment?

There, in the ICU, day after dreary day, night after sleepless night, alternating between pacing up and down cold sterile hallways and sitting on hard plastic chairs in the waiting room listening to the moans and prayers of other grieving souls, I took a knee and bowed my head and asked God to heal Maureen's broken back. I prayed with deepest sincerity. I cried out to God to overlook my doubts in the name of Maureen. I willingly suspended all disbelief. At that time and in that place, I was once again a believer. I believed because I wanted to believe that if there was any justice in the universe—any at all—this sweet, loving, smart, responsible, devoted, caring spirit did not deserve to be in a shattered body. A just and loving God who had the power to heal would surely heal Maureen. He didn't. He didn't, I now believe, not because "God works in mysterious ways" or "He has a special plan for Maureen"—the nauseatingly banal comforts believers sometimes offer in such trying and ultimately futile times—but because there is no God.

The Principle of Principled Values

If it turns out that I am wrong and that there is a God, and it is the Judeo-Christian God more preoccupied with belief than behavior, then I'd rather not spend eternity with him and would joyfully go to the other place where I suspect most of my family, friends, and colleagues will be, since we share most of the same principled values.

Whether or not there is a God, however, the principles that I hold and try to live by should stand on their own. In philosophy this is known as "Euthyphro's dilemma," first delineated 2,500 years ago by the Greek philosopher Plato in his dialogue *Euthyphro*. Plato's protagonist Socrates asks a young man named Euthyphro the following question: "The point which I should first wish to understand is whether the pious or holy is beloved by the gods because it is holy, or holy because it is beloved of the gods?" That is, do we judge some actions to be pious or holy because the gods happen to love those actions, or do the gods love those actions because they are inherently pious or holy? The dilemma stands in monotheism today just as it did for the polytheism of the ancient Greeks: Does God embrace moral principles naturally occurring and external to him because they are sound ("holy"), or are these moral principles sound only because God says that they are sound?[3]

If moral principles hold value only because we believe that God created them, then what is their value if there is no God? The principle of truth telling and honesty in human interactions, for example, is the foundation of trust and is absolutely essential for human relations; this is true whether or not there is a source outside of our world to validate such principles. Do we really need God to tell us that murder is wrong? Isn't breaking a promise immoral because it destroys trust between people, and not because the creator of the universe says it is immoral? Thus it is that most of the principles I have inculcated along my belief journey—including my political, economic, and social attitudes—turn out to be shared even by my theist and conservative friends and colleagues, and thus I do not fit the traditional labels of either liberal or conservative. It is to this part of my belief journey we turn to now.

A Radical for Liberty

I cannot say for certain whether it was the merits of free market economics and fiscal conservatism that convinced me of their veracity, or if it was my temperament and personality that reverberated so well with their cognitive style. As it is for most belief systems we hold, it was probably a combination of both. I was raised by parents who could best be described as fiscally conservative and socially liberal, which today would be called *libertarian*, but there was no such label when they were coming of age in the 1940s and 1950s. Throughout my childhood I was inculcated with the fundamental principles of economic conservatism: hard work, personal responsibility, self-determination, financial autonomy, small government, and free markets.

It was in this state of economic preparedness that I first encountered *Atlas Shrugged* by the novelist-philosopher Ayn Rand when I was a senior at Pepperdine University. I was unfamiliar with the book and the author, and I was not a big reader of fiction, but I managed to drag myself through the first hundred pages until I was finally hooked. Millions of readers have managed the hurdle themselves, and her followers boast that a survey of books that "made a difference in readers' lives" in 1991 conducted by the Library of Congress and Book of the Month Club found that *Atlas Shrugged* was rated second only to the Bible (although it appears that the "survey" was more of a promotional campaign to entice readers to purchase copies of books carried by the Book of the Month Club).[4] Rand's popularity and influence continue to this day. In 2009, on the heels of the trillion-dollar bailout with its accompanying program of government intervention into the free market that could have been ripped from the pages of *Atlas*, readers turned to Rand as never before. Tea parties posterized *Atlas* with such memorable Randenalia as "Atlas is Shrugging," "Who is John Galt?" and the über-cool "The Name is Galt. John Galt." Sales of *Atlas* approached half a million copies that year alone, putting it in competition for sales with the top new novels of the year—not bad for a half-century-old thousand-plus-page novel chockablock with lengthy speeches about philosophy, metaphysics, economics, politics, and even sex and money.[5]

What is the appeal of Rand's characters and her plotlines that makes people want to read her books and inveigle others to do so as well? It is, I suspect, because in this postmodern age of moral relativism Ayn Rand stood for something clearly, unequivocally, unreservedly, and with passion. Her characters are *Homo economicus* on steroids: ultra-rational, utility-maximizing, freely choosing übermensches. According to Rand's recent biographer Jennifer Burns in *Goddess of the Market: Ayn Rand and the American Right*, the ultimate appeal of Rand is her almost messianic vision of the world: "Rand intended her books to be a sort of scripture, and for all her emphasis on reason it is the emotional and psychological sides of her novels that make them timeless."[6] Indeed, even though Rand called her philosophy *Objectivism*, which she said is grounded in four central tenets—objective reality, reason, self-interest, and capitalism—the pull of her gravity comes out of her passion for life and values.

Of course, the shortcomings of Rand and her movement were not lost to my skeptical scrutiny. In my 1997 book, *Why People Believe Weird Things*, I devoted a chapter to the cultlike following that developed around Rand ("The Unlikeliest Cult in History," I called it), in an attempt to show that extremism of any kind, even the sort that eschews cultish behavior, can become irrational. Many of the characteristics of a cult, in fact, seemed to fit what the followers of Objectivism believed, most notably veneration of the leader, belief in the inerrancy and omniscience of the leader, and commitment to the absolute truth and absolute morality as defined by the belief system. To wit, I cited the description of Rand's inner circle by Nathaniel Branden—Rand's chosen intellectual heir—in which he listed the other central tenets (besides the four above) to which followers were to adhere, including:

> Ayn Rand is the greatest human being who has ever lived. *Atlas Shrugged* is the greatest human achievement in the history of the world. Ayn Rand, by virtue of her philosophical genius, is the supreme arbiter in any issue pertaining to what is rational, moral, or appropriate to man's life on earth. No one can be a good Objectivist who does not admire what Ayn Rand admires and condemn what Ayn Rand condemns. No one can be a fully consistent individualist who disagrees with Ayn Rand on any fundamental issue.[7]

Nevertheless, any discussion of Rand's followers or her salacious personal life must carry this disclaimer: *Criticism of the founder of a philosophy does not, by itself, constitute a negation of any part of the philosophy.* By most accounts, Sir Isaac Newton was a narcissistic, misogynistic, egocentric curmudgeon, and yet his theories about light, gravity, and the structure of the cosmos stand on their own and would be no more or less true if he were a saintly gentleman. Rand's critique of communism may have been energized and animated by the horrific experiences she and her family endured under the brutal Communist regime in Russia (including the confiscation of her father's business), but her criticisms of communism would be just as true or false (they're true) had she been raised a farm girl in Iowa.

Most of what Rand taught either gelled with what I already believed or reinforced the belief pathway I had already started down, so I have no problem identifying myself as a fan of Ayn Rand and a proponent of her work, as long as it is clear that where scientific data conflict with political and economic philosophy, I am going with the data. For example, I am most troubled by Rand's theory of human nature as wholly selfish and competitive, defined in *Atlas* through the famous "oath" pronounced by the novel's heroes: "I swear—by my life and my love of it—that I will never live for the sake of another man, nor ask another man to live for mine." Evolutionary psychologists and anthropologists have now demonstrated unequivocally that humans have a dual nature of being selfish, competitive, and greedy as well as altruistic, cooperative and charitable. And in *The Science of Good and Evil* and *The Mind of the Market*, I built a case for evolutionary ethics and evolutionary economics that most Randians would find quite palatable with free market economics. Reading Rand, and absorbing the logic of her case for economic freedom and political liberty—she referred to herself as a "radical for capitalism"—led me to the extensive body of work on the science of markets and economies and the philosophy of liberty and freedom, all of which resonated deeply with my personality and temperament. I am a radical for liberty.

One source of influence on my political and economic thought was a retired physicist named Andrew Galambos, who taught private courses through his own Free Enterprise Institute. He called his field *volitional science*, and I took the introductory course, V-50. It was a combination of philosophy of science, economics, politics, and history, the likes of which I never heard in college. This was free market capitalism on performance-enhancing drugs. It was also a very black-and-white worldview in which Adam Smith is good, Karl Marx is bad; individualism is good, collectivism is bad; free economies are good, mixed economies are bad. Rand advocated limited government, but even that was too much for Galambos, whose theory outlined a society in which everything would be privatized until the government simply withers away. How could this work? It is based on Galambos's definition of freedom as "the societal condition that exists when every individual has full (i.e. 100%) control over his own property." Thus, a free society is one where "anyone may do anything that he pleases—with no exceptions—so long as his actions affect only his own property; he may do nothing which affects the property of another without obtaining consent of its owner." Galambos identified three types of property: *primordial* (one's life), *primary* (one's thoughts and ideas), and *secondary* (derivatives of primordial and primary property, such as the utilization of land and material goods). Capitalism, then, is "that societal structure whose mechanism is capable of protecting all forms of private property completely." To realize a truly free society, then, we have merely "to discover the proper means of creating a capitalist society."[8]

This was capitalism no economist would recognize, but Galambos had the chutzpah to sell it with passion, and many of us carried his ideas out into the world—to the extent that we were allowed, anyway; we all had to sign a contract promising that we would not disclose his ideas to anyone, while we were also encouraged to solicit others to enroll. As in the case with Rand, some of my politics and economics were shaped by Galambos, but my skepticism kicked in after the inchoate enthusiasm waned—most notably the translation of theory into practice. Property definitions are all well and good, but what happens when we cannot agree on property rights infringements? The answer was inevitably something like this: "In a truly free society all such disputes will be peacefully resolved through private arbitration." Such counterfactual fantasies reminded me of my Marxist professors who answered challenges along the same lines ("in a truly communist society, X would not be a problem").

Through the people who recommended Galambos to me I met one of his protégés named Jay Stuart Snelson, who taught courses under his own Institute for Human Progress after he had a falling out with Galambos. To distance himself from his mentor, Snelson built his theory of a free market society on the Austrian School of Economics, most notably the work of the Austrian economist Ludwig von Mises and his 1949 magnum opus *Human Action*. Outlining the countless and varied governmental actions that attenuate freedom, Snelson explained that "Freedom exists where the individual's discretion to choose is not confiscated by interventionism. The free market exists where people have the unrestricted freedom to buy and sell." Although thieves, thugs, muggers, and murderers confiscate our freedoms, Snelson continued, congressmen, senators, governors, and presidents restrict our freedoms on a scale orders of magnitude greater than all private criminals combined. And they do so, Snelson showed, with the best of intentions, because they believe that the "confiscation of the people's freedom to choose will achieve the greatest satisfaction for the greatest number." With such good intentions, and the political power to enforce them, states have intervened in business, education, transportation, communications, health services, environmental protection, crime prevention, free trade overseas, and countless other areas.

How these services could all be successfully privatized was the primary thrust of Snelson's work. He believed that the social system that optimizes peace, prosperity, and freedom is one "where anyone at any time can choose to produce or provide any product or service, hire any employee, choose any production, distribution, or sales site, and offer to sell products or services at any price." The only allowable restrictions are from the market itself. So employed, systematically throughout the world, a free market society would "open the world to all people."[9]

These were heady words for a heady time in my life, before formal commitments to career and family were congealed. For several years I taught Snelson's principles course, along with my own courses on the history of science and the history of war. I also developed a monthly discussion group I named the "Lunar Society"—after the famous eighteenth-century Lunar Society of Birmingham—centered on books such as *Human Action*. As a social scientist in search of a research project, I accepted Ludwig von Mises's challenge: "One must study the laws of human action and social cooperation as the physicist studies the laws of nature."[10] We were going to build a new science, and out of that science we would build a new society. I even penned a "Declaration of Freedom" and a speech entitled "I Have a Dream II."[11] What could be grander?

Well, as Yogi Berra once said, "In theory, there is no difference between theory and practice. In practice there is." I soon discovered that Berra's principle applies in spades to the economic sphere. We live in a world dramatically different from that envisioned by my visionary mentors, so I turned my attention to the writings of economists from the Austrian

School and their protégés at the University of Chicago, who were decidedly becoming more mainstream in the 1980s as the country began a systematic shift toward the right. Through these writings I found a scientific foundation for my economic and political preferences. The founders of the Austrian and Chicago schools of economics—of which I consider myself a member even today—penned a number of books and essays whose ideas burned into my brain a clear understanding of right and wrong human action.

I read Friedrich A. von Hayek's *The Constitution of Liberty* and *The Road to Serfdom*; I absorbed Henry Hazlitt's *Economics in One Lesson*, an exceptional summary of free market economics; and I found Milton Friedman's *Free to Choose* to be one of the clearest expositions of economic theory ever penned. His PBS documentary series by the same name—introduced by the most muscular libertarian in history, Arnold Schwarzenegger—was so powerful that I purchased the series on video and watched the episodes several times.[12] In the giants of libertarian thought who most shaped my thinking, Ludwig von Mises was first among equals; he taught me that interventionism leads to more interventionism, and that if you can intervene to protect individuals from dangerous drugs, what about dangerous ideas?[13]

It is this link between freedom and ideas that brings together my passion for science and my love of liberty, and has led to the type of science that I practice today.

An Unauthorized Autobiography of Science

Over the past three decades I have noted two disturbing tendencies in both science and society: first, to rank the sciences from "hard" (physical sciences) to "medium" (biological sciences) to "soft" (social sciences); second, to divide science writing into two forms, technical and popular. As such rankings and divisions are wont to be, they include an assessment of worth, with the hard sciences and technical writing respected the most, and the soft sciences and popular writing esteemed the least. Both of these prejudices are so far off the mark that they are not even wrong.

I have always thought that if there must be a rank order (which there mustn't), the current one is precisely reversed. The physical sciences are hard, in the sense that calculating differential equations is difficult, for example. The number of variables within the causal net of the subject matter, however, is comparatively simple to constrain and test when contrasted with, say, computing the actions of organisms in an ecosystem or predicting the consequences of global climate change. Even the difficulty of constructing comprehensive models in the biological sciences, however, pales in comparison to that of the workings of human brains and societies. By these measures, the social sciences are the hard disciplines, because the subject matter is orders of magnitude more complex and multifaceted with many more degrees of freedom to control and predict.

Between technical and popular science writing, there is what I call *integrative science*, a process that blends data, theory, and narrative. Without all three of these metaphorical legs, the seat upon which the enterprise of science rests will collapse. Attempts to determine which of the three legs has the greatest value is on a par with debating whether π or r^2 is the most important factor in computing the area of a circle. I classify two types of narrative. Formal science writing—what I call the *narrative of explanation*—presents a neat and tidy step-by-step process of introduction-methods-results-discussion grounded in a nonexistent "scientific method" of observation-hypothesis-prediction-experiment followed in a linear fashion. This type of science writing is like autobiography, and as the comedian Steven Wright said, "I'm writing an unauthorized autobiography." Any other kind is fiction. It is also a type of Whiggish history—the conclusion draws the explanation toward it, forcing facts and events to fall neatly into a causal chain where the final outcome is an inevitable result of a logical sequence.

Informal science writing—what I call the *narrative of practice*—presents the actual course of science as it is sewn through with periodic insights and subjective intuitions, random guesses and fortuitous findings. Science, like life, is messy and haphazard, full of quirky contingencies, unexpected bifurcations, serendipitous discoveries, unanticipated encounters, and unpredictable outcomes. Where a narrative of explanation might read something like "the data lead me to conclude . . ." a narrative of practice reads more like "Huh, that's weird."

The rest of this particular integrative work of science appears in the style of the narrative of practice and is, in a manner of speaking, an unauthorized autobiography of the science of belief.

What If I'm Wrong? What I Would Say to God

I am old enough now to have learned the hard way that there is always the possibility I could be wrong. I have been wrong about many things, so it is possible that I am wrong about God.

Maybe what Chick D'Arpino experienced that early morning in 1966 was the real deal: an intentional agent outside of our world—call it God, an Intelligent Designer, ET, or the source—spoke to Chick and delivered a message that by most people's judgment would be a welcome one: there is an entity out there who cares for us. That is most certainly what Chick believes to this day, despite the fact that he knows all about the neuroscience of such experiences. Perhaps Francis Collins is right in his reasoning that there had to be a first cause and prime mover of the cosmos, an actual (not imaginary) intentional agent who arranged the laws of nature to give rise to stars, planets, life, intelligence, and us.

Maybe all those other mystics and sages and regular folks in history and today who have touched the spirit world or encountered the paranormal are simply more attuned to another dimension, their skepticism reduced enough to allow their minds to connect to such a source. This is, in fact, what the great Institute for Advanced Study physicist Freeman Dyson believes. In a 2004 essay on the paranormal, Dyson concludes with a "tenable" hypothesis that "paranormal phenomena may really exist" because, he says, "I am not a reductionist" and "that paranormal phenomena are real but lie outside the limits of science is supported by a great mass of evidence." That evidence is entirely anecdotal, he admits, but because his grandmother was a faith healer and his cousin edits a journal on psychic research, and because anecdotes gathered by the Society for Psychical Research and other organizations suggest that under certain conditions (for example, stress), some people sometimes exhibit some paranormal powers, "I find it plausible that a world of mental phenomena should exist, too fluid and evanescent to be grasped with the cumbersome tools of science."[14]

Maybe there is mind outside of the brain, maybe God is mind or some manifestation thereof, and if so maybe the mind transcends the body and continues after death and this is how we may ultimately connect to the divine. What if it is mind itself that brought the universe into existence in the first place? In this scenario, maybe God is the universal mind and the afterlife is where minds go without their brains.

Maybe. But I doubt it. I believe I have outlined a reasonable explanation for Chick D'Arpino's experience as a stress-induced auditory hallucination, not unlike the sensed-presence effect experienced by climbers, explorers, and ultraendurance athletes, which I describe at length in chapter 5. As for Dyson's endorsement of the paranormal, he is one of the greatest minds of our time and thus whatever he says is worthy of serious consideration. But even a mind of this staggering genius cannot override the cognitive biases that favor anecdotal thinking. The only way to find out if anecdotes represent real phenomena is controlled tests. Either people can read other people's minds (or ESP cards), or they can't. Science has unequivocally demonstrated that they can't. And being a holist instead of a reductionist, being related to a psychic, or reading about weird things that befall people does not change this fact.

On the matter of the God question, either God exists or he does not, regardless what I think on the matter, so I'm not particularly worried about it, even if the afterlife turns out to be what Christians think it is with a heaven and a hell, and with belief in God and his Son as the requisite criteria for entry. Why?

First of all, why would an all-knowing, all-powerful, all-loving God care whether I *believed* in him? Shouldn't he know this ahead of time in any case? Even assuming that he has granted me free will, since God is said to be omniscient and outside of time and space, shouldn't he know everything that happens? In either case, why would "belief" matter at all, unless God were more like the Greek and Roman gods who competed with one another for human affections and worship and were filled with such human emotions as jealousy. The Old Testament God Yahweh certainly sounds like this type of deity in the first three of the Ten Commandments (Exodus 20:2–17, King James Version): *"I am the LORD thy God. . . . Thou shalt have no other gods before me. Thou shalt not make unto thee any graven image, or any likeness of any thing that is in heaven above, or that is in the earth beneath, or that is in the water under the earth. Thou shalt not bow down thyself to them, nor serve them: for I the LORD thy God am a jealous God, visiting the iniquity of the fathers upon the children unto the third and fourth generation of them that hate me."*

Yikes! The sins of the fathers are to be borne by their children's children's children? What sort of justice is that? What kind of God is this? This just sounds so . . . well . . . ungodly to my ears. Most *people* have learned to get over jealousy, and I've even managed to keep it in check much of the time myself, and I'm no god, that's for sure.[15] Wouldn't an omniscient, omnipotent, omniphilic deity be more concerned with how I comported myself in *this world*, rather than obsessing over whether I believe in him and/or his Son in hopes of getting to the right place in the *other world*? I would think so. Behavioral comportment dines at the high table of morality and ethics; jealousy feasts on the empty calories of baser human emotions.

In any case, if there is an afterlife and a God who resides over it, I intend to make my case along these lines:

Lord, I did the best I could with the tools you granted me. You gave me a brain to think skeptically and I used it accordingly. You gave me the capacity to reason and I applied it to all claims, including that of your existence. You gave me a moral sense and I felt the pangs of guilt and the joys of pride for the bad and good things I chose to do. I tried to do unto others as I would have them do unto me, and although I fell far short of this ideal far too many times, I tried to apply your foundational principle whenever I could. Whatever the nature of your immortal and infinite spiritual essence actually is, as a mortal finite corporeal being I cannot possibly fathom it despite my best efforts, and so do with me what you will.

PART II

THE BIOLOGY OF BELIEF

The first principle is that you must not fool yourself—and you are the easiest person to fool.

—RICHARD FEYNMAN,
SURELY YOU'RE JOKING, MR. FEYNMAN, 1974

4

Patternicity

IMAGINE THAT YOU ARE A HOMINID WALKING ALONG THE SAVANNA of an African valley three million years ago. You hear a rustle in the grass. Is it just the wind or is it a dangerous predator? Your answer could mean life or death.

If you assume that the rustle in the grass is a dangerous predator but it turns out that it is just the wind, you have made what is called a *Type I error* in cognition, also known as a *false positive*, or believing something is real when it is not. That is, you have found a nonexistent pattern. You connected (A) a rustle in the grass to (B) a dangerous predator, but in this case A was not connected to B. No harm. You move away from the rustling sound, become more alert and cautious, and find another path to your destination.

If you assume that the rustle in the grass is just the wind but it turns out that it is a dangerous predator, you have made what is called a *Type II error* in cognition, also known as a *false negative*, or believing something is not real when it is. That is, you have missed a real pattern. You failed to connect (A) a rustle in the grass to (B) a dangerous predator, and in this case A was connected to B. You're lunch. Congratulations, you have won a Darwin Award. You are no longer a member of the hominid gene pool.

Our brains are belief engines, evolved pattern-recognition machines that connect the dots and create meaning out of the patterns that we think we see in nature. Sometimes A really is connected to B; sometimes it is not. The baseball player who (A) doesn't shave and (B) hits a home run forms a false association between A and B, but it is a relatively harmless one. When the association is real, however, we have learned something valuable about the environment from which we can make predictions that aid in survival and reproduction. We are the descendants of those who were most successful at finding patterns. This process is called *association learning* and is fundamental to all animal behavior, from *C. elegans* to *H. sapiens*. I call this process *patternicity*, or *the tendency to find meaningful patterns in both meaningful and meaningless noise.*

Unfortunately, we did not evolve a baloney-detection network in the brain to distinguish between true and false patterns. We have no error-detection governor to modulate the pattern-recognition engine. The reason has to do with the relative costs of making Type I and Type II errors in cognition, which I describe in the following formula:

$$P = C_{TI} < C_{TII}$$

Patternicity (P) will occur whenever the cost (C) of making a Type I error (TI) is less than the cost (C) of making a Type II error (TII).

The problem is that assessing the difference between a Type I and Type II error is highly problematic—especially in the split-second timing that often determined the difference between life and death in our ancestral environments—so the default position is to assume that all patterns are real; that is, assume that all rustles in the grass are dangerous predators and not the wind.

This is the basis for the evolution of all forms of patternicity, including superstition and magical thinking. There was a natural selection for the cognitive process of assuming that all patterns are real and that all patternicities represent real and important phenomena. We are the descendants of the primates who most successfully employed patternicity.

Note what I am arguing here. This is not just a theory to explain why people believe weird things. It is a theory to explain *why people believe things*. Full stop. Patternicity is the process of seeking and finding patterns, connecting the dots, linking A to B. Again, this is nothing more than association learning, and all animals do it. It is how organisms adapt to their ever-changing environments when evolution is too slow. Genes are selected for and against in changing environments, but this takes time—generations of time. Brains learn, and they can learn almost instantaneously—time is not an issue.

In a 2008 paper entitled "The Evolution of Superstitious and

Superstition-Like Behaviour,"[1] Harvard biologist Kevin R. Foster and University of Helsinki biologist Hanna Kokko tested an earlier version of my theory through evolutionary modeling, a tool used to assess the relative costs and benefits of different relationships between organisms. For example, to whom should you offer help? In evolutionary theory, altruistically helping others seems problematic because in a selfish gene model, shouldn't we hoard all resources and never help anyone? No. Hamilton's rule—named for the renowned British evolutionary biologist William D. Hamilton—states that $br > c$: a positive social interaction between two individuals may occur when the benefit (b) of the genetic relatedness (r) exceeds the cost (c) of the social action. A sibling, for example, may make an altruistic sacrifice for another sibling when the cost of doing so is surpassed by the genetic benefits derived from getting its genes into the next generation through the surviving sibling. That is, you are more likely to help a full brother than you are a half brother, and a half brother more than a complete stranger.[2] Blood really is thicker than water.

Of course, organisms do not consciously make such calculations. Natural selection made them for us and imbued us with moral emotions that guide behavior. In *The Science of Good and Evil* I worked out the evolutionary advantages of being pro-social, cooperative, and altruistic not only to blood relatives, but to fellow group members and even strangers who have become honorary friends or relatives through positive social interactions. Examples include food redistribution and tool sharing among members of a tribe. In this context, evolution endowed us with a rule of thumb that says "be generous and helpful to our blood relatives and those who are nice and generous to us." Even unrelated members of a clan who exhibit such positive attributes trigger in our brains a moral pattern: (A) Og was nice to me, so (B) I should be nice to Og; and (C) if I help Og, (D) Og will return the favor. In *The Mind of the Market* I demonstrated that this effect can be seen between clans and tribes when they participated in mutually beneficial exchanges, also known as trade. Even in the modern world, opening trade borders between two countries tends to lower tensions and aggressions between them, and closing trade borders—imposing trade sanctions—increases the likelihood that two nations will fight. These are both good examples of moral patternicities that have worked for and against our species.[3]

Foster and Kokko used Hamilton's rule to derive their own formula to demonstrate that whenever the cost of believing that a false pattern is real is less than the cost of not believing a real pattern, natural selection will favor the patternicity.[4] Through a series of complex formulas that included additional stimuli (wind in the trees) and prior events (past experience with predators and wind), the authors demonstrated that "the inability of individuals—human or otherwise—to assign causal probabilities to all sets of events that occur around them will often force them to lump causal associations with non-causal ones. From here, the evolutionary rationale for superstition is clear: natural selection will favour strategies that make many incorrect causal associations in order to establish those that are essential for survival and reproduction." In other words, we tend to find meaningful patterns whether they are there or not, and there is a perfectly good reason to do so. In this sense, patternicities such as superstition and magical thinking are not so much errors in cognition as they are natural processes of a learning brain. We can no more eliminate superstitious learning than we can eliminate all learning. Although true pattern recognition helps us survive, false pattern recognition does not necessarily get us killed, and so the patternicity phenomenon endured the winnowing process of natural selection. Because we must make associations in order to survive and reproduce, natural selection favored all association-making strategies, even those that resulted in false positives. With this evolutionary perspective we can now understand that *people believe weird things because of our evolved need to believe nonweird things.*

The Evolution of Patternicity

Anecdotal association is a form of patternicity that is all too common and that leads to faulty conclusions. I heard that Aunt Mildred's cancer went into remission after she imbibed extract of seaweed. Hey, maybe it works. Then again, maybe it doesn't. Who can tell? There is only one surefire method of proper pattern recognition, and that is science. Only when a group of cancer patients taking seaweed extract is compared to a control group can we draw a valid conclusion (and not always then).

As I write this, there is a major brouhaha over a form of anecdotal association involving vaccinations and autism, with some parents of autistic children claiming that shortly after they took their children in for (A) the MMR (measles, mumps, rubella) vaccine they were (B) diagnosed with autism. This is patternicity where it really counts. On National Autism Awareness Day in 2009, Larry King hosted a debate on his show in which he had on one side of his table a couple of medical researchers and experts on autism and vaccines who explained that no connection between the two has ever been made, that the allegedly toxic chemical thimerosal was removed from vaccines in 1999, and that children born after thimerosal was removed are still being diagnosed with autism. On the other side of the table were the actor Jim Carrey and his ex–*Playboy* bunny partner Jenny McCarthy, with videos of her adorable son exhibiting obvious signs of autism. Who are you going to believe—a couple of nerdy brainiacs with expertise, or a couple of glamorous maniacs with celebrity? It was a classic case of the emotional brain running roughshod over the rational brain, as McCarthy tugged on the heartstrings of viewers while the scientists struggled to elucidate how proof is established in science through careful controlled experiments and epidemiological studies. Once again, the rational bit was in the emotional horse's mouth, but the reins gave no direction that day.

The problem we face is that superstition and belief in magic are millions of years old whereas science, with its methods of controlling for intervening variables to circumvent false positives, is only a few hundred years old. Anecdotal thinking comes naturally, science requires training. Any medical huckster promising that A will cure B has only to advertise a handful of successful anecdotes in the form of testimonials.

B. F. Skinner was the first scientist to systematically study superstitious behavior in animals, noting that when food was presented to pigeons at random intervals instead of more predictable schedules of reinforcement—for which pecking a key inside a box in which the pigeon was placed would result in delivery of the food through a small food hopper (see figure 1)—the pigeons exhibited an odd assortment of behaviors, such as side-to-side hopping or twirling around counterclockwise before pecking the key. It was an avian rain dance of sorts. The pigeons did this because they were put on something called a *variable interval* (VI) schedule of reinforcement, in which the time interval between getting the food reward for pecking a key varied. In that interval of time between pecking the key and the hopper delivering the food, whatever the pigeons happened to be doing was scored in their little brains as a pattern.

Supporting my thesis that such patternicities were important in the

Figure 1. Patternicity in Pigeons
Inside a Skinner box in Douglas Navarick's laboratory at California State University–Fullerton, where I conducted research on learning in the 1970s, one of our pigeons has learned to peck at the two keys above to receive grain through a food hopper below. Skinner discovered that if he randomly delivered the food reinforcement, whatever the pigeon happened to be doing just before the delivery of the food would be repeated the next time, such as spinning around once to the left before pecking at the key. This is pigeon patternicity, or the learning of a superstition. PHOTO BY THE AUTHOR.

evolution of response behaviors to changing environments, Skinner noted, "each response was almost always repeated in the same part of the cage, and it generally involved an orientation toward some feature of the cage. The effect of the reinforcement was to condition the bird to respond to some aspect of the environment rather than merely to execute a series of movements." These superstitious behaviors were intensely repeated, typically five or six times in a matter of fifteen seconds or so, as Skinner concluded: "The bird behaves as if there were a causal relation between its behavior and the presentation of food, although such a relation is lacking."[5] In the bird's brain, (A) twirling around once and pecking the key was connected to (B) food. That is basic patternicity. If you doubt its potency as a force in human behavior, just visit a Las Vegas casino and observe people playing the slots with their varied attempts to find a pattern between (A) pulling the slot machine handle and (B) the payoff. Pigeons may have bird brains, but when it comes to such basic patternicities, our brains are little different.

Inspired by Skinner's classic experiments, Koichi Ono of Komazawa University in Japan ran human subjects through the equivalent of a Skinner box by having them sit in a booth in which there were three levers.[6] Independent of their pulling the levers (but unknown to them) the subjects were then exposed to a number counter that granted them one point at a time, which was followed by a flashing light and buzzer (a scaled-down slot machine, as it were). The points were delivered in a VI schedule of reinforcement (just like the pigeons) of, on average, either 30 seconds (with a range of 3 to 57 seconds) or 60 seconds (with a range of 25 to 95 seconds). Before the experiment began the subjects were instructed, "The experimenter does not require you to do anything specific. But if you do something, you may get points on the counter. Now try to get as many points as possible."

Since the subjects could not predict when the points would be delivered (because the schedule of delivery was variable), and people just seem to have a natural propensity to pull levers, some of them inferred a connection between (A) pulling the handles and (B) getting points. Patternicity. And there were some doozies. Subject 1 happened to get a point after pulling the levers in the order of left, middle, right, right, middle, left, and so repeated that pattern three more times. Subject 5 began the session with short pulls of all the levers, with the points accumulating quite independently of his pulls, but then by chance he happened to be holding the middle lever when a point was delivered, so thereafter he performed the superstitious ritual of three short pulls followed by holding the middle lever. Of course, the longer he held the lever the greater the chance that he would get another point (because they were delivered on a variable time schedule). After minute nine of the thirty-minute session, Subject 5 had his ritual down pat. Subject 15 developed the strangest rite of all. Five minutes into her session a point was delivered the moment she happened to touch the point counter. Thereafter she started touching anything and everything within reach, and, of course, since the points continued to be delivered, this odd touching behavior was reinforced. At the ten-minute mark she got a point just as she happened to jump on the floor, whereby she promptly abandoned touching and took up jumping as her new strategy, climaxing in a point being scored when she touched the ceiling, leading her to end the session early from ceiling-touching exhaustion.

Technically speaking, in Ono's words, "superstitious behavior is defined as behavior produced by response independent schedules of reinforcer delivery, in which only an accidental relation exists between responses and delivery of reinforcers." That's a fancy way of saying that superstitions are just an accidental form of learning. This is patternicity. Can such learned superstitious patternicities be unlearned? They can. In 1963, Skinner's Harvard colleagues Charles Catania and David Cutts put humans through the pigeon paces by instructing each of twenty-six undergraduate subjects to press one of two different buttons on a box whenever a yellow light flashed and to try to accumulate as many points as possible on a counter. Whenever the subject gained a point a green light flashed. A red light indicated that the session was over, which was when the subject reached one hundred points. Unbeknownst to the subjects, only the right button could generate points, and those points were delivered on a VI schedule of reinforcement, with an average time between point delivery of thirty seconds. The results were revealing in that human brains are no less superstitious than bird brains: most of the subjects quickly developed superstitious button-pushing patterns between the left and right buttons, because if they pressed the left button just before the right button happened to deliver a point, that particular pattern was reinforced. Once subjects established a superstitious button-pushing pattern, they stuck with that pattern throughout the session because they continued to be reinforced for it.

To extinguish the Type I false-positive pattern, Catania and Cutts introduced what is called a *changeover delay* (COD), which added a period of time between presses on the left button and subsequent reinforced presses on the right button, thereby untangling them from any meaningful pattern. That is, where (A) the left button was incorrectly associated with (B) a point, a superstitious pattern was established; but by separating A and B in time the association link was disconnected. As you might expect (and certainly hope), humans needed a longer COD than pigeons because, presumably, we have a greater cognitive capacity for holding associations in memory than birds do. But this is a double-edged sword. Our greater capacity for learning is often offset by our greater capacity for magical thinking. Superstition in pigeons can be easily extinguished; in humans it is much more difficult.[7]

Hardwired Patternicity

Patternicity is common across the animal kingdom. Early studies in the 1950s by Niko Tinbergen and Konrad Lorenz, who pioneered the study of ethology—the evolutionary origins of animal behavior—demonstrated the capacity of many organisms to rapidly form lasting patterns. Lorenz, for example, documented *imprinting*, a type of phase-dependent learning whereby the youth of a species at a critical period in their development will form a fixed and lasting pattern of memory for whoever or whatever appears before them during that brief span of time. In the baby greylag geese that Lorenz studied, for example, the object of gaze in the critical period of thirteen to sixteen hours old is normally a mother, and thus she becomes imprinted in their brains. To test this hypothesis, the mischievous Lorenz made certain that it was he who was in the ducklings' visual field at the critical moment, and thereafter "momma" Konrad led his flock around the grounds of his research station.[8]

A form of reverse imprinting can be found in humans in the incest taboo. Two people growing up in close proximity to each other during a critical period in childhood are unlikely to find each other sexually attractive as adults. Evolution has programmed within us a rule of thumb: don't mate with those with whom you've grown up because they are very likely your siblings and are thus too genetically similar.[9] Again, we don't make genetic calculations. Natural selection did the calculating for us and endowed us with emotions, in this case incest disgust. Our brains are developmentally sensitive to forming incest patternicities, and that happens even with people we grow up with who are stepsiblings or others not genetically related to us. This is a Type I error, a false positive, and it evolved because in our Paleolithic past the other people in our childhood homes were most likely blood relatives.

In his studies of herring gulls, Niko Tinbergen observed that when the chick perceived the mother gull's yellow beak with a red dot, it promptly began pecking at it, which triggered the mother to regurgitate some food for her chick to eat. Further experimental studies of this phenomenon

Figure 2. The SS-IRM-FAP System of Patternicity

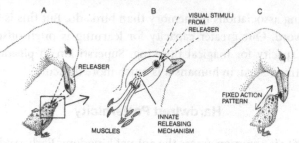

a. Niko Tinbergen discovered that when a herring gull chick sees its mother gull's yellow beak with a red dot, it promptly begins pecking at it, which causes the mother to regurgitate food for her chick to eat. This is the Sign Stimulus (SS)—Innate Releasing Mechanism (IRM)—Fixed Action Pattern (FAP) process. FROM JOHN ALCOCK, *ANIMAL BEHAVIOR: AN EVOLUTIONARY APPROACH* (SUNDERLAND, MASS.: SINAUER ASSOCIATES, 1975), P. 164. ORIGINALLY APPEARED IN NIKO TINBERGEN AND A. C. PERDECK, "ON THE STIMULUS SITUATION RELEASING THE BEGGING RESPONSE IN THE NEWLY HATCHED HERRING GULL CHICK," *BEHAVIOUR* 3 (1950): 1–39.

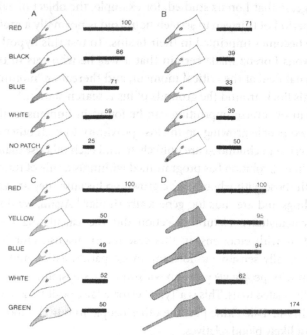

b. Further experimental studies of the SS-IRM-FAP patternicity phenomenon revealed that yellow bills with a red dot receive four times as many pecks from hatchlings over uniformly yellow beaks with no red dot, and that some bill shapes act as superstimuli, triggering excessive begging. FROM NIKO TINBERGEN AND A. C. PERDECK, *BEHAVIOUR* 3 (1950): 1–39. REPRINTED IN JOHN ALCOCK, *ANIMAL BEHAVIOR: AN EVOLUTIONARY APPROACH* (SUNDERLAND, MASS.: SINAUER ASSOCIATES, 1975), P. 150.

revealed that yellow bills with a red dot receive more than three times as many pecks from hatchlings over uniformly yellow beaks with no red dots. Tinbergen found that isolated hand-reared birds would also sometimes peck at cherries or the red bottoms of tennis shoes. This suggests that extremely young birds have an unlearned preference for the color red, especially when placed on a beak. (See figure 2.) Tinbergen codified this sequence as follows: a *sign stimulus* triggers an *innate releasing mechanism* in the brain that leads to a *fixed action pattern* of behavior, or *SS-IRM-FAP*. In the case of the herring gull chick, the red dot in contrast with the yellow beak of its mother acted as a sign stimulus that triggered an innate releasing mechanism in its brain to deliver a fixed action pattern of pecking at the red dot. In turn, the chick's pecking acted as a sign stimulus for the mother that triggered an innate releasing mechanism in its brain to deliver a fixed action pattern of food regurgitation.[10]

Facial Recognition Patternicity

Face recognition in humans is another form of the SS-IRM-FAP system of patternicity, and it begins shortly after birth. When an infant observes the cooing happy face of its mother or father, the face acts as a sign stimulus that initiates the innate releasing mechanism in its brain to trigger the fixed action pattern of smiling back, thereby setting up a symphony of parent-child staring and cooing and smiling—and bonding attachment. It need not even be a real face. Two black dots on a cardboard cutout elicit a smile in infants, although one dot does not, indicating that the newborn brain is preconditioned by evolution to look for and find the simple pattern of a face represented by two to four data points: two eyes, a nose, and a mouth, which may even be represented as two dots, a vertical line, and a horizontal line.

Facial-recognition software was built into our brains by evolution because of the importance of the face in establishing and maintaining relationships, reading emotions, and determining trust in social interactions. We notice the whites of someone's eyes for directionality of their gaze. We detect the dilation of another's pupils as a sign of arousal (anger, sexual, or otherwise). We scan others' faces for emotional leakage: sadness, disgust, joy, surprise, anger, and happiness. We subtly notice the difference between a real and a fake smile in the upturn of the outer eyelids for the genuine article. Faces are important to a social primate species such as ourselves.

This is why we are so inclined to see faces in random patterns in nature: the face on Mars that is an eroded mountain is my favorite example, but there are many others. (See figure 3.)

The location in the brain where faces are recognized and processed has now been established by neuroscientists. In general, inside the temporal lobes of the brain (just above your ears) there is a structure called the *fusiform gyrus* that we know is actively involved in facial recognition because damage to it makes it difficult or impossible to recognize the face of someone you know, even your own in a mirror! More specifically, there are two separate neural pathways: one for processing faces in general and another for processing facial characteristics in particular. This is done through two different types of neurons: large (*magno*) cells that comprise the relatively rapid-firing *magnocellular pathway* that processes large receptive fields and carries low-spatial-frequency (coarse-grained data) information (the general face), and smaller cells that comprise the relatively slower firing *parvocellular pathway* that processes small receptive fields and carries high-spatial-frequency (fine-grained data) information (facial details such as eyes, nose, and mouth).

Further, it appears that the brain first processes the global shape of a face, such as the general outline with two eyes and a mouth, and then processes the details of facial features, such as the eyes, nose, and mouth. This is why when you examine the upside-down photograph of President Obama (in figure 3) you recognize him immediately; but if you stare at it awhile you will see that there is something odd about his eyes and mouth in one of the pictures. Turn the book upside down and you will see what it is. This is the effect of your two different facial-recognition networks operating at different rates and granularity. First there is the rapid assessment that it is a face, and then the recognition that it is the face of someone you know; then there is the processing of the details of that face, which takes a bit longer. The former happens quickly and unconsciously while the latter happens slowly and consciously.[11]

This difference between slow and rapid processing of information is interesting because in the search for the neural correlates of consciousness, most theories hold that rapid unconscious processing happens before slower conscious awareness. In a famous 1985 study by neuroscientist Benjamin Libet, he took EEG readings of subjects sitting in front of a screen in which a dot was moving about a circle (like the second hand on a clock face). The subjects were asked to do two things: (1) note

Figure 3. Faces Everywhere

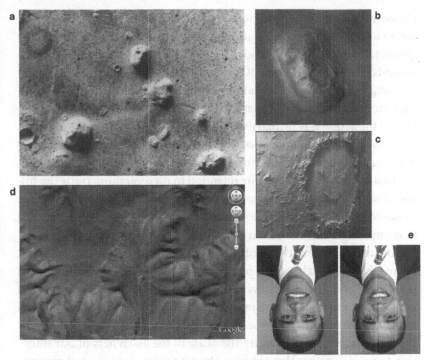

The human face is so important in the expression of emotions that we have evolved facial-recognition networks in our brains (see details in the text), to the point where we see faces everywhere we look. Here are a few examples.

a. The face on Mars, original grainy photograph from 1976 *Viking* spacecraft mission. COURTESY OF NASA.

b. The face on Mars, closer detailed photograph from 2000 *Mars Surveyor* mission. COURTESY OF NASA.

c. The happy face on Mars. COURTESY OF NASA.

d. Indian chief head or random configuration of hills and valleys? Configuration is in Cypress County, Alberta, Canada, southeast of Calgary just north of the U.S. border. Turn the book upside down to view the image from a different perspective or enter the coordinates (+50° 0' 38.20", −110° 6' 48.32") into Google Maps and zoom in on the image and rotate it yourself to see the face pattern appear and disappear. COURTESY OF GOOGLE MAPS.

e. Which upside-down photo of President Barack Obama looks odd? Turn the book upside down to find out (see the text for explanation). Original illusion was discovered by Peter Thompson of York University and published in 1980: PETER THOMPSON, "MARGARET THATCHER: A NEW ILLUSION," *PERCEPTION* 9, NO. 4 (1980): 483–84. THE OBAMA ILLUSION MAY BE FOUND AT MIGHTY OPTICAL ILLUSIONS: http://www.moillusions.com/2008/12/who-says-we -dont-have-barack-obama.html.

the position of the dot on the screen when they first became aware of the desire to act, and (2) press a button that also recorded the position of the dot on the screen. The difference between 1 and 2 was two hundred milliseconds. That is, two-tenths of a second lapsed between thinking about pressing the button and actually pressing the button. The EEG recordings for each trial revealed that the brain activity involved in the initiation of the action was primarily centered in the secondary motor cortex, and that part of the brain became active three hundred milliseconds *before subjects reported their first awareness of a conscious decision to act.*

That is, the awareness of our intention to do something trails the initial wave of brain activity associated with that action by about three hundred milliseconds—three-tenths of a second lapsed between the brain making a choice and our awareness of the choice. Add to this processing time the other two-tenths of a second to act on the choice, and it means that a full half second passes between our brain's intention to do something and our awareness of the actual act of doing it. The neural activity that precedes the intention to act is inaccessible to our conscious mind, so we experience a sense of free will. But it is an illusion, caused by the fact that we cannot identify the cause of the awareness of our intention to act.[12] Together these studies show how deeply ingrained patternicity is in our brains, hardwired into our unconscious and generating patterns beneath our awareness.

A final example in our facial-recognition patternicity is the now well-documented facial greeting found in nearly every human group around the world (except where it is culturally suppressed, as in Japan). When greeting over a distance people smile and nod, and if friendly they raise their eyebrows in a rapid movement for approximately one-sixth of a second. In the 1960s, the Austrian ethologist Irenäus Eibl-Eibesfeldt traversed the globe filming people with an ingeniously devised camera equipped with an angle lens, in which the camera appeared to be pointed in one direction but the filming was actually taking place at a ninety-degree angle from where it was pointing. Thus, the facial expressions of people from urban Europe to rural Polynesia were "unobtrusively measured" and later analyzed in slow motion. There is an innate pattern of greeting everywhere in the world that people are born understanding without any cultural training. The pattern is not just for happy greetings. Eibl-Eibesfeldt also recorded remarkable similarities across radically different cultures in other emotional expressions, such as anger, characterized

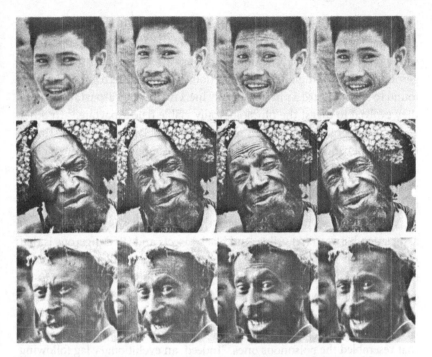

Figure 4. The Innate Pattern of Face Greetings Around the World
The Austrian ethologist Irenäus Eibl-Eibesfeldt traversed the globe filming
people with a hidden lens as they greeted one another. He discovered that
when greeting over a distance people smile and nod, and if friendly they
raise their eyebrows in a rapid movement for approximately one-sixth of a
second. This is an example of innate facial patternicity. FROM IRENÄUS EIBL-
EIBESFELDT, *ETHOLOGY* (NEW YORK: HOLT, RINEHART AND WINSTON), 1970.

by opening the corners of the mouth, frowning, clenching the fists, stamp-
ing on the ground, and even hitting at objects.[13] Eibl-Eibesfeldt's research
has since been corroborated by Paul Ekman, and together they have presented a body of uncontestable evidence for the evolutionary origins of
facial patternicities.[14] (See figure 4.)

Mimicking Patterns

Mimicry is another form of patternicity. In their paper on the evolution
of patternicity discussed above, Foster and Kokko presented three such
examples: (1) predators who normally avoid eating dangerous yellow and

black insects also avoid harmless insects with similar yellow and black markings;[15] (2) predators of snakes who normally avoid preying upon poisonous species also avoid the nonpoisonous varieties that mimic the dangerous types;[16] (3) single-celled *E. coli* (found in the human gut) have been found to swim toward a physiologically inert methylated aspartate because they evolved to digest the physiologically viable true aspartate.[17] In other words, these organisms formed meaningful associations between stimuli (visual, taste) and their effects (dangerous, poisonous) because such associations are vital to survival; as such, the ability to make such associations was selected for and could therefore be exploited by other organisms by tricking the system.

What happens with mimicry, as in the first example, is that the original association between (A) yellow and black insects and (B) dangerous is that nondangerous insects resembling the dangerous ones will also be avoided by predators and thus are more likely to survive and pass on the genes for coloration that more closely match the dangerous species. The second example illustrates the same principle of mimicry and exploitation of an A-B association in which evolution favored those nonpoisonous snakes that resembled the poisonous ones. "Indeed, an evolutionary lag following a changed environment provides another route to superstitious behaviours," Foster and Kokko explained, "whereby an organism associates two events that once were, but are no longer causally related, e.g. a predator goes extinct but the prey still hides at night."

The third example of the *E. coli* swimming toward the taste of a substance chemically similar to aspartate because of its original preference for the real thing has obvious parallels with the human enjoyment of artificial sweeteners as well as with our modern problem of obesity. In the natural environment, (A) sweet and rich foods are strongly associated with (B) nutritious and rare. Therefore, we gravitate to any and all foods that are sweet and rich, and because they were once rare we have no satiation network in the brain that tells us to shut off the hunger mechanism, so we eat as much as we can of them. On the other end of the taste spectrum, there is the well-known taste aversion effect—one-trial learning—where the pairing of a food or drink with severe nausea and vomiting often results in a long-term aversion for that food or drink. In my case it was a graduate school pairing of (A) too much cheap red wine with (B) a night of vomiting that made it difficult for me for decades to enjoy red wines, even expensive labels. The evolutionary significance is clear: foods that can kill you (but don't) should never be tried a second time, so one-trial learning evolved as an important adaptation.

Supernormal Patternicities

Supernormal stimuli combine the principles of mimicry and the SS-IRM-FAP system and are another example of an innate form of patternicity. Niko Tinbergen, for example, discovered that gull chicks peck even more fervently at a fake bill that is longer and narrower than the real beak of their mother. He also studied a species of bird that normally nests upon small pale blue eggs with gray specks on them and found that he could get them to prefer sitting on giant bright blue eggs speckled with black polka dots. It's a form of tricking a brain preprogrammed by evolution to expect certain patterns by exposing it to exaggerated forms of the same.[18]

Harvard University evolutionary psychologist Deirdre Barrett, in her 2010 book *Supernormal Stimuli*, documented numerous instances of ancient innate human patternicities hijacked by the modern world.[19] In addition to the pattern of sweet and rich foods leading to obesity mentioned above, Barrett outlined how modernity has commandeered our ancient propensities for patterns of sexual preferences, leading to expectations of women's faces and figures to match the supernormal stimuli seen in perfect (and perfectly modified) supermodels with long legs, hourglass figures, 0.7 waist-to-hip ratios, enlarged breasts, perfectly symmetrical faces with blemish-free complexions, full lips, large alluring eyes with dilated pupils, and full, thick heads of hair. In the environment of our Paleolithic ancestors, the "normal" dimensions of these physical characteristics were proxies for genetic health, and thus there was a natural selection for emotional preference for women who approximated such physicality. Like food that is nutritionally rich and environmentally rare, such physical characteristics are both strongly desired and without satiation, so our brains can be tricked into feeling that more is better.

Today, of course, no one walks into a nightclub with calipers to measure waist-to-hip ratios or facial symmetries. Evolution has done the measuring for us, leaving us with such essential emotions as sexual desire. In the SS-IRM-FAP system, such "normal" features act as a sign stimulus to initiate the innate releasing mechanism in the brain of arousal that leads to the fixed action pattern of soliciting contact for sex. Thus, "super-normal" stimuli, such as silicone-enhanced breasts, lip implants, makeup to enhance the eyes, rouge to blush the cheeks, high heels to extend the legs, and the like, all trigger an even stronger emotional and behavioral response.

What women prefer in men is just as real and natural, of course: women are attracted to men who are taller than them, with narrow waists and broad shoulders, lean and muscular builds, symmetrical faces and clear complexions, and strong jaws and chins. These are all characteristics related to a good balance of testosterone and other hormones, and they serve as proxies for genetic health in terms of selecting a mate with whom to have children. Because sexuality is so much more visually attended to by men, however, pornography as a supernormal stimulus is almost entirely a guy thing. Porn for women—actually the title of a parody in which fully clothed men are performing domestic chores ("I just vacuumed the whole house!")—is mainly found in soap operas, chick flicks, and especially romance novels in which the plot concerns the heroine "finding and capturing the heart of the one right man," wrote Barrett. "Sex may be explicit, implied, or not destined to occur until after a proposal of marriage, which constitutes the end of the book."[20]

There are many other forms of preprogrammed patternicities in super-normal stimuli. There is, for example, our natural "territorial imperative," in which we have a strong desire to protect what is ours, especially literal territory in the form of land, community, and nation. This, too, has been usurped by modernity. As Barrett notes, there is "a compelling instinct to provide for one's offspring; this is practically synonymous with whose genes will survive." In the modern world, however, territory has taken on supernormal dimensions. "Now the powerful and rich can direct these instincts at supernormal family estates, trust funds that endure for genera-tions, and, in the case of monarchies, permanent rulership for the family."[21]

Most territorial animals resolve land disputes with threat gestures, vocal cries, and—if worse comes to worst—a brief physical attack in which someone may actually get pushed, shoved, or even bitten. In fact, in laboratory "eye gaze" experiments, primatologists triggered male rhesus monkeys to make threatening gestures and displays, and even aggressive motions toward them, by simply staring at the monkeys with an open mouth. Once again returning to the SS-IRM-FAP system, the closed eye-lid and open mouth serve as a sign stimulus to set off an innate releasing mechanism of anger and thereby release the fixed action pattern of aggression or reciprocal threat display. In this research we also find direct evidence for the IRM in single-cell recording from the brain stem of monkeys, in which there is a significant increase in neuronal activity when the experimenter stares at the monkey; the breaking of the gaze decreases neuronal activity, along with aggressive responses.[22]

Patternicity and Control

Patternicities do not occur randomly but are instead related to the context and environment of the organism, to what extent it believes that it is in control of its environment. Psychologists call this *locus of control*. People who rate high on *internal* locus of control tend to believe that they make things happen and that they are in control of their circumstances, whereas people who score high on *external* locus of control tend to think that circumstances are beyond their control and that things just happen to them.[23] The thinking here is that having a high internal locus of control leads you to be more confident in your personal judgment, more skeptical of outside authorities and sources of information, and have a lower tendency to conform to external influences. In fact, people who consider themselves "skeptics" about the paranormal and supernatural tend to score high in *internal locus of control*, whereas self-reported "believers" in ESP, spiritualism, reincarnation, and mystical experiences in general tend to rate high in *external locus of control*.[24]

Locus of control is also mediated by levels of certainty or uncertainty in physical and social environments. Bronislaw Malinowski's famous studies of superstitions among the Trobriand Islanders in the South Pacific demonstrated that as the level of uncertainty in the environment increases so, too, does the level of superstitious behavior. Malinowski noted this in particular among the Trobriand fishermen—the farther out to sea they sailed the more uncertain the conditions grew, along with the uncertainty of success at a catch. Their levels of superstitious rituals rose with their levels of uncertainty. "We find magic wherever the elements of chance and accident, and the emotional play between hope and fear have a wide and extensive range," Malinowski explained. "We do not find magic wherever the pursuit is certain, reliable, and well under the control of rational methods and technological processes. Further, we find magic where the element of danger is conspicuous."[25]

I have made a similar observation on superstitions among athletes, most notably baseball players. As fielders succeeding over 90 percent of the time, they exhibit almost no superstitious rituals, but when they pick up a bat and go to the plate—where they are sure to fail at least seven out of ten times—they suddenly become magical thinkers employing all manner of bizarre ritualistic behaviors in order to cope with the uncertainty.[26]

Risk and control were tested in a 1977 study that found that if you show parachute jumpers about to leap out of a plane a photographic representation of noise (such as the "snow" on a television screen) they are far more likely to see a nonexistent embedded figure than if you presented it to them earlier. Uncertainty makes people anxious, and anxiety is related to magical thinking. A 1994 study, for example, showed that anxious first-year MBA students are far more conspiratorially minded than their more secure second-year colleagues. Even such base emotions as hunger can influence your perceptual patternicity. A 1942 study found that when ambiguous images are shown to both hungry and satiated people, the former are more likely to see food. And apropos the current recession, economic environments may lead to misperceptions where, in one experiment, children from poor neighborhoods and working-class families tend to overestimate the size of coins compared to the estimates made by children from wealthy neighborhoods and families.[27]

The relationship between personality, belief, and patternicity was explored by experimental psychologist Susan Blackmore, famous for her dramatic reversal from believer to skeptic of the paranormal after years of conducting research trying to find the elusive effects of ESP. What she discovered was that people who believe in ESP tend to look at data sets and see evidence of the paranormal, whereas skeptics do not. In one study, for example, Blackmore and her colleagues had subjects complete a paranormal belief scale, then presented them with photographs of common objects with varying degrees of degeneration into noise (0 percent, 20 percent, 50 percent, and 70 percent) and asked them if they could recognize and identify each object. The results revealed that believers were significantly more likely than nonbelievers to see objects in the noisiest images but to misidentify them. (See figure 5.)[28] In other words, they saw more patterns but made more Type I false-positive errors.

A similar effect was found in an experiment in which subjects were asked to determine the probability of the roll of a die. Try it yourself. Imagine that you have a die in your hand that you roll three consecutive

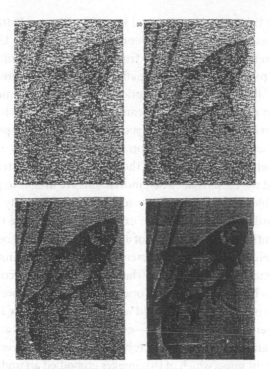

Figure 5. Patternicity and Belief
Psychologist Susan Blackmore discovered that believers in ESP and other forms of the paranormal were more likely to see an object in the maximally degraded image in the upper left corner than were skeptics of the paranormal, but the believers made more identification mistakes. ILLUSTRATIONS COURTESY OF SUSAN BLACKMORE.

times and note the outcome. Which of the following sequences is more likely: 2-2-2 or 5-1-3? Most people say that the second outcome is more likely than the first, because it seems like a streak of 2s is more improbable. In fact, both are equally likely because dice have no memory, and for each roll a 2 is as likely to come up as a 5 or 1 or 3. This psychological effect is called *repetition avoidance* and it affects believers and skeptics differently. When believers in ESP are given this choice they tend to rate 5-1-3–type sequences as significantly more probable than skeptics do. That is, they find greater meaning in randomness.[29]

An even more direct link between patternicity and perceived levels of control over the environment was demonstrated in a 2008 study descriptively titled "Lacking Control Increases Illusory Pattern Perception," by management researchers Jennifer Whitson at the University of Texas–Austin and Adam Galinsky from Northwestern University, who studied how psychological states are affected by corporate environments. Defining an "illusory pattern perception" (a form of patternicity) as "the identification of a coherent and meaningful interrelationship among a set of random or unrelated stimuli (such as the tendency to perceive false correlations, see imaginary figures, form superstitious rituals, and embrace conspiracy beliefs, among others)," the researchers conducted six experiments to test the thesis that "when individuals are unable to gain a sense of control objectively, they will try to gain it perceptually."[30] Why do people do this? "Because," Whitson explained to me as she tried to gain a sense of control in a quiet corner of a busy airport jetting between conferences, "feelings of control are essential for our well-being—we think clearer and make better decisions when we feel we are in control. Lacking control is highly aversive, and one fundamental way we can bolster our sense of control is to understand what's going on. So we instinctively seek out patterns to regain control—even if those patterns are illusory."

Whitson and Galinsky sat subjects before a computer screen, telling one group they must guess which of two images embodied an underlying concept the computer had selected. For example, they might see a capital *A* and a lowercase *t*, colored, underlined, or surrounded by a circle or square. Subjects would then guess at an underlying concept, such as all capital *A*s are red. There was no actual underlying concept—the computer was programmed to randomly tell the subjects they were either "correct" or "incorrect." Consequently, they developed a sense of lacking control. Another group did not receive randomized feedback and so felt more in control. In the second part of the experiment subjects were shown twenty-four "snowy" photographs, half of which contained hidden images such as a hand, horses, a chair, or the planet Saturn, whereas the other half just consisted of grainy random dots. (See figure 6 for an example of Saturn dots versus random dots.) Although nearly every subject correctly identified the hidden figures, subjects in the lack-of-control group found more patterns in the photographs that had no embedded images compared to subjects in the baseline group.

In a second experiment, Whitson and Galinsky had subjects vividly recall an experience in which they either had full control or lacked control over a situation. The subjects then read stories in which outcomes of

Figure 6. Find the Hidden Pattern
Most people can see the hidden figure of Saturn in the photograph on the left. Can you pick out the hidden figure in the photograph on the right? If not, then you are probably feeling in control of your life because subjects who are put into a situation where they feel out of control are more likely to see a pattern in this random series of dots. ILLUSTRATIONS COURTESY OF JENNIFER WHITSON.

situations for the characters were preceded by unconnected and superstitious behaviors (such as foot stomping before entering a meeting) that led to success (such as having one's idea approved in the meeting). The subjects were then asked whether they thought the characters' behavior was related to the outcome. Those who had recalled an experience in which they lacked control perceived a significantly greater connection between the two unrelated events than those who recalled an experience in which they felt in control. Interestingly, the low-control subjects who read a story about an employee who failed to receive a promotion tended to believe that a behind-the-scenes conspiracy was the cause.

"Consider 9/11," Whitson suggested when I mentioned the time spent by skeptics in debunking conspiracy theories. "There we saw an unstable environment caused by the terrorist attacks that led directly and almost instantly to the generation of hidden conspiracy theories." But 9/11 *was* a conspiracy, I reminded her, only it was a conspiracy by nineteen members of al-Qaeda to fly planes into buildings, not an "inside job" by the Bush administration. What's the difference between these two conspiracies? "It may be that even though we were told immediately that it was al-Qaeda, there was a terrible uncertainty about the future, a sense of loss of control," Whitson conjectured, "leading to the search for hidden patterns, which the 9/11 'truthers' think they found."

Maybe. I suspect this is partially true, but there is another factor that

I call *agenticity* that comes into play with conspiracy theories that I will explore in the next chapter. For now, keep in mind that research consistently shows that once people have established what they think is the cause of an event they just observed—(in other words, they have formed a link between A and B)—they will then continue to gather information to support that causal link over other possibilities—if they can even think of alternatives once the first causal link is established, which they usually cannot.

Interestingly, it appears that a negative event, such as a sporting game loss or a failure to achieve a goal, produces even faster causal links and support for those links, especially if it is an unexpected event. Observers (especially fans) produce more causal explanations when a winning team unexpectedly loses to a vastly inferior opponent (an "upset" loss), or vice versa, than if the event went as expected.[31] As a lifelong observer of the usually successful Los Angeles Lakers, for example, I can attest to the fact that long winning streaks are notched up to such simple explanations as smooth teamwork, hard work, and the natural talent of the players, whereas the occasional loss generates dozens of column inches and hours of radio talk time in the endless search for this, that, and the other cause— Kobe and Shaq's feud, Phil's bad back, payroll disputes, too much travel, too many Hollywood distractions, and so on, anything but the fact that the other team just outplayed them.

The most intriguing and practical finding by Whitson and Galinsky came when they tested the relationship between lack of control and pattern perception in the stock market. Control was manipulated by describing the market environment as either volatile (one group of subjects was shown a headline that read "Rough Seas Ahead for Investors" with a short paragraph description that included the line that investing in the stock market was "like walking through a minefield") or stable (the other group was shown a headline that read "Smooth Sailing Ahead for Investors" with a short paragraph description that included the line that investing in the stock market was like "walking through a field of flowers"). Subjects were then exposed to uncorrelated sets of information about stocks; they read a series of twenty-four statements about the finances of two companies, some positive and some negative. Company A had sixteen positive and eight negative statements while Company B had eight positive and four negative statements. Even though the ratio of positive to negative statements was the same for both companies (2:1), subjects exposed to the "volatile market" condition ("Rough Seas Ahead") were significantly less likely to invest in Company B compared to those subjects exposed to the "stable market" condition ("Smooth Sailing Ahead"). Why? Because those subjects in the "volatile market" condition remembered more negative statements made about Company B, whereas those in the "stable market" condition accurately remembered the number of negative statements. Why should this be?

This is the result of something called *illusory correlation*, the perception of a causal relationship between two sets of variables where none exists, or the overestimation of a connection between two variables. The illusory correlation effect is strongest when people form false associations between (X) membership in a statistically small group and (Y) rare and usually negative traits or behaviors. Trivially, for example, people tend to recall the days when they (X) washed their car and (Y) it rained; nontrivially, white Americans typically overestimate the rate that (X) African Americans are (Y) arrested.[32]

What can we do about illusory correlation and the broader problem of illusory pattern detection? In their final experiment, Whitson and Galinsky created a sense of lacking control in two groups of subjects, then asked one group to contemplate and affirm their most important values in life—a proven technique for reducing learned helplessness. The researchers then presented those same snowy pictures, finding that those who lacked control but had no opportunity for self-affirmation saw more nonexistent patterns than did those in the self-affirmation condition.

Interestingly, Whitson confessed to me, she originally devised this research protocol when she was going through a particularly stressful time in her life and feeling rather out of control herself. Call it therapeutic science. It seems to work. "Before undergoing surgery," Whitson reflected, "people given details about what is going to happen have less anxiety and may even recover faster. Knowledge is another form of control." This is reminiscent of a 1976 study by Harvard psychologist Ellen Langer and her colleague Judith Rodin, now president of the Rockefeller Foundation, in a New England nursing home. Residents were given plants and the opportunity to see weekly films but with some variation of control. Residents on the fourth floor, who were in charge of watering the plants and could choose the night of the week they wanted to view the film, lived longer and healthier lives than the other residents, even those given plants that were watered by the staff. It was the sense of control that had the apparent effect on health and well-being.[33] Perhaps this is what Voltaire meant at the end of *Candide*, in the title character's rejoinder to Dr. Pangloss's proclamation that "all events are linked up in this best of all possible worlds": "'Tis well said," replied Candide, "but we must cultivate our gardens."

The Power and Perils of Patternicity

Occasionally I am challenged about the harm of people embracing superstitions, along the lines of: "Oh, come on, Shermer, let people have their delusions. What's the harm?" Setting aside for the moment the playful reading of one's astrology chart in the newspaper or one's fortune in an after-dinner cookie, my general answer is that it is better to live in a real world than a fantasy world. The harm, in fact, can be deadly serious when our patternicities are of the Type I false-positive type.

What's the harm? Ask the victims of John Patrick Bedell, the gunman who attacked guards at the entrance of the Pentagon in March 2010, who now appears to have been a right-wing extremist and 9/11 "truther." In an Internet posting, he said that he intended to expose the truth behind the 9/11 "demolitions." Apparently the delusional Bedell intended to shoot his way into the Pentagon to find out what really happened on 9/11. Death by conspiracy.

Death by theory provides another case in point. In April 2000, a ten-year-old girl named Candace Newmaker began treatment for something called *attachment disorder* (AD). Candace's adoptive mother of four years, Jeane Newmaker, was having trouble handling what she considered to be Candace's disciplinary problems. When Jeane sought help from a therapist affiliated with the so-called Association for Treatment and Training in the Attachment of Children,[34] she was told Candace needed attachment therapy (AT), based on the theory that if a normal attachment is not formed during the critical first two years, then reattachment can be done later. This is a little like arguing that if imprinting in a baby duckling does not happen in the early critical period it can be done at a later time (it can't).

According to the theory behind AT, in order for this later attachment process to be successful, the child must first be subjected to physical "confrontation" and "restraint" in order to release supposedly repressed abandonment anger. The process repeats for as long as is necessary—

hours, days, even weeks—until the child is physically exhausted and emotionally reduced to an "infantile" state. Then the parents cradle, rock, and bottle-feed the child, implementing a "reattachment." This would be like taking a full-grown duck and attempting to reduce it back to its duckling stage through physical and emotional constraints, and then seeing if it will attach to its mother. That's the theory anyway. The practice resulted in something rather different . . . and deadly.

Candace was taken to Evergreen, Colorado, where she was treated by Connell Watkins, a nationally prominent attachment therapist and past clinical director for the Attachment Center at Evergreen, along with her associate Julie Ponder, a recently licensed family counselor from California. The treatment was conducted in Watkins's home and videotaped. According to trial transcripts, Watkins and Ponder conducted more than four days of "holding therapies," in which they grabbed or covered Candace's face 138 times, shook or bounced her head 392 times, and shouted into her face 133 times. When this failed to break her, they put the tiny sixty-eight-pound Candace inside a flannel sheet and covered her with sofa pillows, while several adults (with a combined weight of nearly seven hundred pounds) lay on top of her so that she could be "reborn." Ponder told Candace that she was "a teeny little baby" in the womb, commanding her to "come out head first" and "push with your feet." In response, Candace screamed, "I can't breathe, I can't do it! Somebody's on top of me. I want to die now! Please! Air!"

According to AT theory, Candace's reaction was a sign of her emotional resistance; she needed more confrontation to reach the rage necessary to "break through" the wall and achieve emotional healing. Putting theory into practice, Ponder admonished her: "You're gonna die." Candace begged, "Please, please, I can't breathe." Ponder instructed the others to "press more on top," on the premise that AD children exaggerate their distress. Candace vomited, then cried "I gotta poop." Her mother entreated, "I know it's hard but I'm waiting for you."

After forty minutes of this torture Candace went silent. Ponder rebuked her "Quitter, quitter!" Someone joked about performing a C-section, while Ponder patted a dog that meandered by. After thirty more minutes of silence, Watkins sarcastically remarked, "Let's look at this twerp and see what's going on—is there a kid in there somewhere? There you are lying in your own vomit—aren't you tired?"

Candace Newmaker was not tired; she was dead. "This ten-year-old child died of cerebral edema and herniation caused by hypoxic-ischemic encephalopathy," the autopsy report clinically stated. The proximate cause of Candace's death was suffocation, and her therapists received the minimum sentence of sixteen years for "reckless child abuse resulting in death." The ultimate cause was pseudoscientific quackery masquerading as psychological science. In their penetrating analysis of the case, *Attachment Therapy on Trial*, Jean Mercer, Larry Sarner, and Linda Rosa write: "However bizarre or idiosyncratic these treatments appear—and however ineffective or harmful they may be to children—they emerge from a complex internal logic, based, unfortunately, on faulty premises."[35]

These therapists killed Candace not because they were evil, but because they were in the grip of a pseudoscientific belief grounded in superstition and magical thinking. Hence, an extreme example of the power and the peril of patternicity, and the deadly force of belief-dependent realism.

5

Agenticity

LET US RETURN TO OUR ERSTWHILE HOMINID ON THE PLAINS OF Africa who hears a rustle in the grass, and the crucial matter of whether the sound represents a dangerous predator or just the wind. This is an important distinction on a number of levels, not the least of which is life or death, but take note that there is another difference: "wind" represents an *inanimate force* whereas "dangerous predator" indicates an *intentional agent*. There is a big difference between an inanimate force and an intentional agent. Most animals can make this distinction on the superficial (but vital) life-or-death level, but we do something other animals do not do.

As large-brained hominids with a developed cortex and a "theory of mind"—the capacity to be aware of such mental states as desires and intentions in both ourselves and others—we practice what I call *agenticity: the tendency to infuse patterns with meaning, intention, and agency.* That is, we often impart the patterns we find with agency and intention, and believe that these intentional agents control the world, sometimes invisibly from the top down, instead of bottom-up causal laws and randomness that makes up much of our world.[1] Souls, spirits, ghosts, gods, demons, angels, aliens, intelligent designers, government conspiracists, and all manner of invisible agents with power and intention are believed to haunt our world and control our lives. Combined with our propensity to find meaningful patterns in both meaningful and meaningless noise, patternicity and agenticity form the cognitive basis of shamanism, paganism, animism, polytheism, monotheism, and all modes of Old and New Age spiritualisms.[2] And much more. The Intelligent Designer is said to be an invisible agent who created life from the top down. Extraterrestrial intelligences are often portrayed as powerful beings coming down from on high to warn us of our impending self-destruction. Conspiracy theories predictably include hidden agents at work behind the scenes, puppet masters pulling political and economic strings as we dance to the tune of the Bilderbergers, the Rothschilds, the Rockefellers, or the Illuminati. Even the belief that the government can impose top-down measures to rescue the economy is a form of agenticity, with President Obama being hailed with almost messianic powers as "the one" who will save us.

There is now substantial evidence from cognitive neuroscience that humans readily find patterns and impart agency to them. In his 2009 book *Supersense*, University of Bristol psychologist Bruce Hood documented the growing body of data that demonstrates our tendency not only to infuse patterns with agency and intention, but to also believe that objects, animals, and people contain an essence—something that is at the core of their being that makes them what they are—and that this essence may be transmitted from objects to people, and from people to people. There are evolutionary reasons for this *essentialism*, rooted in fears about diseases and contagions that contain all-too-natural essences that can be deadly (and hence should be avoided), and thus there was a natural selection for those who avoided deadly diseases by following their instincts about essence avoidance. But we also generalize these essence emotions to both natural and supernatural beings, to any and all objects and people, and to things seen and unseen; we also assume that those seen and unseen objects and people have agency and intention. "Many highly educated and intelligent individuals experience a powerful sense that there are patterns, forces, energies, and entities operating in the world," Hood wrote. "More importantly, such experiences are not substantiated by a body of reliable evidence, which is why they are *supernatural* and *unscientific*. The inclination or sense that they may be real is our supersense."[3]

Examples of agenticity abound. Subjects watching reflective dots move about in a darkened room, especially if the dots take on the shape of two legs and two arms, infer that they represent a person or intentional agent. Children believe that the sun can think and follows them around, and when asked to draw a picture of the sun they often add a smiley face to give agency to it. Genital-shaped foods such as bananas and oysters are often believed to enhance sexual potency. A third of transplant patients believe that the donor's personality or essence is transplanted with the organ. Hood's research team conducted a study among healthy adults in which they first asked them to rate the faces of twenty people for attractiveness, intelligence, and how willing they would be to receive a heart transplant from each person. After these ratings were recorded, Hood then told the subjects that half of the people they had just rated were convicted murderers, then asked them to rerate the pictures. Tellingly, although the ratings of the murderers' attractiveness and intelligence decreased, the biggest drop of all was in the willingness to accept a heart from a murderer, which Hood concluded was due to the fear that some of the essence of evil might be transmitted to the recipient.[4] This finding corroborates the study that reveals most people say they would never wear the sweater of a murderer, showing great disgust at the very thought, as if some of the murderer's evil would rub off in the material of the sweater.[5]

By contrast, in a form of positive agenticity, most people say that they *would* wear the cardigan sweater of the children's television host Mr. Rogers, believing that wearing the sweater would make them a better person.[6] What is the deeper evolutionary basis of this essentialism? "If essences are thought to be transferable, we will not consider ourselves isolated individuals but rather members of a tribe potentially joined to each other through beliefs in supernatural connectedness," Hood suggested. "We will see others in terms of the properties that make them essentially different from us. Such an idea suggests that some essential qualities are more likely to be transmitted than others. Youth, energy, beauty, temperament, strength, and even sexual preference are essential qualities that we attribute to others."[7]

I caught myself in a moment of agenticity during a 2009 trip to Austin for a debate with creationists at the University of Texas. While in town I paid a visit to Lance Armstrong's famous bike shop Mellow Johnny's (so named because Americans butcher the pronunciation of *maillot jaune*, French for "yellow jersey"). In addition to numerous yellow jerseys hanging on the walls, on the showroom floor were several of Armstrong's bikes on which he won seven Tours de France. "People think these are replica bikes," the shop manager told me. "When I explain that these are the actual bikes on which Lance won the tour, they touch them like holy relics." I was amused by the example, but then promptly and without thinking, I purchased an array of Lance Armstrong cycling gear and during my debate that night I donned a pair of Lance Armstrong yellow-rimmed black socks and a "Livestrong" T-shirt underneath my suit. My rational brain does not for a moment believe that the essence of Armstrong's celebrated strength and endurance powered me through the three-hour event, yet for some odd reason I felt more confident. Perhaps, given the influence of belief-dependent realism and the power of placebo, I was a better debater that night. Who knows? There may be natural effects of such supernatural thinking.

We are natural-born supernaturalists, driven by our tendency to find meaningful patterns and impart to them intentional agency. Why do we do this?

Agenticity and the Demon-Haunted Brain

Five centuries ago demons haunted our world, with incubi and succubi tormenting their victims as they lay asleep in their beds. Two centuries ago spirits haunted our world, with ghosts and ghouls harassing their sufferers all hours of the night. For the past century aliens have haunted our world, with grays or greens molesting people in their sleep, delivering messages to them as they lay awake, or abducting them out of their beds and whisking them away to the mother ship for prodding and probing. Today people are undergoing out-of-body experiences (OBEs), floating above their beds, out of their bedrooms, and even off the planet into space.

What is going on here? Are these elusive creatures and mysterious phenomena in our world or in our minds? By now you know that I will argue that they are entirely in our heads, even while they are modified and tweaked by the culture into which we happened to be born. The evidence that brain and mind are one is now overwhelming. Consider the research by the Laurentian University neuroscientist Michael Persinger, who in his laboratory in Sudbury, Ontario, induces all of these events in volunteers by subjecting their temporal lobes to patterns of magnetic fields. Persinger uses electromagnets inside a modified motorcycle helmet (sometimes called the *God Helmet*) to produce *temporal lobe transients*— increases and instabilities in the neuronal firing patterns in the temporal lobe region just above the ears—in the brains of subjects. Persinger believes that the magnetic fields stimulate "microseizures" in the temporal lobes, often producing what can best be described as "spiritual" or "supernatural" episodes: the sense of a presence in the room, an out-of-body experience, bizarre distortion of body parts, and even profound religious feelings of being in contact with God, gods, saints, and angels. Whatever we call them, the process itself is an example of agenticity.

Why does this happen? Because, said Persinger, our "sense of self" is maintained by the left hemisphere temporal lobe. Under normal brain functioning this is matched by the corresponding systems in the right hemisphere's temporal lobe. When these two systems are out of sync, the left hemisphere interprets the uncoordinated activity as "another self" or a "sensed presence," because there can only be one self. Two "selves" are reconfigured as one self and one something else, which may be labeled as an angel, demon, alien, ghost, or even God. When the amygdala is involved in the transient events, said Persinger, emotional factors significantly enhance the experience, which, when connected to spiritual themes, can be a powerful force for intense religious feelings.[8]

Having read about Persinger's research I was naturally curious to know if his helmet would work its magic on a skeptic's brain. I had recently tried hypnosis for the first time in nearly two decades for a television series I cohosted for the Fox Family Channel called *Exploring the Unknown*.[9] In my far-less-skeptical early twenties, while training for the three-thousand-mile nonstop transcontinental bicycle Race Across America, I engaged the talents of a former fellow graduate student to teach me self-hypnosis so that I could learn to deal with pain and sleep deprivation. I was easily hypnotized, as evidenced in an ABC *Wide World of Sports* "Up Close and Personal" segment on me in which I was so deeply entranced that my hypnotist colleague had a hard time bringing me out (dramatically revealed on television). But during my *Exploring the Unknown* experience, I was much too anxious about what was going on in my brain during the hypnotic process and thereby negated its effects, leaving me in what was little more than a role-playing mode (which critics of hypnosis think is all that it is anyway). Would the same thing happen in Persinger's lab when they strapped me into the God Helmet, I wondered?

Articulate, smart, and media savvy, Persinger is an interesting character, famous for wearing 1970s-era three-piece suits everywhere he goes (including, allegedly, while mowing the lawn). His jargon-laden descriptions of his research make it hard to know when hypothesis and theory blend into speculation and conjecture. Since the early 1970s, Persinger has devoted his research to testing the hypothesis that paranormal experiences are illusions created by the brain. Tiny changes in brain chemistry or minute alterations in electrical activity can create powerful hallucinations that seem absolutely real. These misfirings of the brain can occur naturally due to external forces. For example, in his "Tectonic Strain Theory," Persinger speculates that earthquake activity may generate excessive electromagnetic fields that influence brains, which could go a long way to explaining the New Age nuttiness of earthquake-burdened Southern California.

I am skeptical of this hypothesis, given the fact that such fields weaken by the square of the distance: double the distance from the source and you receive only a quarter of the energy from it. I live in Southern California. Most earthquake centers are tens to hundreds of miles away from population centers, usually out in the deserts surrounding the Los Angeles basin. This strikes me as dramatically different from wearing a helmet that delivers electromagnetic fields from millimeters away. Whether such natural electromagnetic fields occur in strengths high enough to influence brains in the real world remains to be seen, but Persinger does it artificially in his lab. Data collected from these experiments have formed a foundation for computer simulations of paranormal encounters. "We know that all experience is derived from the brain," Persinger explained in my interview with him. "We also realize that subtle patterns generate complex human experiences and emotions. Thanks to computer technology, we extracted the electromagnetic patterns generated from the brain during these experiences and then re-exposed volunteers to those patterns."

After our interview it was time to run the experiment. A lab assistant strapped me into the helmet, hooked up the leads to my hands, chest, and scalp to measure brain waves, heart rate, and other physiological activity, and sealed me in a soundproof room where I plopped myself into a comfortable chair that could have been Archie Bunker's easy chair from the set of *All in the Family*. Persinger, his assistant, and the camera crew exited the chamber and I settled into cushioned bliss. A voice rang in announcing that the experiment would now begin. Magnetic fields washed over my temporal lobes. My initial reaction was a little giddiness, as if the whole process were a silly exercise that I could easily control, similar to my recent hypnosis experience. I also worried that I might accidentally fall asleep, so I tried to maintain alertness. But remembering how overthinking thwarted my hypnotic efforts, I cleared my head and allowed myself to slump into a state of willful suspension of disbelief. Minutes later, I felt a tug-of-war between the rational and emotional parts of my brain over whether the sense that I wanted to leave my body was real.

"What's happening to Michael now," Persinger explained to my producer during the first set of trials, "is he's being exposed to complex magnetic fields associated with opiate-like experiences such as floating and pleasantness and spinning." Halfway through the experiment Persinger's technicians fiddled with some dials to change the electromagnetic patterns. "At this point there is now another pattern being generated along the right hemisphere which tends to be associated with more terrifying experiences." Indeed, under these patterns volunteers have reported seeing the devil, being grabbed by aliens, and even being transported to hell. As I told Persinger in a postexperiment debriefing for the show, "In the first one, it felt like something went by me. . . . I wasn't sure if it was me leaving or somebody or something coming by me. It was very strange. Then, in the second round, there was the feeling of being in waves and that I wanted to come out of my body but I kept going back in. I can really see how if somebody was slightly more fantasy prone and tends to interpret environmental stimuli in a paranormal way, this kind of experience would be a real wild trip."[10]

Temporal lobe stimulation may not account for every encounter with the paranormal, but Persinger's research may be the first step toward demystifying a number of centuries-old puzzles. As he summed up for our show, "Four hundred years ago the paranormal included what in large part is science today. That's the fate of the paranormal—it becomes science, it becomes normal." Or, it simply disappears under the scrutiny of the scientific method.

Agents Who Stare at Goats

Belief in the paranormal is itself an extension of agenticity, as hidden powers are thought to emanate from powerful agents. During my graduate stint in experimental psychology in the 1970s, I saw on television the Israeli psychic Uri Geller bend cutlery and reproduce drawings using, so he said, psychic powers alone. For a while I kept an open mind to the possibility that such phenomena could be real, until I saw James "The Amazing" Randi on Johnny Carson's *Tonight Show*, where Randi used magic tricks to duplicate Geller's effects. (As Randi likes to say, "If Geller is bending spoons with psychic power he's doing it the hard way.") Randi bent spoons, replicated drawings, levitated tables, and even performed a psychic surgery. When asked about Geller's ability to pass the tests of professional scientists, Randi explained that scientists are not trained to detect trickery and intentional deception, the very art of magic.

Randi is right. I vividly recall a seminar that I attended in 1980 at the Aletheia Foundation in Grants Pass, Oregon, in which a holistic healer named Jack Schwarz impressed us by shoving a ten-inch sail needle through his arm with no apparent pain and only a drop of blood. Years later, and to my chagrin, Randi performed the same feat with the simplest of magic. I attended that seminar at the behest of a woman I was dating named Allison, an Oregonian brunette attractive in a New Ageish way, before the New Age fully blossomed in the 1980s. She wore natural-fiber dresses, flowers in her hair, and nothing on her feet. But what most intrigued me in our year of distance dating were Allison's spiritual gifts. I knew she could see through me metaphorically, but Allison also saw things that she said were not allegorical: body auras, energy chakras, spiritual entities, and light beings. One night she closed the door and turned off the lights in my bathroom and told me to stare into the mirror until my aura appeared. I stared vacuously into space. During a drive through the Oregon countryside late one cold night she pointed out spiritual beings dotting the landscape. I stared blankly into the dark. I tried to see the world as Allison did, but I couldn't. She could see invisible intentional agents but I could not. She was a believer and I was a skeptic. The difference doomed our relationship.

By 1995, just as the heyday of New Age codswallop was winding down, a story broke that for the previous quarter century the Central Intelligence Agency, in conjunction with the U.S. Army, had invested $20 million in a highly secret psychic spy program called *Stargate* (also *Grill Flame* and *Scanate*). Stargate was a Cold War project intended to close the "psi gap" (the psychic equivalent of the missile gap) between the United States and Soviet Union. The Soviets were training psychic spies, so we would as well. The story of Stargate—a form of agenticity at the CIA—reemerged while I was writing this chapter in the form of a feature film based on the book *The Men Who Stare at Goats* by British investigative journalist Jon Ronson. This is a *Through the Looking Glass*-

like story of what the CIA—operating through something called *Psychological Operations* (PsyOps)—was researching: invisibility, levitation, telekinesis, walking through walls, and even killing goats just by staring at them, with the ultimate goal of killing enemy soldiers telepathically. In one project, psychic spies attempted to use "remote viewing" to identify the location of missile silos, submarines, POWs, and MIAs from a small room in a run-down Maryland building. If these skills could be honed and combined, it was believed, perhaps military officials could zap remotely viewed enemy missiles in their silos.[11]

Initially, the Stargate story received broad media attention—including a special investigative report on ABC's *Nightline*—and made minor celebrities out of a few of the psychic spies, such as Ed Dames and Joe McMoneagle. As regular guests on Art Bell's pro-paranormal radio talk show *Coast to Coast*, the former spies spun tales that, had they not been documented elsewhere, would have seemed like the ramblings of paranoid delusionists. For example, Ronson connects some of the bizarre torture techniques used on prisoners at Guantánamo Bay, Cuba, and Iraq's Abu Ghraib prison with similar techniques employed during the FBI siege of the Branch Davidians in Waco, Texas. FBI agents blasted the Branch Davidians all night with such obnoxious sounds as screaming rabbits, crying seagulls, dentist drills, and (I'm not making this up) Nancy Sinatra's "These Boots Are Made for Walking." The U.S. military employed the same technique on Iraqi prisoners of war, replacing Sinatra's ballad with the theme song from the PBS kids' television series *Barney and Friends*—a tune many parents concur does become torturous with repetition.

One of Ronson's sources, none other than Uri Geller (of bent-spoon fame), led him to Major General Albert Stubblebine III, who directed the psychic spy network from his office in Arlington, Virginia. Stubblebine thought that with enough practice he could learn to walk through walls, a belief encouraged by Lieutenant Colonel Jim Channon, a Vietnam vet whose postwar experiences at such New Age meccas as the Esalen Institute in Big Sur, California, led him to found the "first earth battalion" of "warrior monks" and "Jedi knights." These warriors, according to Channon, would transform the nature of war by entering hostile lands with "sparkly eyes," marching to the mantra of "ohm," and presenting the enemy with "automatic hugs" (acts colorfully carried out by George Clooney's character in the film version of *The Men Who Stare at Goats*).

Disillusioned by the ugly carnage of modern war, Channon envisioned a battalion armory of machines that would produce "discordant sounds" (Nancy and Barney?) and "psycho-electric" guns that would shoot "positive energy" at enemy soldiers.

As entertaining as all this is, can anyone actually levitate, turn invisible, walk through walls, or view a hidden object remotely? No. Under controlled conditions, remote viewers have never succeeded in finding a hidden target with greater accuracy than random guessing. The occasional successes you hear about are due either to chance or suspect experimental conditions, such as when the person who subjectively assesses whether the remote viewer's narrative description matches the target already knows the target location and its characteristics. When both the experimenter and the remote viewer are blinded to the target, psychic powers vanish.

Herein lies an important lesson that I have learned in many years of paranormal investigations: what people remember happening rarely corresponds to what actually happened. Case in point: Ronson interviewed a martial arts teacher named Guy Savelli, who claimed that he was involved in the psychic spy program and had witnessed soldiers killing goats by staring at them, and that he himself had done so as well. But as the details of the story unfold we discover that Savelli was recalling, years later, what he remembered about a particular "experiment" with thirty numbered goats. Savelli randomly chose Goat 16 and gave it his best death stare. But he couldn't concentrate that day, so he quit the experiment, only to be told later that Goat 17 had subsequently died. End of story. No autopsy or explanation of the cause of death. No information about how much time had elapsed between the staring episode and death; the conditions of the room into which the thirty goats had been placed (temperature, humidity, ventilation, and so forth); how long the goats were in the room, and so forth. When asked for corroborating evidence of this extraordinary claim, Savelli triumphantly produced a videotape of another experiment where someone else supposedly stopped the heart of a goat. But the tape showed only a goat whose heart rate dropped from sixty-five to fifty-five beats per minute.

That was the extent of the empirical evidence of goat killing, and as someone who has spent decades in the same fruitless pursuit of phantom goats, I conclude that the evidence for the paranormal in general doesn't get much better than this. They shoot horses, don't they?

Telephoning Dead Agents

In the fall of 2008, I attended a paranormal conference in Pennsylvania where I was to deliver the keynote address, an odd juxtaposition if ever there was one—a skeptic of the paranormal lecturing about the nonexistence of ESP to a room full of self-proclaimed psychics, mediums, astrologers, tarot-card readers, palm readers, and spiritual gurus of all stripes. I figured the experience of hanging out with paranormal believers was worth the transcontinental trip, if for no other reason than to collect more data on why people believe in invisible powers and agents. I wasn't disappointed. The first session I attended was on talking to the dead. Of course, anyone can talk to the dead—it's getting the dead to talk back that is the hard part. And yet right there at the front of the room that is what appeared to be happening—the dead were talking back, through a small box on a table.

"Is Matthew there?" asked Cheyenne, an attractive blonde who was directing her voice toward the box, clearly assuming that her brother would come through from the other side.

"Yes," the speaker in the box squawked.

With a connection "validated," Cheyenne shakily continued: "Was the suicide a mistake?"

A voice crackled, "My death was a mistake."

With tears now cascading down her cheeks, Cheyenne asked to speak with her mother, and with the matrilineal connection made, Cheyenne sputtered out, "Do you see my children, your beautiful grandchildren?"

Mom replied, "Yes. I see the children."

Cheyenne's life-affirming messages were coming out of Thomas Edison's "telephone to the dead," or at least a facsimile of a rumored machine that by all accounts the great inventor never built. It was just one of many readings that day (at ninety dollars a pop) conducted by Christopher Moon, the ponytailed senior editor of *Haunted Times* magazine and HauntedTimes.com, a clearinghouse for all things paranormal.

I couldn't hear Cheyenne's brother, mother, or any other incorporeal spirits, until Moon interpreted the random noises emanating from the machine that, he explained to me, was created by a Colorado man named Frank Sumption. "Frank's Box," according to its inventor, "consists of a random voltage generator, which is used to tune an AM receiver module rapidly. The audio from the tuner (raw audio) is amplified and fed to an echo chamber, where the spirits manipulate it to form their voices." (See figure 7.) Apparently this is difficult for the dead to do, so Moon employs the help of "Tyler," a spirit "technician" on the "other side" who he calls upon to corral wayward spirits to within earshot of the receiver. What it sounds like to the untrained ear (that is to say, anyone not within earshot of Moon's interpretative voice) is the rapid twirling of a radio dial so that only noises and word and sentence fragments are audible.

"Are the dead in that little box?" I asked Moon.

"I don't know where the dead are. Another dimension probably," Moon conjectured unhelpfully.

"Well, since we know how easy it is for our brains to find meaningful patterns in meaningless noise," I continued, "how can you tell the difference between a dead person's real words and random radio noises that just sound like words?"

Surprisingly, Moon agreed with me: "You have to be very careful. We record the sessions and get consistency in what people hear."

I persist: "Consistency . . . as in what, ninety-five percent, fifty-one percent?"

"A lot," Moon rejoins.

Figure 7. Telephone to the Dead
"Frank's Box," also known as the "telephone to the dead," rumored to have been first invented by Thomas Edison, is today constructed by a Colorado man named Frank Sumption. PHOTOGRAPH BY THE AUTHOR.

"A lot, as in . . . ?" Our impromptu Q&A ended there, as the next session was about to start and I didn't want to miss the lecture on "Quantum Mechanics: Is It Proving the Existence of the Paranormal?" by another ponytailed speculator with the uni-name Konstantinos.

That evening in my keynote address I explained how "priming" the brain to see or hear something increases the likelihood that the percepts will obey the concepts. I played a portion of Led Zeppelin's "Stairway to Heaven," first forward with the words on the screen: *If there's a bustle in your hedgerow / Don't be alarmed now / It's just a spring clean for the May queen / Yes, there are two paths you can go by / But in the long run / There's still time to change the road you're on.* I joked that I'm not sure what the lyrics mean forward, but that when I was in high school they were deeply meaningful. Then I played this portion of the song backward with no words on the screen, and almost everyone heard "Satan," with some also hearing "sex" or "666." Finally, I played it again after priming their brains with the alleged lyrics on the screen. The auditory data jumped out of the visual primes, and everyone could now clearly hear: *Oh, here's to my sweet Satan / The one whose little path will make me sad / whose power is Satan / He'll give you / Give you 666 / There was a little tool shed where he made us suffer, sad Satan.*[12] The effect is stunning to audiences who, with their unprimed ears, can hear one or maybe two words, but when primed can make out the entire lyrical score.[13]

These are all examples of patternicity and agenticity, and the next day I put them to the test when Moon gave me a personal demo. With the telephone to the dead squawking away I tried to connect to my deceased father and mother, asking for any "validation" of a connection—name, cause of death . . . anything. I coaxed and cajoled. Nothing. Moon asked Tyler to intervene. Nothing. Moon said he heard something, but when I pressed him he came up with nothing. I willingly suspended my disbelief in hopes of talking to my parents whom I miss dearly. Nothing. I searched for any pattern I might find. Nothing. And that, I'm afraid, is my assessment of the paranormal. Nothing.

Agenticity and the Sensed-Presence Effect

One of the most effective means we have of understanding how the brain works is when it doesn't work well, when something goes wrong, or under extreme stress or conditions. As an example of the latter, there is a phenomenon well known among mountain climbers, polar explorers, isolated sailors, and endurance athletes called the "third-man factor," but what I call the *sensed-presence effect.* The sensed presence is sometimes described as a "guardian angel" that appears in extreme and unusual environments.[14] Particularly in life-and-death struggles for survival in these exceptionally harsh climes, or under unusual strain or stress, the brain apparently conjures up help for physical guidance or moral support. The descriptive phrase *third man* comes from T. S. Eliot's poem "The Waste Land":

> *Who is the third who always walks beside you?*
> *When I count, there are only you and I together.*
> *But when I look ahead up the white road*
> *There is always another one walking beside you*
> *Gliding wrapt in a brown mantle, hooded.*

In his footnotes to these lines, Eliot explained that they "were stimulated by the account of one of the Antarctic expeditions (I forget which, but I think one of Shackleton's): it was related that the party of explorers, at the extremity of their strength, had the constant delusion that there was *one more member* than could actually be counted."[15] In fact, in Sir Ernest Henry Shackleton's account it was a fourth man who accompanied the remaining three in the party: "It seemed to me often that we were four, not three." No matter, whether it is a third man, fourth man, angel, alien, or extra man, it is the sensed presence that interests us here because this is yet another example of the brain's capacity for agenticity; I shall refer to such companions as *sensed presences* and the process as the *sensed-presence effect.*

In his book *The Third Man Factor,* John Geiger lists the conditions that are associated with the generation of a sensed presence: monotony, darkness, barren landscapes, isolation, cold, injury, dehydration, hunger, fatigue, and fear.[16] To this list we can add sleep deprivation, which probably accounts for Charles Lindbergh's sensed presence during his transatlantic flight to Paris. During the historic journey, Lindbergh became aware that he had company in the cockpit of the *Spirit of St. Louis*: "The fuselage behind me becomes filled with ghostly presences—vaguely outlined forms, transparent, moving, riding weightless with me in the plane. I feel no surprise at their coming. There's no suddenness to their appearance." Most critically, these were not aberrations of the cockpit environment such as fog or reflected light because, as Lindbergh recounts, "Without turning my head, I see them as clearly as though in my normal field of vision." Lindbergh even heard "voices that spoke with authority and clearness," yet after the flight reported, "I can't remember a single word they said." What were these phantom beings doing there? They were there to help, "conversing and advising on my flight, discussing problems of my navigation, reassuring me, giving me messages of importance unattainable in ordinary life."[17]

The famous Austrian mountaineer Hermann Buhl, the first person to summit the 26,660-foot Nanga Parbat—the ninth highest peak in the world and known as "Killer Mountain" because of the number of climbers (thirty-one) who have perished there—suddenly found himself with company on his way back down, even though he was climbing alone: "Out on the Silbersattel I see two dots. I could shout with joy; now someone is coming up. I can hear their voices too, someone calls 'Hermann,' but then I realize that they are rocks on Chongra Peak that rises up behind. It is a bitter disappointment. I set off again subdued. This realization happens frequently. Then I hear voices, hear my name really clearly— hallucinations." Throughout the ordeal, in fact, Buhl said, "I had an extraordinary feeling, that I was not alone."[18]

Such accounts have become legion in climbing lore. The most famous climbing soloist in history (and the first to summit Mount Everest without bottled oxygen), Reinhold Messner, recalls having many conversations with imaginary companions during his expeditions into the thin air of the Himalayas. Linking the sensed-presence effect to beliefs more broadly, I was intrigued to read the account of climber Joe Simpson about what happened to him during the descent from the 20,814-foot summit of Siula Grande in the Peruvian Andes after an accident threatened his survival. As Simpson struggled to make it back to base camp, a second mind suddenly materialized in his head to give him aid and comfort. After determining that the voice was not emanating from his Walkman cassette player, Simpson decided that it was something else entirely: "The voice was clean and sharp and commanding. It was always right, and I listened to it when it spoke and acted on its decisions. The other mind rambled out a disconnected series of images, and memories and hopes, which I attended to in a daydream state as I set about obeying the orders of the voice."[19]

Consistent with belief-dependent realism and my thesis that belief comes first, explanation second, the self-declared atheist Simpson attributed his experience to a "sixth sense" that he figured was probably an evolutionary remnant from the distant past that he simply called "the voice." By contrast, in William Laird McKinlay's classic survival memoir, *The Last Voyage of the Karluk*, the deeply religious Arctic explorer described a sensed-presence experience that "filled me with an exultation beyond all earthly feeling. As it passed, and I walked back to the ship, I felt wholly convinced that no agnostic, no sceptic, no atheist, no humanist, no doubter, would ever take from me the certainty of the existence of God."[20] Indeed, as the psychologist James Allan Cheyne, an expert on the study of preternatural experiences, observed: "There is often a dual consciousness associated with the presence in which a hard-nosed realist is simultaneously aware that the presence is not real in the normal sense of the term, yet utterly compelling; so compelling, and persistent, that food may be offered to the presence in a casual and automatic manner."[21] That's the power of agenticity.

I've had many such experiences myself in association with the three-thousand-mile nonstop transcontinental bicycle Race Across America (RAAM), which in 1993 was ranked by *Outside* magazine as "the world's toughest sporting event" (based on such criteria as distance, course difficulty, pain and suffering, environmental conditions, dropout rate, recovery time, and other factors).[22] RAAM starts on the West Coast and ends on the East Coast, with competitors sleeping only when necessary and stopping as little as possible. The top cyclists complete the three-thousand-mile distance in eight and a half to nine days, averaging 325 to 350 miles a day and sleeping only about ninety minutes a night. Weather conditions vary from 120 degrees Fahrenheit in the California deserts to the low 30s and freezing over the Colorado Rockies. The pain from saddle sores and pressure points and the agony of fatigue are almost unbearable. There is no time to recover. The dropout rate of about two-thirds is a staggering testimony to the difficulty of this ultramarathon event, and in nearly three decades of racing fewer than two hundred people have earned the coveted RAAM ring. The Race Across America is a rolling experiment in physical exhaustion and psychological deterioration, which when coupled with sleep deprivation has produced some wild and wacky stories from the highways and byways of America. I know because I cofounded the race with three other men in 1982 and competed in it five times.

All RAAM riders have stories to tell about bizarre experiences they have had under these extraordinary conditions. I would often perceive clusters of mailboxes on the side of the road in the Midwest as cheering fans come out to root us on. Blotches in the pavement from minor road repairs looked like animals and mythical creatures. In the 1982 race Olympic cyclist John Howard told the ABC television camera crew: "The other day I saw about fifty yards of Egyptian hieroglyphics spread along the highway—craziest thing I've ever seen, but it was there!" In that same race John Marino recalled, "In the fog of Pennsylvania I was riding along and I visualized myself riding sideways in a fog tunnel. I put my hand down, stopped, got off the bike and sat down, then got back on the bike." In the 1986 race, Gary Verrill recounted his out-of-body-like experience: "After day three my consciousness was in a dream state. I was alert enough to carry on a conversation, but at the same time was viewing myself from another plane. The sensation was exactly like dreaming—the only difference was in the disappointment of not being able to wake up or control the dream."[23]

When I was the race director in the 1990s, I would routinely come across blurry-eyed cyclists in the middle of the night blathering on about guardian angels, mysterious figures, and assorted cabals and conspiracies against them. One night in Kansas (where Dorothy had her vision quest to Oz) I came across a RAAM rider standing next to some railroad tracks. When I asked him what he was doing he explained that he was waiting to take the train to see God. More recently, five-time winner Jure Robic witnessed asphalt cracks morph into coded messages, and hallucinated bears, wolves, and even aliens. A member of the Slovenian army, Robic once dismounted his bike to engage in combat a gaggle of mailboxes he was convinced were enemy troops, and another year he found himself being chased by a howling band of black-bearded horsemen. "Mujahedeen, shooting at me," Robic recalled. "So I ride faster."[24]

A sister event to RAAM is the one-thousand-mile nonstop Iditarod sled dog race from Anchorage to Nome, Alaska, in which mushers go for nine to fourteen days on minimum sleep, are alone except for their dogs, rarely see other competitors, and hallucinate horses, trains, UFOs, invisible airplanes, orchestras, strange animals, voices without people, and occasionally phantom people on the side of the trail or imaginary friends hitching a ride on their sleds and chatting them up during long lonely stretches. Four-time winner Lance Mackey recalled a day when he was riding the sled and saw a girl sitting by the side of the trail knitting. "She laughed at me, waved, and I went by her and she was gone. You just laugh."[25] A musher named Joe Garnie became convinced that a man was riding in his sled bag. He politely asked the man to leave, but the man didn't move. Garnie tapped him on the shoulder and insisted he depart his sled, and when the stranger refused Garnie swatted him.[26]

What is happening in the brain during an agent-filled sensed-presence experience? Because they happen under such differing environments, I strongly suspect that there is more than one cause. If it happened only at high altitude, for example, we might finger hypoxia as the suspect, but arctic explorers experience it at low altitudes. Perhaps it is freezing cold temperatures, but solo sailors and RAAM riders in warm climes sense presences. I suspect that extreme environmental conditions are a necessary but not sufficient explanation of the sensed-presence experience. Whatever its immediate cause (temperature, altitude, hypoxia, physical exhaustion, sleep deprivation, starvation, loneliness, fear), a deeper cause of the sensed-presence effect is to be found in the brain. I suggest four explanations: (1) an extension of our normal sense of presence of ourselves and others in our physical and social environments; (2) a conflict between the high road of controlled reason and the low road of automatic emotion; (3) a conflict within the body schema, or our physical sense of self, in which your brain is tricked into thinking that there is another you; or (4) a conflict within the mind schema, or our psychological sense of self, in which the mind is tricked into thinking that there is another mind.

1. *An extension of our normal sense of presence of ourselves and others in our physical and social environments.* This process of sensing a presence is probably just an extension of our normal expectations of others around us because we are such a social species. We have all lived with others, particularly in our formative childhood and teenage years, and we develop a sense of their presence whether they are there or not. Under normal conditions, you come home from school or work expecting your fellow family members to either be home or to arrive soon. You scan for telltale cues of cars or keys or coats. You listen for their familiar sounds of welcome. Their presence is either sensed or anticipated. For years after my mother died, whenever I visited my father at the home where I grew up I had this overwhelming feeling that she would come around the corner at any moment, even though my rational brain kept

correcting my emotional expectations. For eight years after my mother's passing my stepfather kept close company with his demonstrative black Lab Hudson, and whenever I stopped by the house Hud would always come running to greet me; even after he was gone I still felt like he'd come running to the door. So ingrained are these sensed-presence expectations that even years later, whenever I was in my ancestral home, I had the eerie feeling that my stepdad and I were not alone.

2. *A conflict between the high road of controlled reason and the low road of automatic emotion.*[27] Brain functions can be roughly divided into two processes: *controlled* and *automatic*. Controlled processes tend to use linear step-by-step logic and are deliberately employed, and we are aware of them when we use them. Automatic processes operate unconsciously, nondeliberately, and in parallel. Controlled processes tend to occur in the front (orbital and prefrontal) parts of the brain. The prefrontal cortex (PFC) is known as the executive region because it integrates the other regions for long-term planning. Automatic processes tend to occur in the back (occipital), top (parietal), and side (temporal) parts of the brain. The amygdala is associated with automatic emotional responses, especially fear. During extreme and unusual events there may be a competition between these controlled and automatic brain systems. As in the fight-or-flight response—in which blood flow is shunted toward the center of the body and away from the periphery, where cuts and gashes could result in death through blood loss—the high road of controlled reason begins to shut down due to oxygen deprivation, sleep deprivation, extreme temperatures, starvation, exhaustion, and the like. The body powers down higher functions in order to preserve the lower functions necessary for basic survival. In the course of normal, day-to-day living, these controlled circuits of reason keep our automatic circuits of emotions in check, and we do not just give in to every whim and impulse. But remove the rational governor and the emotional machinery begins to spin out of control.

Research shows, for example, that at low levels of stimulation, emotions appear to play an advisory role, carrying additional information to the decision-making areas of the brain along with inputs from higher-order cortical regions of the brain. At medium levels of stimulation, conflicts can arise between high-road reason centers and low-road emotion centers. At high levels of stimulation (as in extreme environmental conditions and physical and mental exhaustion), low-road emotions can so overrun high-road cognitive processes that people can no longer reason their way to a decision; they report feeling "out of control" or "acting against their own self-interest."[28] Perhaps this is when the brain calls forth the sensed-presence companion.

3. *A conflict within the body schema, or our physical sense of self, in which your brain is tricked into thinking that there is another you.* Remember, the primary function of the brain is to run the body, which mostly involves sending and receiving signals from muscles, tendons, tissues, and organs. What we think of as our exalted mind capable of higher-order functions, such as aesthetic appreciation, mathematical computation, or philosophical speculation, is a result of the cerebral cortex that sits atop the massive structure of the brain that mostly concerns itself with countless other mundane and subconscious processes that make a living body possible. As such, your brain develops an overarching portrait of your body, from your toes and fingers through your legs and arms and right into your torso and up your back to the top of your head. This is your body schema, and it extends beyond the body into the world when your thinking engages with other people through language, when you write something down on paper or type it into a computer, or perform any other extended reach from inside your head to outside your body. This is sometimes called *embodied cognition*, the *extended mind*, or, in the philosopher Andy Clark's apt descriptor, "supersizing the mind."[29] Physically touching someone is a mind extension, and if they touch you back it creates a feedback loop. Language was the first evolved form of extended mind, and the written word extended language even further, as did the printing press, printed books, and newspapers. Most recently, radio, television, and especially the Internet have supersized the brain and extended the mind throughout the globe and even into space.

This body schema is you, and there is only one of you.[30] If for any reason your brain is tricked (or altered or damaged) into thinking that there is another you—an internal doppelgänger—this inevitably conflicts with your single-body schema. To adjust for this anomaly, your brain constructs a plausible explanation for this other you: it is actually someone or something else: a noncorporeal entity or soul coming out of your body (as in the out-of-body experience), or that there is another person—a sensed presence—nearby.

The mismatch between your body schema and the artificially induced doppelgänger probably occurs between the parietal lobe and the temporal lobe in your brain. Specifically, it is the job of the *posterior superior parietal lobe* to orient your body in physical space (the back and upper regions of this lobe sit above and to the rear of the temporal lobe above your ears). This is the part of the brain that can tell the difference between you and not-you, which is to say everything else outside of your body. When this part of the brain is quiescent during deep meditation and prayer (as witnessed in brain scan studies), subjects (Buddhist monks and Catholic nuns) report feeling at one with the world or in deep contact with the transcendent.[31] In a manner of speaking, meditation and prayer have created a mismatch between the body schema and the world, and it is possible that something like this happens under extreme and unusual conditions.

Phantom limbs are another perceptual mismatch. At the University of California–San Diego, the neuroscientist V. S. Ramachandran ("Rama") has used the concept of the body schema to treat phantom pain in patients who have lost an arm. Essentially, these patients are suffering from a limb schema mismatch, with their eyes reporting that there is no limb while their body schema still maintains the limb image. Why this should result in pain is not clear. Rama suggests several explanations, including irritation of the nerve endings, central remapping (leading to referred sensations) in which "some low threshold touch input might cross-activate high threshold pain neurons," and a "mismatch between motor commands and the 'expected' but missing visual and proprioceptive input" that "may be perceived as pain."[32] Whatever the cause, the patient's brain sends a signal to the phantom arm to move, but the signal sent back to the brain is that it can't move (patients report feeling as if their arm is "stuck in cement" or "frozen in a block of ice"), and thus there is a "learned paralysis." To correct the mismatch, Rama constructed a mirror box. The patient inserted his left phantom arm into one side of the box behind the mirror and his right intact arm into the other side. The mirror reflected the right whole arm as a mirror image of the left phantom arm. Rama then had the man wiggle the fingers of the right arm, which sent signals back to his brain that the phantom arm was moving, thereby overriding the learned paralysis and leading to a dramatic reduction of phantom pain.[33]

Phantom limbs, body schemas, and visual and auditory hallucinations are all neural correlates of the dualistic stance that mind and body exist as separate agents both in ourselves and others, and we thereby attribute intentional agency not only to real others but to phantom others as well.

4. *A conflict within the mind schema, or our psychological sense of self, in which the mind is tricked into thinking that there is another mind.* Our brains consist of many independent neural networks that at any given moment are working away at various problems in daily living. And yet we do not feel like we're a bundle of networks. We feel like a single mind in one brain. The neuroscientist Michael Gazzaniga thinks that we have a neural network that coordinates all the other neural networks and weaves them together into a whole. He calls this the *left-hemisphere interpreter,* the brain's storyteller that puts together countless inputs into a meaningful narrative story. Gazzaniga discovered this network while studying split-brain patients whose hemispheres have been separated to stop the spread of epileptic seizures. In one experiment, Gazzaniga presented the word *walk* to only the right hemisphere of a split-brain patient, who promptly got up and started walking. When asked why, his left-hemisphere interpreter made up a story to explain this behavior: "I wanted to go get a Coke."

We often learn how the brain works from when it doesn't work right. Gazzaniga notes, for example, that patients with "reduplicative paramnesia" believe that there are copies of people or places. They mix these copies up into one experience or story that makes perfect sense to them even if it sounds ridiculous to everyone around them. "One such patient believed the New York hospital where she was being treated was actually her home in Maine," Gazzaniga recalled. "When her doctor asked how this could be her home if there were elevators in the hallway, she said, 'Doctor, do you know how much it cost me to have those put in?' The interpreter will go to great lengths to make sure the inputs it receives are woven together to make sense—even when it must make great leaps to do so. Of course, these do not appear as 'great leaps' to the patient, but rather as clear evidence from the world around him or her."[34] This is, in part, what I mean by *patternicity* and *agenticity,* although these are just descriptive terms for a cognitive process. What we really want to know is what the neural correlates are for this process, and for the generation of sensed presences and other forms of ephemeral agenticities. This left-hemisphere interpreter is a good candidate for where it happens.

My brother-in-law Fred Ziel, who has climbed many of the highest and most dangerous peaks of the Himalayas, tells me that he has twice experienced a sensed presence. The first time was when he was frost-bitten and without oxygen at the limit of physical effort above the Hillary Step, the last hurdle on Mount Everest's southeast ridge. The second time was on Everest's north ridge after he collapsed from dehydration and hypoxia (oxygen deprivation) at twenty-six thousand feet. Both times he was alone and wishing he had a companion, which his brain obligingly provided. Tellingly, when I asked his opinion as a medical doctor on possible hemispheric differences to account for such phenomena, Fred noted, "Both times the sense was on my right side, perhaps related to being left-handed." Neuroscientists believe that our "sense of self" is located primarily in the temporal lobe of the left hemisphere, and that our divided brain means that left- and right-brain circuits are criss-crossed so that, for example, the right visual field is registered in the left hemisphere's visual cortex. Perhaps the oxygen deprivation at twenty-six thousand feet, or the bitter cold, or the pain of frostbite, or the feeling of being abandoned and alone—or some combination thereof—triggered the left temporal lobe in Fred's brain to generate "another self." Since the brain has only one body and one mind schema—one self—a second self can be perceived only as another being outside the body, a sensed presence nearby.

The sensed presence may be the left-hemisphere interpreter's explanation for right-hemisphere anomalies. Or there may be neural network conflicts in body or mind schemas. Or it may be loneliness and fear extending our normal sensed presence of real others into imagining ephemeral companions. Whatever its cause, the fact that it happens under so many different conditions tells us that the presence is inside the head and not outside the body.

⚬

These examples of and explanations for superstition and magical thinking, rooted in association learning, theory of mind, sensed presences, the supersense, and the like—under the rubric of patternicity and agenticity—are not causal explanations per se. Labeling a cognitive process is a heuristic to help us get our minds around a problem to be solved or a mystery to be explained, but they are only labels, in the same way that calling a set of hallucinatory symptoms *schizophrenia* explains the cause of those symptoms. We need to bore deeper into the brain itself to understand the ultimate nature of belief and the true cause of our tendency to find meaningful patterns in meaningful and meaningless noise, and to infuse those patterns with meaning, intention, and agency. The actions of neurons in the brain is where we will find the ultimate causal explanation.

6

The Believing Neuron

ALL EXPERIENCE IS MEDIATED BY THE BRAIN. THE MIND IS WHAT THE brain does. There is no such thing as "mind" per se, outside of brain activity. *Mind* is just a word we use to describe neural activity in the brain. No brain, no mind. We know this because if a part of the brain is destroyed through stroke or cancer or injury or surgery, whatever that part of the brain was doing is now gone. If the damage occurs in early childhood when the brain is especially plastic, or in adulthood in certain parts of the brain that are conducive to rewiring, then that brain function—that "mind" part of the brain—may be rewired into another neural network in the brain. But this process just further reinforces the fact that without neural connections in the brain there is no mind. Nevertheless, fuzzy explanations for mental processes are still commonly employed.

Force Mental: A Nonexplanation for Mind

When I was a psychology undergraduate at Pepperdine University we were required to take a course called *physiological psychology*, which today would be called *cognitive neuroscience*. It turned out to be a real eye-opener for me in the study of the mind because our professor— Darrell C. Dearmore, one of the clearest expositors of science I've ever had—bore deep into the core of the brain to reveal the foundational structure of all thought and action: the neuron. Before I understood how the neuron works, I was satisfied with fuzzy-word explanations for what was going on inside people's heads, such as "thinking" or "processing" or "learning" or "understanding" together collected under the rubric of "mind," as if these were causal accounts for brain processes. They are not. They are just words to describe a process, which itself needs a deeper explanation.

In the early twentieth century the British biologist Julian Huxley parodied the French philosopher Henri Bergson's fuzzy explanation for life as being caused by an *élan vital* (vital force), which Huxley said was like explaining a railroad steam engine as being driven by its *élan locomotif* (locomotive force). Richard Dawkins brilliantly employed a similar analogy to parody intelligent design explanations for life. To say that the eye, or the bacteria flagellum, or DNA are "designed" tells us nothing. Scientists want to know *how* they were designed, what *forces* were at work, how the *process* of development unfolded, and so on. Dawkins imagined a counterfactual history in which Andrew Huxley and Alan Hodgkin, winners of the Nobel Prize for figuring out the molecular biophysics of the nerve impulse, in a creationist worldview attributed it to "nervous energy."[1]

Inspired by Dawkins's satirical dialogue, imagine if David Hubel and Torsten Wiesel—winners of the 1981 Nobel Prize for their pioneering research in brain circuitry and determining the neurochemistry of vision—had, instead of spending years getting down to the cellular and molecular level of understanding how the brain converts photons of light into neural impulses, simply attributed the process to *force mental*.

"Now see here, Hubel, this business about how photons of light are transduced into neural activity is a dreadfully thorny problem. I just can't understand how it works, can you?"

"No, my dear Wiesel, I can't, and implanting those electrodes into monkeys' brains is truly unpleasant and messy, and I have the hardest time getting the electrode into the right spot. Why don't we just say that the light is converted into a nerve impulse by *force mental*?"

What would *force mental* explain? Nothing. It would be like describing your automobile's engine as operating by *force combustion*, which fails to capture what is actually going on inside the cylinders of an internal combustion engine: a piston compresses a vaporous mixture of gasoline and air that is ignited by a spark plug causing an explosion that drives the piston down thereby turning a crank arm that is connected to a drive shaft that is linked to a differential that rotates the wheels.

This is what I mean when I say that the mind is what the brain does. The neuron and its actions are to psychology what the atom and gravity are to physics. To understand belief we have to understand how neurons work.

Synaptic States and Believing Neurons

The brain consists of about a hundred billion neurons of several hundred types, each of which contains a cell body, a descending axon cable, and numerous dendrites and axon terminals branching out to other neurons in approximately a thousand trillion synaptic connections between those hundred billion neurons. These are staggering numbers. A hundred billion neurons is 10^{11}, or a 1 followed by 11 zeros: 100,000,000,000. A thousand trillion connections is a quadrillion, or 10^{15}, or a 1 followed by 15 zeros: 1,000,000,000,000,000. The number of neurons in a human brain is about the same number of stars in the Milky Way galaxy—literally an astronomical number! The number of synaptic connections in the brain is equivalent to the number of seconds in 30 million years. Think about that for a moment. Start counting seconds as "one one thousand, two one thousand, three one thousand. . . ." When you get to 86,400 that is the number of seconds in a day; when you reach 31,536,000 that is the number of seconds in a year; and when you finally reach one trillion seconds you will have been counting for about 30,000 years; now, do that 30,000-year counting block one thousand more times and you will have counted the number of synaptic connections in your brain.

Large neuronal counts do generate greater computational power to be sure (like adding more processor chips or memory cards to your computer), but the action is in the individual neurons themselves. Neurons are elegantly simple and yet beautifully complex electrochemical information-processing machines. Inside a resting neuronal cell there is more potassium than there is sodium, and a predominance of anions—negatively charged ions—gives the inside of the cell a negative charge. Depending on which type of neuron it is, if you put a tiny electrode inside the neuronal cell body in a resting state it would read −70 mv (a millivolt is equal to one-thousandth of a volt). In this resting state the cell wall of the neuron is impermeable to sodium but permeable to potassium. When the neuron is stimulated by the actions of other neurons (or the electrical machinations of curious neuroscientists with electrodes), the permeability of the cell wall changes, allowing sodium to enter and thereby shift the electrical balance from −70 mv toward 0. This is called the *excitatory postsynaptic potential*, or EPSP. The *synapse* is the tiny gap between neurons, so *postsynaptic* means the neuron on the receiving end of the signal that travels across the synaptic cleft is the one being excited to reach its potential to fire. By contrast, if the stimulation comes from inhibitory neurons it causes the voltage to shift downward from −70 mv to −100 mv, making the neuron less likely to fire, and this is called the *inhibitory postsynaptic potential*, or IPSP. Although there are hundreds of different types of neurons, we can classify most of them as either excitatory or inhibitory in their actions.

If there are enough EPSPs built up (from numerous neuronal firings in sequence or from multiple connections from many other neurons) for the permeability of the neuron cell wall to reach a *critical point*, sodium rushes in, causing an instant spike in voltage to +50 mv, which spreads throughout the cell body and cascades down the axon into the terminals. Just as quickly, the neuron's voltage collapses back down to −80 mv, then returns to the −70 mv resting state. This process of the cell wall becoming permeable to sodium with a corresponding shift in voltage from negative to positive that travels down the axon to the dendrites and their synaptic connections to other neurons is called an *action potential*. More colloquially, we say that the cell "fired." The buildup of EPSPs is called *summation*, and there are two types: (1) *temporal summation*, in which two EPSPs from a single neuron are enough for the receiving neuron to reach its critical point and fire; and (2) *spatial summation*, in which two EPSPs from two different neurons arrive at the same time and are enough for the receiving neuron to reach its critical point and fire. This electrochemical change of voltage spike and sodium permeability propagates down the axon sequentially from the cell body to the axon terminals, and this is called, appropriately enough, *propagation*. The speed of propagation depends on two conditions: (1) the diameter of the axon (the bigger the faster), and (2) the myelination of the axon (the more myelin sheath there is covering and insulating the axon the faster the propagation of the impulse down it).[2]

Note that if the critical point for the neuron to fire is not reached, then it does not fire; if the critical point is reached then the neuron does fire. It's an on-or-off, all-or-nothing system. Neurons do not fire "soft" in response to a weak stimulus, nor do they fire "hard" in response to a strong stimulus. They either fire or they do not fire. Therefore, neurons communicate information in one of three ways: (1) *firing frequency* (the number of action potentials per second), (2) *firing location* (which neurons fire), and (3) *firing number* (how many neurons fire). In this way, it is said that neurons are binary in action, analogous to the binary digits of a computer—1 and 0—which correspond to an "on" or "off" signal being passed along a neural pathway or not. If we consider these neuronal on-or-off states as a type of mental state, with one neuron giving us two mental states (on or off), then there are 2×10^{15} possible choices available to the brain in processing information about the world and the body it is running. Since we take in only a tiny fraction of this number, the brain—for all intents and purposes—is an infinite information-processing machine.

How do individual neurons and their action potentials create complex thoughts and beliefs? It begins with something called neural *binding*. A "red circle" would be an example of two neural network inputs ("red" and "circle") bound into one percept of a red circle. Downstream neural inputs, such as those closer to muscles and sensory organs, converge as they move upstream through *convergence zones*, which are brain regions that integrate information coming from various neural inputs (eyes, ears, touch, and so forth) so that what you end up experiencing is a whole object instead of countless fragments of an image. In the upside-down image of President Obama presented in chapter 4, we initially see the integrated face holistically, and only later do we begin to notice that there is something wrong with the eyes and mouth; as explained, this is due to two different neural networks operating at different speeds—the whole face percept first, then the parts of the face second.

Binding involves so much more than this, however. There may be hundreds of percepts streaming into the brain from the various senses that must be bound together for higher brain regions to make sense of it all. Large brain areas such as the cerebral cortex coordinate inputs from smaller brain areas such as the temporal lobes, which themselves collate neural events from still smaller brain modules such as the fusiform gyrus (for facial recognition). This reduction continues all the way down to the single neuron level, where highly selective neurons—sometimes described as "grandmother" neurons—fire only when subjects see someone they know. There are neurons that fire only when an object moves left to right across your visual field. There are other neurons that fire only when an object moves right to left across your visual field. And there are still other neurons that have an action potential only when they receive EPSP inputs from other neurons that fire in response to diagonal-moving objects in your visual field. And so on up the networks goes the binding process. There are even neurons that fire only when you see someone you recognize. Caltech neuroscientists Christof Koch and Gabriel Kreiman, in conjunction with UCLA neurosurgeon Itzhak Fried, for example, have even found a single neuron that fires when the subject is shown a photograph of Bill Clinton and no one else. Another one fires only when shown Jennifer Aniston, but not a photograph of her and Brad Pitt.[3]

Of course, we are not aware of the workings of our own electrochemical systems. What we actually experience is what philosophers call *qualia*, or subjective states of thoughts and feelings that arise from a concatenation of neural events. But even qualia is itself a type of neural binding effect, integrating inputs from countless other neural networks downstream. It really does all come down to the electrochemical process of neuronal action potentials, or neurons firing and communicating with one another, passing information along the way. How do they do this? More chemistry.

Communication between neurons happens in that impossibly tiny synaptic cleft between neurons. When the action potential of a neuron rushes down the axon and reaches the terminals it triggers the release of tiny packets of *chemical transmitter substances* (CTS) into the synapse. When taken up by the connecting neuron, the CTS act as an EPSP to change the voltage and permeability of the postsynaptic neuron, thereby causing it to fire and propagate its action potential down its axon to its terminals to release its CTS into the next synaptic gap, and so on down the line in a neural network. When you stub your toe the pain signal travels along a circuitry from the pain receptors in the tissues in your toe all the way up to the brain, which registers the pain and processes the signal to other areas of the brain that send additional signals to contract muscles to pull your foot away from the offending object, all at a speed that feels almost instantaneous.

There are many types of CTS. The most common are known as the *catecholamines* and include *dopamine, norepinephrine (noradrenaline)*, and *epinephrine (adrenaline)*. The CTS act like keys for the locks on the postsynaptic neuron. If the key fits and turns, the neuron fires; if it doesn't the door remains closed and the postsynaptic neuron quiet. After the firing process occurs, most unused CTS go back to the presynaptic neuron where it is either reused or destroyed by monoamine oxidase (MAO) in a process called *Uptake I.* If there are too much CTS floating around in the synaptic gap, then the rest gets sucked up into the post-synaptic neuron in a process called *Uptake II.*

Drugs act on synapses and the release of CTS and subsequent uptake processes. Amphetamines, for example, speed up the release of CTS into the synapse, thereby accelerating the neural communication process—that's why they're called *speed.* Reserpine, once commonly prescribed for psychosis, breaks up CTS vesicles in the presynaptic neuron so that the MAO destroys them before they are used, thereby slowing neural networks and controlling mania, hypertension, and other symptoms of an overactive nervous system. Cocaine blocks Uptake I so that the CTS just stay in the synapse and keep the neurons firing away at an accelerated rate, cranking up neural networks into a frenzied state—think Robin Williams with a microphone and an audience; in point of fact, Williams attributes much of his manic comedy in the 1980s to his cocaine addiction. As one of the most common CTS, dopamine is critical to the smooth communication between neurons and muscles, and when there isn't enough of it patients lose motor control and shake uncontrollably. This is called Parkinson's disease, one treatment for which is L-dopa, a dopa-mine agonist that stimulates the production of more dopamine.

How do we build a system from the bottom up, starting with a chemical transmitter substance such as dopamine, and bind the inputs into an integrated belief system? Through behavior. Remember, the primary function of the brain is to run the body and help it survive. One way it does that is through association learning, or patternicity. This is the link from neuronal action potentials to human action.

Dopamine: The Belief Drug

Of all the chemical transmitter substances sloshing around in your brain, it appears that dopamine may be the most directly related to the neural correlates of belief. Dopamine, in fact, is critical in association learning and the reward system of the brain that Skinner discovered through his process of operant conditioning, whereby any behavior that is reinforced tends to be repeated. A reinforcement is, by definition, something that is rewarding to the organism; that is to say, it makes the brain direct the body to repeat the behavior in order to get another positive reward. Here is how it works.

In the divided brain stem—one of the most evolutionarily ancient parts of the brain shared by all vertebrates—there are pockets of roughly fifteen thousand to twenty-five thousand dopamine-producing neurons on each side that shoot out long axons connecting to other parts of the brain. These neurons stimulate the release of dopamine whenever it is determined that a received reward is more than expected, which causes the individual to repeat the behavior. The release of dopamine is a form of information, a message that tells the organism "Do that again." Dopamine produces the sensation of pleasure that accompanies mastering a task or accomplishing a goal, which makes the organism want to repeat the behavior, whether it is pressing a bar, pecking a key, or pulling a slot machine lever. You get a hit (a reinforcement) and your brain gets a hit of dopamine. *Behavior—Reinforcement—Behavior. Repeat sequence.*

The dopamine system, however, has its pluses and minuses. On the positive side, dopamine has been linked to a peanut-sized bundle of neurons in the middle of the brain called the *nucleus accumbens* (NAcc), which is known to be associated with reward and pleasure. In fact, dopamine appears to fuel this so-called pleasure center of the brain that has been implicated in the "high" derived from both cocaine and orgasms. This "pleasure center" was discovered in 1954 by James Olds and Peter Milner of McGill University, when they accidentally implanted an electrode into the NAcc of a rat and discovered that the rodent became very energized. They then set up an apparatus so that whenever a rat pressed a bar it generated a small electrical stimulation to the area. The rats pressed the bar until they collapsed, even to the point of forgoing food and water.[4] The effect has since been found in all mammals tested, including people who have undergone brain surgery and had their NAcc stimulated. The word they used to describe the effect was *orgasm*.[5] Now *that* is the type specimen of a positive reinforcement!

Unfortunately, there's a downside to the dopamine system, and that is addiction. Addictive drugs take over the role of reward signals that feed into the dopamine neurons. Gambling, pornography, and drugs such as cocaine cause the brain to flood itself with dopamine in response. So, too, do addictive ideas, most notably addictive *bad* ideas, such as those propagated by cults that lead to mass suicides (think Jonestown and Heaven's Gate), or those propagated by religions that lead to suicide bombing (think 9/11 and 7/7).

An important caveat about dopamine: neuroscientists make a distinction between "liking" (pleasure) and "wanting" (motivation), and there is a lively debate about whether dopamine acts to stimulate pleasure or to motivate behavior. A positive reinforcement may lead to behavioral repetition because it feels good (liking, or the pure pleasure of getting the reward) or because it will feel bad if the behavior isn't repeated (wanting, or motivation to avoid the anxiety of not getting the reward). The first reward is related to the pure pleasure of, say, an orgasm, whereas the second is related to the anxiety addicts feel when their next fix is in doubt. The research I cited above supports the pleasure thesis, but new research has scientists leaning toward the motivation position.[6] The UCLA neuroscientist Russell Poldrack told me that, based on this new data, he suspects "the role of dopamine is in motivation rather than in pleasure per se, whereas the opioid system appears to be central to pleasure." He points out, for example, "you can block the dopamine system in rats and they will still enjoy rewards, but they just won't work to get them."[7] It is a subtle but important distinction, but for our purposes in understanding the neural correlates of belief, the central point is that dopamine reinforces behaviors and beliefs and patternicity, and thus it is one of the primary belief drugs.

The connection between dopamine and belief was established by experiments conducted by Peter Brugger and his colleague Christine Mohr at the University of Bristol in England. Exploring the neurochemistry of superstition, magical thinking, and belief in the paranormal, Brugger and Mohr found that people with high levels of dopamine are more likely to find significance in coincidences and pick out meaning and patterns where there are none. In one study, for example, they compared twenty self-professed believers in ghosts, gods, spirits, and conspiracies to twenty self-professed skeptics of such claims. They showed all subjects a series of slides consisting of people's faces, some of which were normal while others had their parts scrambled, such as swapping out eyes or ears or noses from different faces. In another experiment, real and scrambled words were flashed. In general, the scientists found that the believers were much more likely than the skeptics to mistakenly assess a scrambled face as real, and to read a scrambled word as normal.

In the second part of the experiment, Brugger and Mohr gave all forty subjects L-dopa, the drug used for Parkinson's disease patients that increases the levels of dopamine in the brain. They then repeated the slide show with the scrambled or real faces and words. The boost of dopamine caused both believers and skeptics to identify scrambled faces and real and jumbled words as normal. This suggests that patternicity may be associated with high levels of dopamine in the brain. Intriguingly, the effect of L-dopa was stronger on skeptics than believers. That is, increased levels of dopamine appear to be more effective in making skeptics less skeptical than in making believers more believing.[8] Why? Two possibilities come to mind: (1) perhaps the dopamine levels of believers are already higher than those of skeptics and so the latter will feel the effects of the drug more; or (2) perhaps the patternicity proclivity of believers is already so high that the effects of the dopamine are lower than those of skeptics. Additional research shows that people who profess belief in the paranormal—compared to skeptics—show a greater tendency to perceive "patterns in noise,"[9] and are more inclined to attribute meaning to random connections they believe exist.[10]

Finding the Signal in the Noise

What is it that dopamine does, exactly, when it enhances belief? One theory—promulgated by Mohr, Brugger, and their colleagues—is that dopamine increases the signal-to-noise ratio (SNR), that is, the amount of signal your brain will detect in background noise.[11] This is the error-detection problem associated with patternicity. The signal-to-noise ratio is, in essence, a problem in patternicity—finding meaningful patterns in both meaningful and meaningless noise. The SNR is the proportion of patterns that your brain detects in the background noise, whether or not the patterns are real. How does dopamine affect this process?

Dopamine enhances the ability of neurons to transmit signals between one another. How? By acting as an agonist (as opposed to antagonist), or a substance that enhances neural activity. Dopamine binds to specific receptor molecule sites on the synaptic clefts of the neurons, as if it were the CTS that normally bind there.[12] It increases the rate of neural firing in association with pattern recognition, which means that synaptic connections between neurons are likely to increase in response to a perceived pattern, thereby cementing those perceived patterns into long-term memory through the actual physical growth of new neural connections and the reinforcement of old synaptic links.

Increasing dopamine increases pattern detection; scientists have found that dopamine agonists not only enhance learning but in higher doses can also trigger symptoms of psychosis, such as hallucinations, which may be related to that fine line between creativity (discriminate patternicity) and madness (indiscriminate patternicity). The dose is the key. Too much of it and you are likely to be making lots of Type I errors—false positives—in which you find connections that are not really there. Too little and you make Type II errors—false negatives—in which you miss connections that are real. The signal-to-noise ratio is everything.

Patternicity in the Brain

In his Pulitzer Prize–winning book *The Dragons of Eden*, Carl Sagan conjectured where in the brain superstition and magical thinking are likely to be found: "There is no doubt that right-hemisphere intuitive thinking may perceive patterns and connections too difficult for the left hemisphere; but it may also detect patterns where none exist. Skeptical and critical thinking is not a hallmark of the right hemisphere."[13] In an extension of the experiment by Susan Blackmore discussed in chapter 4, in which she found a difference between believers and skeptics on the propensity to find meaningful patterns in meaningless noise, Peter Brugger presented random dot patterns in a divided visual field paradigm so that either the left hemisphere (via the right visual field) or the right hemisphere (via the left visual field) of the brain was exposed to the image. (Recall that our brains are split down the middle and divided into two hemispheres connected in the middle at the *corpus callosum*; inputs from the left side of the body go to the right hemisphere and inputs from the right side of the body to the left hemisphere.) Brugger found that his subjects perceived significantly more meaningful patterns in the right hemisphere than in the left hemisphere, and this happened for both believers and skeptics.[14]

Subsequent studies found hemispheric differences between believers and skeptics. In one study, Brugger's team had blindfolded subjects hold a rod in their hands and physically estimate its middle point. Subjects were also given the Magical Ideation Scale questionnaire, which measures paranormal beliefs and experiences. What the scientists found is bizarre: believers in the paranormal estimated the middle point of the rod more to the left of center, which means that their right hemispheres were influencing their perception of space and distance. Brugger's lab then ran another experiment in which strings of letters forming either a word or nonsense were presented to the left visual field and the right visual field, instructing the subjects to respond when they recognized a word. The subjects also rated their belief in ESP on a six-point scale. Results: skeptics had greater left hemispheric dominance compared to believers, and believers had superior right hemispheric performances compared to skeptics. Adding EEG measures to the experiment revealed that believers had more right hemisphere activity compared to disbelievers in ESP.[15]

What does all this mean? Split-brain studies show that there are many distinct differences between the left brain and the right brain, but that the differences are far more subtle and nuanced than originally believed (thereby discounting most of the claims made in the endless stream of self-help books published on how to, for example, improve your right brain by using your left hand more, or improve your left brain through certain right-handed exercises). Nevertheless, there are dissimilar tendencies between the hemispheres, with the left cortex dominant in verbal tasks such as writing and speaking, and the right cortex dominant in nonverbal and spatial tasks. It's too simple to say that the left hemisphere is your literal, logical, rational brain and your right hemisphere is your metaphorical, holistic, intuitive brain, but it is a good first-order approximation of the division of labor in your head.

This is not to say that the dominance (however slight) of one hemisphere over the other is good or bad. It depends on the task. Creativity in all fields (art, music, literature, and even science), for example, appears to be related to right-brain dominance, and this makes sense given that the ability to find new and interesting patterns in both meaningful and meaningless noise is what creativity is all about. Were we only logic machines churning out products that were the result of strictly defined cognitive algorithms, nothing new would ever be created or discovered. At some point we must think outside the box and connect the dots into new patterns. Of course, the rub is in striking the right balance between finding a few new and interesting patterns within the background noise and finding nothing but patterns and leaving no noise. Perhaps this is the difference between creativity and madness.

Patternicity, Creativity, and Madness

In a sense, creativity involves a process of patternicity, of finding novel patterns and generating original products or ideas from them. Of course, the products or ideas must be useful or appropriate for a given context or environment for us to label them as creative, or else every amateur scientist and *American Idol* contestant would be indistinguishable from Einstein or Mozart. The connection between patternicity, creativity, and madness comes from a thinking style that is too all inclusive and that indiscriminately sees patterns everywhere. "When I was investigating the neuroscience of creativity," the clinical psychologist Andrea Marie Kuszewski explained, "one of the things I came across was the trait of 'lack of latent inhibition,' or as Hans Eysenck described it, an 'all-inclusive thinking style.' People on the schizophrenic spectrum tend to have an all-inclusive thinking style, which means they see patterns where no meaningful patterns exist, and cannot tell the difference between a meaningful or a non-meaningful pattern."[16]

This is, in fact, what was found by Max Planck Institute cognitive neuroscientist Anna Abraham and her colleagues, in a 2005 study designed to explore the link between creativity and a personality trait called *psychoticism*, one of three traits that the psychologist Hans Eysenck included in his P-E-N model of personality (the other two being extraversion and neuroticism). Eysenck was the first to suggest a possible correlation of psychoticism with creativity, and that too much of it can lead to psychoses and schizophrenia because of its characteristic "overinclusive cognitive style," which can lead to seeing patterns where none exist. We might think of this as patternicity on steroids. Abraham explored two dimensions of personality in eighty healthy subjects: the originality/novelty dimension and the practicality/usefulness dimension. She and her colleagues predicted that "higher levels of psychoticism would accompany a greater degree of conceptual expansion and elevated levels of originality in creative imagery, but would be unrelated to the practicality/usefulness of an idea." This is indeed what they found. Subjects with higher levels of psychoticism were more creative but in less practical ways, and Abraham and her colleagues concluded that this was due to their capacity for "associative thinking" (finding associations between random things) instead of "goal-related thinking."[17] That is, finding new and useful patterns is good, finding new patterns everywhere and being unable to discriminate between them is bad.

The next step in the causal chain to understanding patternicity and false pattern detection is to determine where in the brain this would happen. "People like this tend to have a prefrontal cortex (PFC) that does not process dopamine properly (the PFC is the area of cognitive control)," Kuszewski hypothesized, "and also have a less than optimally functioning *anterior cingulate cortex* (ACC). This area is activated when given options between multiple choices, and having to decide which option is the correct one. I like to think of it as the area in the brain that helps you to notice the details that distinguish two near-identical pictures, with only a few minor details that are different. You rely on the ACC to notice what the difference is (or the 'error') in picture A that makes it vary from picture B. Or more simply, the area in the brain that helps you locate Waldo in those *Where's Waldo?* puzzle books."[18]

So we might consider the ACC as the *Where's Waldo? Detection Device*. But what has that got to do with creativity and madness? "If you think of noticing patterns, a person with schizophrenia picks up on ridiculous patterns and draws conclusions based on them," Kuszewski continued. "For example, a stranger across the room looked at you, then made a phone call, and then looked at you again, therefore the false conclusion is that the person is stalking you, and was calling conspirators to come and hunt you down."

Right, that's what we call conspiratorial thinking, but just because you're paranoid doesn't mean they're not after you, so how can we tell the difference?

"Schizophrenics who are delusional see patterns like this all the time and think they are relevant. Their PFC and their ACC are not functioning to weed out the unlikely patterns, but instead see all patterns and give them equal weight for relevance."[19] In a way, there's a fine line between the creative genius of finding novel patterns that change the world and the madness or paranoia of seeing patterns everywhere and being unable to pick out the important ones. "A person who is successfully creative will see many patterns also (because creative people have an overinclusive thinking style) but will have a superior functioning

PFC and ACC that tell him which patterns make no sense, and which ones are useful, relevant, yet original ideas," Kuszewski concluded.

An instructive example would be a comparison between the Nobel Prize-winning physicist Richard Feynman, who did top-secret government work on the Manhattan Project to build an atomic bomb (and whose quirkiness extended no further than playing bongo drums, sketching nudes, and cracking safes), and the Nobel Prize-winning mathematician John Nash, who was diagnosed schizophrenic and portrayed in the film *A Beautiful Mind* as a man struggling with paranoid delusions about top-secret government work on a code-breaking project to detect enemy information patterns. Both Feynman and Nash were creative geniuses who made novel discoveries of unique patterns worthy of a Nobel Prize— Feynman in quantum physics and Nash in game theory—but Nash's cognitive style was all inclusive. He saw patterns everywhere, including complex conspiracies with nonexistent government agents and no basis in reality.

Someone in between Feynman and Nash on the patternicity scale is the Nobel Prize-winning geneticist Kary Mullis, the scientist behind the development of the polymerase chain reaction (PCR), the idea for which he says came to him late one night when he was driving through the mountains of northern California: "Natural DNA is a tractless coil, like an unwound and tangled audio tape on the floor of the car in the dark. I had to arrange a series of chemical reactions, the result of which would represent and display the sequence of a stretch of DNA. The odds were long. Like reading a particular license plate out on Interstate 5 at night from the moon."[20] Mullis's insight was that he could use a pair of chemical primers to bracket a desired DNA sequence and copy it using DNA polymerase, which would make it possible for a small strand of DNA to be copied an almost infinite number of times. By most accounts Mullis is a creative genius who loves to surf. He has an eccentric zeal for California counterculture with its propensity for artificially altering one's states of consciousness. His work has revolutionized biochemistry, molecular biology, genetics, medicine, and even forensics—those cheek-swab tests for DNA that you see on various crime television shows, for example, use the PCR method.

I first met Mullis at a social gathering after a conference several years ago. After a few beers loosened both of our tongues, he was only too happy to regale me with stories about his close encounter with an extraterrestrial (a "glowing raccoon" he says), his belief in astrology, ESP, and the paranormal (he says he doesn't "believe" but he "knows" they are real), his skepticism about global warming, HIV, and AIDS (he doesn't believe that humans cause global warming or that HIV causes AIDS), and his unadulterated endorsement of just about any claim that is routinely debunked in *Skeptic* magazine—claims that 99 percent of all scientists reject. I remember sitting there, thinking, "I can't believe this guy won a Nobel Prize! Are they just giving those things away to anyone these days?"

Well, now I think I know why Kary Mullis is a creative genius who also believes weird things: he has his pattern-detection filter dialed up to *wide open*, thereby availing himself to a wide variety of patterns, most of which are nonsense. But every now and again . . . It may be that 99 percent of scientists are skeptical of what Kary Mullis believes, but 99 percent of scientists never win a Nobel Prize.[21]

I documented a similar effect in my biography of Alfred Russel Wallace, the codiscoverer (with Charles Darwin) of natural selection.[22] Wallace was a brilliant synthesizer of masses of biological data into a few core principles that revolutionized ecology, biogeography, and evolutionary theory. In addition to being a pathbreaking scientist, Wallace was also a firm believer in phrenology, spiritualism, and psychic phenomena. He routinely attended séances and wrote serious scientific papers defending the paranormal against the skepticism of his fellow scientists as vociferously as he proffered natural selection over the views of his creationist colleagues. In hindsight, Wallace was ahead of his time in defending women's rights and in wildlife preservation, but he was on the wrong side in the anti-vaccination campaign that he helped lead in the late nineteenth century. He got himself into a legal entanglement with a flat-Earth defender—after proving to the lunatic that the earth really was round Wallace spent years in court trying to collect the prize money that was offered for the debate. Wallace fell for a scam surrounding a "lost poem" of Edgar Allan Poe's (allegedly written to cover a hotel bill in California), and even eventually broke with Darwin over the evolution of the human brain, which Wallace believed could not be the product of natural selection. He had what I call a *heretic personality*, or "the unique pattern of relatively permanent traits that makes an individual open to subjects at variance with those considered authoritative."

Wallace's patternicity filter was porous enough to let through both revolutionary and ridiculous ideas at the same time. Perhaps, we might speculate, the gain on the anterior cingulate cortex of Mullis and Wallace was turned down, thereby enabling their creative genius to emerge, along with their propensity for paranormal piffle.[23]

There is, in fact, good evidence to support the hypothesis that the anterior cingulate cortex is our error-detection network. Studies show, for example, that the ACC in particular becomes very active during the famous Stroop task, in which the names of colors are presented to subjects in either the same or a different color than the name denotes. The task is to identify the color of the letters only. When the name of the color and the color of the letters are the same then identifying the color of the letters is easy, but when the name of the color and the color of the letters are different, identifying the color of the letters is greatly slowed by the cognitive conflict inherent in the task. This is, in essence, an error-detection task.[24] Another example is a go/no-go task in which subjects are to press a button when an *A* appears on a screen in conjunction with an *X* but not in conjunction with other letters. When a letter combination similar to *AX* is used—such as *AK*—the error-detection difficulty increases, and along with it activity in the ACC.[25] Interestingly, research comparing schizophrenic patients with healthy subjects on such tasks reveals that the detection errors are higher for the schizophrenics, who often (although not always) also show less activity in their ACC.[26]

Here a plausible explanation for the link between patternicity, creativity, and madness presents itself. We are all pattern seekers, but some people find more patterns than others, depending on how indiscriminately they connect the dots between random events and how much meaning they put into such patterns. For most of us, most of the time, our error-detection networks (the ACC and the PFC) weed out some but not all of the false patterns we pick up through association learning, and we lead moderately creative (but not world-changing) lives, dealing with our various superstitions that come from false patterns that slipped through our pattern-detection filters. Some people are ultraconservative in their patternicity, see very few patterns, and are not very creative, while others are indiscriminate in their patternicity and find patterns everywhere they look; this may lead to creative genius or conspiratorial paranoia.

The Neuroscience of Agenticity

This process of explaining the mind through the neural activity of the brain makes me a *monist*. Monists believe that there is just one substance in our head—brain. *Dualists*, by contrast, believe that there are two substances—brain and mind. This is a very old problem in philosophy dating back to the seventeenth century when the French philosopher René Descartes put it on the intellectual landscape, with *soul* the preferred term of the time (as in "body and soul" instead of "brain and mind"). Broadly speaking, monists assert that body and soul are the same, and that the death of the body—particularly the disintegration of DNA and neurons that store the informational patterns of our bodies, our memories, and our personalities—spells the end of the soul. Dualists contend that body and soul are separate entities, and that the soul continues beyond the existence of the body. Monism is counterintuitive. Dualism is intuitive. It just seems like there is something else inside of us, and our thoughts really do feel like they are floating around up there in our skulls separate from whatever it is our brains are doing. Why?

We are natural-born dualists, argued Yale University psychologist Paul Bloom in his book *Descartes' Baby*. Children and adults alike, for example, speak of "my body," as if "my" and "body" are two different entities. We revel in films and books that take such dualisms as their themes. In Kafka's *Metamorphosis* a man falls asleep and wakes up as a cockroach, but his personality is intact inside the insect. In the film *All of Me*, the soul of Lily Tomlin battles with the soul of Steve Martin for control of his body. In *Freaky Friday*, mother and daughter (Jamie Lee Curtis and Lindsay Lohan) trade bodies with their essences unbroken. In *Big* and *13 Going on 30*, the characters' essences leapfrog ages with, respectively, Tom Hanks getting immediately younger and Jennifer Garner growing instantly older.

"In fact most people around the world believe that an even more radical transformation actually takes place," Bloom explained. "Most people believe that when the body is destroyed, the soul lives on. It might ascend to heaven, or descend to hell, go off into some sort of parallel world, or occupy some other body, human or animal. Even those of us who do not hold such views have no problems understanding them. But they are only coherent if we see people as separate from their bodies."[27]

In one among many experiments Bloom recounted, for example, young children are told a story about a mouse that gets munched by an alligator. The children agree that the mouse's body is dead—it does not need to go to the bathroom, it can't hear, and its brain no longer works. However, they insist that the mouse is still hungry, concerned about the alligator, and wants to go home. "This is the foundation for the more articulated view of the afterlife you usually find in older children and adults," Bloom explained. "Once children learn that the brain is involved in thinking, they don't take it as showing that the brain is the source of mental life; they don't become materialists. Rather, they interpret 'thinking' in a narrow sense, and conclude that the brain is a cognitive prosthesis, something added to the soul to enhance its computing power."[28]

The reason dualism is intuitive and monism counterintuitive is that the brain does not perceive the process of binding all the neural networks into one whole self, and so imputes mental activity to a separate source. Hallucinations of preternatural beings such as ghosts, gods, angels, and aliens are perceived as real entities; out-of-body and near-death experiences are processed as external events; and the pattern of information that is our memories, personality, and self is sensed as a soul. The renowned neurologist and author Oliver Sacks, best known for his remarkable work in "awakening" the catatonic brains of encephalitis victims as portrayed in the popular 1990 film *Awakenings* starring Robin Williams, has written a number of books describing the bizarre mental anomalies experienced by his patients—such as the man who mistook his wife for a hat—which are inevitably interpreted by the experiencers as external to their brain.[29]

One elderly patient who suffered from macular degeneration and had completely lost her vision was diagnosed by Sacks with Charles Bonnett syndrome (named for the eighteenth-century Swiss naturalist who first described it), because of her suite of complex visual hallucinations, including and especially faces with distorted teeth and eyes. Another patient developed a tumor in her visual cortex and soon after began hallucinating cartoon characters and even Kermit the Frog that were transparent and covered only half of her visual field. In fact, said Sacks, about 10 percent of visually impaired people experience visual hallucinations; faces (especially distorted faces) are the most common, cartoons are second, and geometric shapes are third. What is going on here?

In the past several years it has been possible to scan the brains of some of these patients inside a functional magnetic resonance imaging (fMRI) machine while they are hallucinating. Not surprising, the visual cortex is activated during these phantasms. During geometric hallucinations it is the primary visual cortex that is most active—the part of the brain that perceives patterns but not images. Hallucinations that include images such as faces are, not surprisingly, associated with more activity in the temporal lobe's fusiform gyrus, which as we saw is involved in the recognition of faces. In fact, people with damage to this area cannot recognize faces, and stimulation of the fusiform gyrus causes people to spontaneously see faces. There is even a tiny portion of the fusiform gyrus dedicated to perceiving eyes and teeth, and during the hallucinations experienced by Charles Bonnett syndrome patients it is this part of the brain that is active. In another part of the brain called the *inferotemporal cortex*, fragments of images—thousands and even millions of fragmentary images—are all stored in individual neurons or small clusters of neurons.

"Normally, this is part of the integrated stream of perception or imagination, and one is not conscious of them," Sacks explained. "If you become visually impaired or blind, the process is interrupted and instead of getting the smoothly organized perception, you are getting an anarchic release of activity from lots of these cells or cell clusters in the inferotemporal cortex and suddenly you start seeing fragments. The mind does its best to organize the fragments and give some coherence to it."[30]

Why does the brain bother to do any of this? As Sacks told one of his patients, who insisted that she was neither crazy nor demented, "As you lose vision, as the visual parts of the brain are no longer getting any inputs from the outside world, they become hyperactive and excitable and they start to fire spontaneously and you start to see things."

In the Charles Bonnett syndrome, we find an example of the foundation for the neural correlates of agenticity. "As Charles Bonnett wondered two hundred and fifty years ago," Sacks concluded, "how is the theater of the mind generated by the machinery of the brain?"[31] We now have a fairly sound understanding of the machinery, thereby rendering the theater of the mind an illusion. There is no theater, and no agent sitting inside the theater watching the world go by on the screen. Yet our intuitions tell us that there is. This is the foundation of agenticity in the brain that further reinforces belief-dependent realism.

Theory of Mind and Agenticity

There is another activity of the brain that I strongly suspect is involved in agenticity, and that is a process called *theory of mind* (ToM), or the fact that we are self-aware of our own beliefs, desires, and intentions, as well as aware that others have beliefs, desires, and intentions. A higher-order ToM allows you to realize that others' intentions may be the same as or different from your own. This is sometimes called *mind reading*, or the process of inferring the intentions of others by projecting yourself into their minds and imagining how you would feel. A still higher-level ToM means that you understand that others also have a theory of mind, and that you know that they know that you know they have a theory of mind. As Jackie Gleason used to growl to Art Carney in the classic 1950s television series *The Honeymooners*, "Norton, you know that I know that you know that I know that...." How does ToM mind reading actually operate in the brain?

In a review of the research on what brain scans have revealed about the location of such mind reading, Glasgow University neuroscientists Helen Gallagher and Christopher Frith concluded that there are three areas consistently activated whenever ToM is needed—located in different areas of the cortex: the *anterior paracingulate cortex*, the *superior temporal sulci*, and the *temporal poles* bilaterally. The first two brain structures are involved in processing explicit behavioral information, such as the perception of intentional behavior on the part of other organisms: "that predator intends to eat me." The temporal poles are essential for the retrieval of personal experiences from memory, such as "the last time I saw a predator it tried to eat me." All three of these structures are necessary for ToM, and Gallagher and Frith go so far as to posit that the anterior paracingulate cortex (located just behind your forehead) is the seat of the theory of mind mechanism.[32]

Theory of mind is a high-road automatic system that kicks in for specified activities involving other people, particularly in social situations. It most likely evolved out of a number of preexisting neural networks used for related activities, such as the ability to distinguish between animate and inanimate objects, to hold the attention of another being or agent by following their eye gaze, the ability to distinguish the actions of self and others, and the ability to represent actions that are goal directed. All of these functions are basic to survival in any social mammal, and thus theory of mind is most likely an *exaptation*, an ex-adaptation (some-times called a preadaptation) or a feature co-opted for a different purpose than the one for which it originally evolved. What might that have been for ToM? Probably imitation, anticipation, and empathy. Enter mirror neurons—specialized neurons that "mirror" the actions of others.

In the late 1980s and early 1990s, Italian neuroscientist Giacomo Rizzolatti and his colleagues at the University of Parma serendipitously discovered mirror neurons when they were recording the activity of single neurons in the *ventral premotor cortex* of macaque monkeys. Pok-ing hair-thin electrodes into individual neurons allows neuroscientists to monitor the rate and pattern of single-cell activity, and in this case the action from the monkey's F5 neurons spiked whenever the monkey reached for a peanut placed in front of it. The serendipity came when one of the experimenters reached in and grabbed one of the peanuts, causing the same neurons in the monkey's brain to fire. Monkey do, monkey see, monkey motor neurons fire. The motor neurons were mirroring the motor activity of others, and thus they became known as mirror neu-rons. As Rizzolatti recalled, "We were lucky, because there was no way to know such neurons existed. But we were in the right area to find them."[33]

Throughout the 1990s neuroscientists scrambled to learn more about mirror neurons, finding them in other parts of the brain, such as the *inferior frontal* and *inferior parietal* regions of the brain, and not only in monkeys but in humans as well through fMRI brain scans.[34] UCLA neuroscientist Marco Iacoboni and his colleagues, for example, imaged the brains of subjects as they watched people make finger movements and then imitated those same finger movements, discovering that the same areas of the frontal cortex and parietal lobe in both conditions were active.[35]

Rizzolatti suggested that mirror neurons are just motor neurons responding to seeing as well as doing. When you see an action it is recorded on your visual cortex, but to more deeply understand what the act means in terms of its consequences the observation must be linked to the motor sys-tem of the brain so that there is an internal check with the external world. With this basic neural network in place, higher-order functions can be lay-ered onto it, such as imitation. In order to imitate someone's actions, you need both a visual memory of how the action looked as well as a motor memory of how the action felt when implemented. There is now consider-able research linking the mirror neural network to imitation learning.

In a 1998 fMRI experiment, for example, people were shown two different hand actions, one without a context and one with a context that revealed the intention of the action. The latter scene activated the subject's mirror neuron network, revealing just where in the brain the perception of another intentional agent is located.[36] A very clever 2005 experiment was conducted in which monkeys watched a person either grasp an object and place it in a cup or grasp an apple and bring it to his mouth—similar action, different intention. Recording forty-one individual mirror neurons in the *inferior parietal lobe* of the monkeys' brains, it was discovered that the "grasp-to-eat" motion triggered fifteen mirror neurons to fire, but these were silent when observing the "grasp-to-place" motion. Interestingly, the neuroscientists concluded, the mirror neurons in this part of the brain "code the same act (grasping) in a different way according to the final goal of the action in which the act is embedded."[37] In other words, there are neurons specialized for discriminating between different intentions: grasping in order to place versus grasping in order to eat. More generally, this implicates mirror neurons in both predicting others' actions and inferring their intentions, which is the very foundation of agenticity.

Belief in the Brain

How is it that people come to believe something that seemingly defies reason? The answer is in the thesis of this book: beliefs come first; reasons for belief follow in confirmation of the realism dependent on the belief. Most belief claims fall somewhere in the fuzzy borderlands between unquestionably true and unmistakably false. How do our brains process such a broad swath of beliefs? To find out, in 2007 neuroscientists Sam Harris, Sameer A. Sheth, and Mark S. Cohen employed fMRI to scan the brains of fourteen adults at the UCLA Brain Mapping Center. They presented their subjects with a series of statements designed to be plainly true, clearly false, or undecidable at the moment. In response, the volunteers were to press a button indicating belief, disbelief, or uncertainty. For example:

MATHEMATICAL

True: $(2+6)+8=16$.

False: *62 can be evenly divided by 9.*

Uncertain: $1.2^{57} = 32608.5153$.

FACTUAL

True: *Most people have ten fingers and ten toes.*

False: *Eagles are common pets.*

Uncertain: *The Dow Jones Industrial Average rose 1.2 percent last Tuesday.*

ETHICAL

True: *It is bad to take pleasure at another's suffering.*

False: *Children should have no rights until they can vote.*

Uncertain: *It is better to lie to a child than to an adult.*

They made four important discoveries:

1. There were significant reaction time differences in evaluating statements. Responses to true (belief) statements were significantly shorter than responses to both false (disbelief) statements and uncertain statements, but there was no difference in reaction time detected between false (disbelief) statements and uncertain statements.
2. Contrasting the reaction to true (belief) statements and false (disbelief) statements yielded a spike in neural activity associated with belief in the *ventromedial prefrontal cortex*, an area of the brain associated with self-representation, decision making, and learning in the context of rewards.
3. Contrasting the reaction to false (disbelief) statements and true (belief) statements showed increased brain activity in the *anterior insula*, associated with responses to negative stimuli, pain perception, and disgust.
4. Contrasting the response to uncertainty statements with both true (belief) statements and false (disbelief) statements revealed elevated neural action in the anterior cingulate cortex—yes, the ACC that is involved in error detection and conflict resolution.

What do these results tell us about belief and the brain? "Several psychological studies appear to support [seventeenth-century Dutch philosopher Baruch] Spinoza's conjecture that the mere comprehension of a statement entails the tacit acceptance of its being true, whereas disbelief requires a subsequent process of rejection," Harris and his collaborators of the study reported. "Understanding a proposition may be analogous to perceiving an object in physical space: We seem to accept appearances as reality until they prove otherwise." Thus, subjects assessed true statements as believable faster than they judged false statements as unbelievable or uncertain statements as undecidable. Further, because the brain appears to process false or uncertain statements in regions linked to pain and disgust, especially in judging tastes and odors, this study gives new meaning to the phrase that a claim has passed the "taste test" or the "smell test."[38] When you hear bullshit, you may know it by its smell.

As for the neural correlates of belief and skepticism, the *ventromedial prefrontal cortex* is instrumental in linking higher-order cognitive factual evaluations with lower-order emotional response associations, and it does so in evaluating all types of claims. Thus, the assessment of the ethical statements showed a similar pattern of neural activation as did the evaluation of the mathematical and factual statements. People with damage in this area have a difficult time feeling an emotional difference between good and bad decisions, and this is why they are susceptible to confabulation—mixing true and false memories and conflating reality with fantasy.

This research supports what I call *Spinoza's conjecture: belief comes quickly and naturally, skepticism is slow and unnatural, and most people have a low tolerance for ambiguity.* The scientific principle that a claim is untrue unless proven otherwise runs counter to our natural tendency to accept as true that which we can comprehend quickly. Thus it is that we should reward skepticism and disbelief, and champion those willing to change their mind in the teeth of new evidence. Instead, most social institutions—most notably those in religion, politics, and economics—reward belief in the doctrines of the faith or party or ideology, punish those who challenge the authority of the leaders, and discourage uncertainty and especially skepticism.

The Brains of Believers and Nonbelievers

In a second fMRI study in search of the neural correlates of religious and nonreligious belief, Sam Harris and his UCLA colleagues scanned the brains of thirty subjects, fifteen self-reported Christians and fifteen self-reported nonbelievers, while they evaluated the truth and falsity of religious and nonreligious propositions. For example, a religious statement was "Jesus Christ really performed the miracles attributed to him in the

Bible." A nonreligious statement was "Alexander the Great was a very famous military leader." The subjects were instructed to push a button indicating that they thought a statement was true (belief) or false (disbelief). Once again, response times were significantly longer for those who perceived statements as false compared to those who interpreted the same statements as true. Tellingly, while both Christians and nonbelievers were faster in responding "true" than "false" on both religious ("Angels really exist") and nonreligious ("Eagles really exist") stimuli (because it's easier for everyone to agree than it is to disagree), nonbelievers were especially quick on the draw to respond to religious statements.

Inside the brain, these scans revealed that for both believers and nonbelievers, for both religious and nonreligious statements, the ventromedial prefrontal cortex, which as noted before is associated with self-relevance, decision making, and learning in the context of rewards, showed an increased signal—that is, more blood delivering oxygen. It's a "dopaminergic system"—remember, dopamine is a neurotransmitter substance associated with pleasure and is involved in the reinforcement of learning. This was the case whether the subjects believed statements about God or statements about ordinary facts. In fact, a direct comparison between belief and disbelief in both believers and nonbelievers showed no difference, leading Harris and his colleagues to conclude "the difference between belief and disbelief appears to be content-independent." That is to say, both believers and nonbelievers appear to evaluate the veracity of both religious and nonreligious claims in the same area of the brain. In other words, there is no "belief" module or "disbelief" module in the brain, no gullibility network or skeptical network.

Subtracting out the response to nonreligious stimuli from the response to religious stimuli revealed a greater BOLD (blood oxygen level–dependent) signal for religious stimuli in the *anterior insula* (associated with pain perception and disgust) and *ventral striatum* (associated with reward), as well as our old friend the ACC, the error-detection and conflict-resolution network. So religious statements provoked more positive and negative effects. Subtracting out the response to religious stimuli from the response to nonreligious stimuli revealed an increase in brain activity in the hippocampus, which is well known to be directly involved in making memories retrievable. Tellingly, this was the case for both believers and nonbelievers, leading Harris and his colleagues to "speculate that both groups experienced greater cognitive conflict and uncertainty while evaluating religious statements," and that "judgments about the nonreligious stimuli presented in our study seemed more dependent upon those brain systems involved in accessing stored knowledge."[39]

Why is this a surprising finding and what's so telling about it? I put the question to Harris, who responded: "I think, given the subject matter, both groups were less certain of their answers. The surprise, of course, is that it was *both* groups. One might have expected Christians to be less certain that 'the Biblical God really exists' than that 'Michael Jordan was a basketball player.' But atheists seem to show the same effect when evaluating a statement like 'The Biblical God is a myth.'"

I also asked Harris about the deeper implications for beliefs and how belief systems work in his discovery that such beliefs appear to be "content-independent." That is, why does it matter that there is only one neural network for belief and disbelief rather than a believing neural network and a skeptical neural network? "It suggests that belief is belief is belief," Harris noted without irony. "In my opinion, this has at least two consequences: (1) It further erodes the spurious distinction between facts and values. If believing that 'torture is wrong' and believing that '2 plus 2 makes 4' are importantly similar, then ethics and science are importantly similar at the level of the brain. (2) It suggests that the validity of a belief depends on how it came to be—on the chains of evidence and reasoning that link it to the world—not merely upon a feeling of conviction." So what? So plenty, Harris continued in his response to my query, because "the feeling of conviction is what we rely upon as consumers of beliefs—but clearly this feeling can become uncoupled from good reasons and good evidence in any domain (mathematical, ethical, etc.)."[40]

Hopefully, what can be decoupled from good reasons and good evidence can be recoupled through counterarguments with even better reasons and better evidence. That is, in any case, what all producers of scientific knowledge hope, which does, after all, spring eternal.[41]

PART III

BELIEF IN THINGS UNSEEN

I worry that . . . pseudoscience and superstition will seem year by year more tempting, the siren song of unreason more sonorous and attractive. Where have we heard it before? Whenever our ethnic or national prejudices are aroused, in times of scarcity, during challenges to national self-esteem or nerve, when we agonize about our diminished cosmic place and purpose, or when fanaticism is bubbling up around us—then, habits of thought familiar from ages past reach for the controls. The candle flame gutters. Its little pool of light trembles. Darkness gathers. The demons begin to stir.

—CARL SAGAN, *THE DEMON-HAUNTED WORLD*

7

Belief in the Afterlife

IN JUNE 2002, BASEBALL LEGEND TED WILLIAMS DIED, A NEWSWORTHY enough story that then got legs when his son whisked the body away to Scottsdale, Arizona, where it was cryogenically frozen at minus 320 degrees, with the hope that someday "Teddy Ballgame" would be resurrected to play again. If Williams's body were one day reanimated would it still be the cranky perfectionist who was the last person in baseball to hit .400? In other words, if future cryonics scientists could bring him back to life, would it still be "him"? Is the "soul" of Ted Williams also in deep freeze along with his brain and body? The answer depends on how *soul* is defined. If by *soul* we mean the pattern of Ted Williams's memories, personality, and personhood, and if the freezing process did not destroy the neural network in the brain where such entities are stored, then yes, the soul of Ted Williams would be resurrected along with his body.

In this sense, the soul is the unique pattern of information that represents a person, and unless there is some medium to retain the pattern of our personal information after we die, our soul dies with us. Our bodies are made of proteins, coded by our DNA, so with the disintegration of DNA our protein patterns are lost forever. Our memories and personality are stored in the patterns of neurons firing in our brains and the synaptic connections between them, so when those neurons die and those synaptic connections are broken, it spells the death of our memories and personality. The effect is similar to the ravages of stroke, dementia, and Alzheimer's disease, but absolute and final. No brain, no mind;

no body, no soul. Until a technology is developed to download our patterns into a more durable medium than the electric meat of our carbon-based protein, the scientific evidence tells us that when we die our pattern of information—our soul—dies with us.

That is the monist position anyway—that there is only one substance. Dualists believe that there is a conscious ethereal substance that is the unique essence of a living being that survives its incarnation in flesh. The ancient Hebrew word for soul is *nephesh*, or "life" or "vital breath"; the Greek word for soul is *psyche*, or "mind"; and the Latin word for soul is *anima*, or "spirit" or "breath." The soul is the essence that breathes life into flesh, animates us, gives us our vital spirit. Given the lack of knowledge about the natural world at the time these concepts were first formed, it is not surprising that ancient peoples reached for such ephemeral metaphors as mind, breath, and spirit. One moment a little dog is barking, prancing, and wagging its tail, and in the next moment it is a lump of inert flesh. What happened in that moment?

In 1907, a Massachusetts physician named Duncan MacDougall tried to find out by weighing six dying patients before and after their death. He reported in the medical journal *American Medicine* that there was a twenty-one-gram difference. Even though his measurements were crude and the weights varied, and no one has been able to replicate his findings, "twenty-one grams" has nonetheless grown to urban legendary status as the weight of the soul, spawning articles, books, and even a feature film of that title.

Death, and the possibility of life continuing beyond it, has spawned countless serious treatises and not a few comedic commentaries. The perpetually anxious Woody Allen has this workaround: "It's not that I'm afraid to die. I just don't want to be there when it happens."[1] Steven Wright thinks he's figured out a solution: "I intend to live forever—so far, so good."[2] Humor aside, since I am a scientist and claims are made that there is scientific evidence for life after death, let us analyze, first, a scientific explanation for why people believe in an afterlife, and, second, what the evidence is for that doubtful future date, and consider what its possibility means for our present state.

Natural-Born Immortalists: Afterlife as Agenticity

In the 2009 Harris Poll of religious beliefs among Americans, respondents were asked to indicate whether they believed in the following:[3]

Belief	Total	Catholic	Protestant	Jewish	Born-Again
God	82 %	94 %	92 %	79 %	97 %
Soul survival	71 %	82 %	85 %	37 %	91 %
Heaven	75 %	86 %	90 %	48 %	97 %
Hell	61 %	70 %	73 %	21 %	89 %
Reincarnation	20 %	19 %	13 %	18 %	14 %

Why do so many people believe in the afterlife? The question can be treated like any other belief question, and science can help illuminate the darkness. I suggest there are at least six solid reasons that lead people to believe there is life after death, based on the causal explanations I proposed for the sensed-presence experience, agenticity, dualism, and especially out-of-body experiences, all of which factor into afterlife accounts.

1. *Belief in the afterlife is a form of agenticity.* In our tendency to infuse the patterns we find in life with meaning, agency, and intention, the concept of life after death is an extension of ourselves as intentional agents continuing indefinitely into the future.

2. *Belief in the afterlife is a type of dualism.* Because we are natural-born dualists who intuitively believe that our minds are separate from our brains and bodies, the afterlife is the logical step in projecting our own mind-agency into the future without our bodies. It may even be a type of sensed-presence effect or third-man factor, with ourselves as that presence continuing on into an imagined ethereal empyrean.

3. *Belief in the afterlife is a derivative of our theory of mind.* We have the ability to understand that others have beliefs, desires, and intentions (we "read their minds") by projecting ourselves into the minds of others and imagining how we would feel. This ToM projection is another form of agenticity and dualism by which we can imagine the intentional minds of both ourselves and others as continuing indefinitely into the future. Since there is good evidence that ToM occurs in the anterior paracingulate cortex just behind the forehead, we might even conjecture that this neural network is integral for belief in the afterlife.[4]

4. *Belief in the afterlife is an extension of our body schema.* Our brains construct a body image out of the myriad inputs from every nook and cranny of our bodies. When this single individual *self* is coupled with our capacity for agenticity, dualism, and theory of mind, we can project that essence into the future, even without a body.

5. *Belief in the afterlife is probably mediated by our left-hemisphere interpreter.* A second neural network that is likely integral for afterlife beliefs is the left-hemisphere interpreter, which integrates inputs from all the senses into a meaningful narrative arc that makes sense of both senseful and senseless data. Tie this process into our body schema, theory of mind, and dualistic agenticity and it becomes clear how easy it is to develop a plot in which we are the lead character whose meaning and importance is central to the story and whose future is eternal.

6. *Belief in the afterlife is an extension of our normal ability to imagine ourselves somewhere else both in space and time, including time immemorial.* Close your eyes and imagine yourself on the warm sands of a tropical beach on a beautiful sunny day. Where are you in this picture? Are you inside your skin looking out from your eyes at the crashing waves in the distance and children playing in the sand? Or are you above yourself looking down on your entire body as if there were a second you hovering overhead? For most people this thought experiment results in the second observational platform. This is called *decentering*, or imagining ourselves somewhere else from an Archimedean point beyond our body. In this same manner we envision ourselves in the afterlife as a decentered image removed from this time and space into an empyreal realm, the literal (and literary) dwelling place of God, the ultimate immortal and eternal agent.

～❧～

In sum, because we so readily impart agency and intention to inanimate objects such as rocks and trees and clouds, and to animate objects such as predators, prey, and our fellow human beings; because we are natural-born dualists who believe in mind beyond body; because we are aware of our own minds and the minds of others; because we are aware of our own bodies as separate from all other bodies; because our brains are naturally inclined to weave all sensory inputs and cognitive thoughts into a meaningful story with ourselves as the central character; and, finally, because we are able to decenter ourselves from our time and space into another time and space, it is natural for us to believe that we have a time-less and eternal essence. *We are natural-born immortalists.*

The Disembodied Mind and the Eternal Soul

Believers in the afterlife, of course, will either reject these lines of evidence that belief in life after death is a product of the brain, or they will argue that their religion is simply reflecting an ontological reality about the universe. They believe in life after death because there really is an afterlife, they will say, and they will offer evidence in support of this claim. But as I have been arguing throughout this book, such rationalization of belief is precisely backward. Belief in the afterlife comes first; rational reasons for the belief come second. Nevertheless, the case for the existence of the afterlife is built around four lines of evidence that may be summarized as follows (from weakest to strongest in evidentiary strength).[5]

1. *Information fields and the universal life force.* According to the theory of morphic resonance, nature preserves data in the form of information fields that exist separately from individual organisms, as evidenced by people who can sense when someone is staring at their backs, by dogs that know when their owners are coming home, and that it is easier to complete the Sunday crossword puzzle later in the day because others have already solved it. These, and many other mysterious psychic phenomena, can be explained by "morphic resonance fields" that connect all living organisms to one another. Information cannot be created or destroyed, only recombined into new patterns, so our personal patterns—our "souls" by my definition—are packages of information that precede birth and survive death.

2. *ESP and evidence of mind.* Experimental research on *psi* (psychic power) and telepathy, in which subjects under controlled conditions can apparently receive images from senders without the use of the five senses, if true would stand as evidence for a disembodied mind that functions independently of the brain and yet can interact with normal matter.

3. *Quantum consciousness.* The study of the actions of subatomic particles through quantum mechanics produces what Einstein called *spooky action at a distance*, where the observation of a particle in one location instantaneously affects a related particle at another location (which could theoretically be in another galaxy), in apparent violation of Einstein's

upper limit of the speed of light. Some scientists take this to mean that the universe is one giant quantum field in which everything (and everyone) is interconnected and can influence one another directly and instantly. For believers in an afterlife, quantum mechanics explain how consciousness arises out of biochemical signals and how our minds may extend into the quantum realm that exists outside the brain.

4. *Near-death experiences.* There are thousands of people who have suffered traumatic accidents, near-drownings, emergency-room collapse, and especially heart attacks who are subsequently resuscitated and report experiencing some aspect of the afterlife—floating out of their bodies, passing through a tunnel or white light, and seeing loved ones or witnessing God, Jesus, or some manifestation of the divine on the other side. If these people were truly dead, then their conscious "self"—their soul or essence—somehow survived the death of the body.

Let's examine each of these carefully.

Information Fields and the Universal Life Force

Have you ever noticed how much easier it is to do a newspaper cross-word puzzle later in the day than it is to do it in the morning? Me neither. But according to British biologist Rupert Sheldrake, it is because the collective wisdom of morning successes resonates throughout the cultural "morphic field." In Sheldrake's theory of morphic resonance, similar forms (morphs, or "fields of information") reverberate and exchange information as extended minds within a universal life force. "As time goes on, each type of organism forms a special kind of cumulative collective memory," Sheldrake wrote in his 1981 book *A New Science of Life.* "The regularities of nature are therefore habitual. Things are as they are because they were as they were." In this and his most popular book, *The Presence of the Past,* Sheldrake, a University of Cambridge trained biologist and onetime research fellow of the Royal Society, explained that morphic resonance is "the idea of mysterious telepathy-type interconnections between organisms and of collective memories within species."[6]

Sheldrake believes that these information fields form a universal life force connecting all organisms and that morphic resonance explains phantom limbs, homing pigeons, how dogs know when their owners are coming home, and how people know when someone is staring at them. "Vision may involve a two-way process, an inward movement of light and an outward projection of mental images," Sheldrake wrote.[7] Thousands of trials conducted by anyone who downloaded the experimental protocol from Sheldrake's Web page "have given positive, repeatable, and highly significant results, implying that there is indeed a widespread sensitivity to being stared at from behind."[8] When someone stares at you it apparently creates something like a ripple in the morphic field that you sense, causing you to turn and look.

Let's examine this claim more closely. First, science is not normally conducted by strangers who happen upon a Web page protocol, so we have no way of knowing if these amateurs controlled for intervening variables and experimenter biases. Second, psychologists dismiss anecdotal accounts of this sense to a reverse self-fulfilling effect: a person suspects being stared at and turns to check; such head movement catches the eyes of would-be starers, who then turn to look at the staree, who thereby confirms the feeling of being stared at. Third, in 2000, John Colwell from Middlesex University, London, conducted a formal test using Sheldrake's suggested experimental protocol, with twelve volunteers who participated in twelve sequences of twenty stare or no-stare trials each, with accuracy feedback provided for the final nine sessions. Results: subjects were able to detect being stared at only when accuracy feedback was provided, which Colwell attributed to the subjects learning what was, in fact, a nonrandom presentation of the experimental trials.[9] When University of Hertfordshire psychologist Richard Wiseman also attempted to replicate Sheldrake's research, he found that subjects detected stares at rates no better than chance. Fourth, there is an experimenter bias problem. Institute of Noetic Sciences researcher Marilyn Schlitz (a believer in psi) collaborated with Wiseman (a skeptic of psi) in replicating Sheldrake's research, and discovered that when *they* did the staring Schlitz found statistically significant results, whereas Wiseman found chance results.[10]

Fifth, the *confirmation bias* may be at work here. In a 2005 special issue of the *Journal of Consciousness Studies* devoted to "Sheldrake and His Critics," I rated the fourteen open-peer commentaries on Sheldrake's target article (on the sense of being stared at) on a scale of 1 to 5 (critical, mildly critical, neutral, mildly supportive, supportive). Without exception, the 1s, 2s, and 3s were all traditional scientists from mainstream institutions, whereas the 4s and 5s were all affiliated with fringe and pro-paranormal institutions.[11] Sheldrake responded that skeptics dampen the morphic field's subtle power, whereas believers enhance it. Of Wiseman,

Sheldrake remarked: "Perhaps his negative expectations consciously or unconsciously influenced the way he looked at the subjects."[12] Perhaps, but how can we tell the difference between negative psi and non-psi? The invisible and the nonexistent look the same.

ESP and Evidence of Mind

For more than a century there have been a number of serious scientists who believed that such epiphenomena were not the products of our tendency to infuse patterns with intentional agents and supernatural forces. They strongly suspected that the brain was tapping into genuine forces not yet measurable through the traditional tools of science. In the late nineteenth century, organizations such as the Society for Psychical Research were founded to employ rigorous scientific methods in the study of psi, and many world-class scientists supported their efforts. In the twentieth century, psi periodically found its way into serious academic research programs, from Joseph Rhine's Duke University experiments in the 1920s to Daryl Bem's Cornell University research in the 1990s. Let's look at this most recent claim of experimental proof more closely, as it is the best argument to date for extrasensory perception.

In January 1994, Bem and his University of Edinburgh parapsychologist colleague Charles Honorton published a paper in the prestigious review journal *Psychological Bulletin* entitled "Does Psi Exist? Replicable Evidence for an Anomalous Process of Information Transfer." Conducting a meta-analysis of forty published experiments, the authors concluded: "The replication rates and effect sizes achieved by one particular experimental method, the ganzfeld procedure, are now sufficient to warrant bringing this body of data to the attention of the wider psychological community." A meta-analysis is a statistical technique that combines the results from many studies to look for an overall effect, even if the results from the individual studies were not significant (that is, they were unable to reject the null hypothesis at the 95 percent confidence level). The ganzfeld procedure places the "receiver" in a sensory isolation room with Ping-Pong ball halves covering the eyes, headphones playing white noise over the ears, and the "sender" in another room psychically transmitting photographic or video images.

Despite finding evidence for psi—subjects had a hit rate of 35 percent when 25 percent was expected by chance—Bem and Honorton lamented:

"Most academic psychologists do not yet accept the existence of psi, anomalous processes of information or energy transfer (such as telepathy or other forms of extrasensory perception) that are currently unexplained in terms of known physical or biological mechanisms."[13]

Why don't scientists accept psi? Daryl Bem has a stellar reputation as a rigorous experimentalist and he has presented us with statistically significant results. Aren't scientists supposed to be open to changing their minds when presented with new data and evidence? The reason for skepticism is that we need both replicable *data* and a viable *theory*, both of which are missing in psi research.

Data. Both the meta-analysis and ganzfeld techniques have been challenged by scientists. Ray Hyman from the University of Oregon found inconsistencies in the experimental procedures used in different ganzfeld experiments, which were lumped together in Bem's meta-analysis as if they used the same procedures. He argued the statistical test employed (Stouffer's Z) was inappropriate for such a diverse data set, and he also found flaws in the target randomization process (the sequence the visual targets were sent to the receiver), resulting in a target selection bias. "All of the significant hitting was done on the second or later appearance of a target. If we examined the guesses against just the first occurrences of targets, the result is consistent with chance."[14] Julie Milton and Richard Wiseman conducted a meta-analysis of thirty more ganzfeld experiments and found no evidence for psi, concluding that psi data are nonreplicable.[15] Bem countered with ten additional ganzfeld experiments he claims are significant, and he has additional research he plans to publish.[16] And so it goes . . . with more to come in the data debate. In general, over the course of a century of research on psi, the tighter the controls on the experimental conditions the weaker the psi effects seem to become until they disappear entirely.

Theory. The deeper reason scientists remain skeptical of psi—and will even if more significant data are published—is that there is no explanatory theory for how psi works. Until psi proponents can explain how thoughts generated by neurons in the sender's brain can pass through the skull and into the brain of the receiver, skepticism is the appropriate response. If the evidence shows that there is such a phenomenon as psi that needs explaining (and I am not convinced that the evidence does support such a conclusion), then we still need a causal mechanism.

Quantum Consciousness

One plausible theory of just such a causal mechanism has been proffered by the American physician Stuart Hameroff and the British physicist Roger Penrose in both technical writings[17] and a popular film improbably titled *What the #$*! Do We Know?!*[18] The film version is artfully edited and features actress Marlee Matlin as a dreamy-eyed photographer trying to make sense of an apparently senseless universe. The film's central tenet is that we create our own reality through consciousness and quantum mechanics. I met the producers of the film the weekend it opened when we were both on a Portland, Oregon, television show, so I got an early screening. I never imagined that a film grounded in an esoteric branch of physics—quantum mechanics—would succeed in the crowded market of popular movies, but it has grossed millions and created a cult following.

The film's avatars are scientists with strong New Age leanings, whose jargon-laden sound bites amount to little more than what Caltech physicist and Nobel laureate Murray Gell-Mann once described as "quantum flapdoodle."[19] University of Oregon quantum physicist Amit Goswami, for example, proclaims with great profundity: "The material world around us is nothing but possible movements of consciousness. I am choosing moment by moment my experience. Heisenberg said atoms are not things, only tendencies." It might prove an interesting experimental test of his theory for Goswami to leap out of a twenty-story building and consciously choose the experience of passing safely through the ground's tendencies.

The work of Japanese researcher Masaru Emoto, author of *The Hidden Messages of Water*, is featured to show how thoughts change the structure of ice crystals—beautiful crystals form in a glass of water with the word "love" taped to it, whereas playing Elvis's "Heartbreak Hotel" causes a crystal to split into two. One can't help but wonder if Elvis's "Burnin' Love" would boil water.

The film's nadir is an interview with "Ramtha," a thirty-five-thousand-year-old spirit channeled by a fifty-eight-year-old woman named J. Z. Knight. In fact, it turns out that many of the film's producers, writers, and actors are members of Ramtha's "School of Enlightenment," where New Age pabulum is dispensed in costly weekend retreats.

The attempt to link the weirdness of the quantum world (such as Heisenberg's uncertainty principle, which states that the more precisely you know a particle's position, the less precisely you know its speed, and vice versa) to mysteries of the macro world (such as consciousness) is based on Penrose and Hameroff's theory of quantum consciousness, which has generated much heat but little light in scientific circles.

Inside our neurons are tiny hollow microtubules that act like structural scaffolding. The conjecture (and that's all it is) is that something inside the microtubules may initiate a wave-function collapse that leads to the quantum coherence of atoms, causing neurotransmitters to be released into the synapses between neurons and thus triggering them to fire in a uniform pattern, thereby creating thought and consciousness. Since a wave-function collapse can come about only when an atom is "observed" (that is, affected in any way by something else), neuroscientist Sir John Eccles, another proponent of the idea, even suggests that "mind" may be the observer in a recursive loop from atoms to molecules to neurons to thought to consciousness to mind to atoms to molecules to neurons to . . . [20]

In reality, the gap between subatomic quantum effects and large-scale macro systems is too large to bridge. In his book *The Unconscious Quantum*,[21] University of Colorado particle physicist Victor Stenger demonstrates that for a system to be described quantum mechanically the system's typical mass m, speed v, and distance d must be on the order of Planck's constant h. "If mvd is much greater than h, then the system probably can be treated classically." Stenger computes that the mass of neural transmitter molecules, and their speed across the distance of the synapse, are about three orders of magnitude too large for quantum effects to be influential. There is no micro-macro connection. Subatomic particles may be altered when they are observed, but the moon is there even if no one looks at it. So what the #$*! is going on here?

Physics envy. The history of science is littered with the failed pipe dreams of ever-alluring reductionist schemes to explain the inner workings of the mind—schemes increasingly set forth in the ambitious wake of Descartes's own famous attempt, some four centuries ago, to reduce all mental functioning to the actions of swirling vortices of atoms, supposedly dancing their way to consciousness. Such Cartesian dreams provide a sense of certainty, but they quickly fade in the face of the complexities of biology. We should be exploring consciousness at the neural level and higher, where the arrow of causal analysis points up toward such principles as emergence and self-organization.

Near-Death Experiences

Since the advent of powerful jet planes capable of such g-force accelera-
tion that pilots can lose consciousness during aerial combat maneuver-
ing, the U.S. Air Force and Navy have undertaken a number of studies
on how to fight what is called *G-LOC*, or g-force-induced loss of consciousness, including special flight suits and training in centrifuges. Dr.
James Whinnery was hired by the military to direct the training and
study of pilots at the Naval Air Warfare Center centrifuge in Warmin-
ster, Pennsylvania. He discovered a remarkable phenomenon: the majority of pilots experienced what Whinnery called "dreamlets," or brief
episodes of tunnel vision, sometimes with a bright light at the end of
the tunnel, as well as a sense of floating, sometimes paralysis, and often
euphoria and a feeling of peace and serenity when they came back to
consciousness.[22]

Sound familiar? These are also the characteristics of a near-death
experience (NDE), first popularized in 1975 by Raymond Moody in his
book *Life After Life*, and now familiar to everyone by the unique set of
signs that include: (1) a floating or flying feeling in which you can look
down and see your body, commonly called an out-of-body experience
(OBE); (2) passing through a tunnel, hallway, or spiral chamber, sometimes with a bright light at the end of it; and (3) perhaps seeing loved
ones who have already passed away, and/or a Godlike image or divine
figure.[23] Whinnery was able to induce the first two of these three more
than a thousand times in sixteen years of study in the controlled conditions of the centrifuge, even videotaping the pilots when they passed out
and noting that this is when they had the experience, leaving no doubt as
to the cause: hypoxia, or oxygen deprivation to the cortex.[24]

Under high g-forces, the blood drains out of the head and pools toward
the center of the torso, rendering these pilots into a gray-out phase followed by a blackout state, all within a matter of fifteen to thirty seconds.
When G-LOC was induced in a gradual fashion by accelerating the centrifuge in a systematic manner, the subjects first experienced tunnel
vision, then blindness, then blackout, which is likely caused by the loss
of oxygen first to the retina then to the visual cortex (producing tunnel
vision as the neurons shut down from the outside to the inside), leading
to total blackout when the majority of the cortex powers down.[25] Dr.
David Comings, a medical doctor and neuroscientist specializing in
altered states of consciousness, notes, "The feelings of serenity and peace

are likely to have been produced by the increased release of various neurotransmitters such as endorphins, serotonin, and dopamine," and that the "NDE proves that when the brain is deprived of oxygen for prolonged periods of time, immediately prior to brain damage a range of physiological events occur that characterize NDE."[26]

Even more directly supportive of my thesis that all such disembodied mental phenomena are the result of brain activity may be found in a 2002 study published in *Nature*, in which Swiss neuroscientist Olaf Blanke and his colleagues reported that they could willfully produce out-of-body experiences through electrical stimulation of the *right angular gyrus* in the temporal lobe of a forty-three-year-old woman suffering from severe epileptic seizures.

With initial mild electrical stimulations of this area of the brain the patient reported "sinking into the bed" or "falling from a height." More intense stimulation led her to "see myself lying in bed, from above, but I only see my legs and lower trunk." Another stimulation induced "an instantaneous feeling of 'lightness' and 'floating' about two meters above the bed, close to the ceiling." The scientists discovered that they could even control the height above the bed that this woman reported by the level of electricity delivered to the temporal lobe. They then asked the patient to stare at her outstretched legs while they stimulated her brain. She reported that she saw her legs "becoming shorter." When they had her bend her legs prior to electrical stimulation, "she reported that her legs appeared to be moving quickly towards her face, and took evasive action." The same thing happened with her arms when the experiment was duplicated.

Blanke's team concluded: "These observations indicate that OBEs and complex somatosensory illusions can be artificially induced by electrical stimulation of the cortex. The association of these phenomena and their anatomical selectivity suggest that they have a common origin in body-related processing, an idea that is supported by the restriction of these visual experiences to the patient's own body." Since the primary function of the brain is to run the body, a displaced body schema may not only help explain the sensed-presence effect, it may generate a sense of the body schema being outside of itself. Blanke and his colleagues conjectured: "It is possible that the experience of dissociation of self from the body is a result of failure to integrate complex somatosensory and vestibular information."[27]

In a related study reported in the 2001 book *Why God Won't Go Away*, neuroscientist Andrew Newberg and his colleague Eugene D'Aquili found that brain scans made when Buddhist monks meditated and Franciscan nuns prayed indicated strikingly low activity in the posterior superior parietal lobe, a region of the brain the authors have dubbed the orientation association area (OAA).[28] The OAA's job is to orient the body in physical space, and people with damage to this area have a difficult time negotiating their way around a house, sometimes even bumping into objects. Even though they can see the obtrusive object, their brain does not process it as something separate from their body. When the OAA is booted up and running smoothly there is a sharp distinction between self and nonself. When OAA is in sleep mode—as in deep meditation and prayer—that division breaks down, leading to a blurring of the lines between reality and fantasy, between feeling in body and out of body. Perhaps this is what happens to monks who experience a sense of oneness with the universe, or with nuns who feel the presence of God, or with alien abductees floating out of their beds up to the mother ship.

This hypothesis was further supported in a 2010 discovery that damage to the posterior superior parietal lobe through tumorous lesions can cause patients to suddenly experience feelings of spiritual transcendence. Italian neuroscientist Cosimo Urgesi and his colleagues at the University of Udine in Italy measured the personalities of eighty-eight patients before and after brain surgery to remove tumors in both the left and right parietal cortex. They specifically noted the change in a relatively stable personality trait called *"self-transcendence,"* which tracks the tendency (or not) to become absorbed in an activity to the point of losing track of time and place, as well as the sense of having a strong spiritual connection with nature. "Damage to posterior parietal areas induced unusually fast changes of a stable personality dimension related to transcendental self-referential awareness," Urgesi explained. "Thus, dysfunctional parietal neural activity may underpin altered spiritual and religious attitudes and behaviors."[29]

Sometimes trauma can trigger such experiences. In a 2001 study published in the British medical journal *Lancet*, Dutch scientist Pim van Lommel and his colleagues reported that of 344 cardiac patients resuscitated from clinical death, 12 percent reported near-death experiences.

These included the full-on out-of-body experience, a light at the end of a tunnel, and so forth. Some of these near-death cardiac patients even described speaking to dead relatives.[30]

Dr. Mark Crisplin, a Portland, Oregon, ER doctor, reviewed the original EEG readings of a number of patients claimed by the scientists as being flatlined or "dead" and discovered that this was not at all the case. "What they showed was slowing, attenuation, and other changes, but only a minority of patients had a flat line, and it [dying] took longer than 10 seconds. The curious thing was that even a little blood flow in some patients was enough to keep EEGs normal." In fact, most cardiac patients were given CPR, which by definition delivers some oxygen to the brain (that's the whole point of doing it). Crisplin concluded: "By the definitions presented in the *Lancet* paper, nobody experienced clinical death. No doctor would ever declare a patient in the middle of a code 99 dead, much less brain dead. Having your heart stop for 2 to 10 minutes and being promptly resuscitated doesn't make you 'clinically dead.' It only means your heart isn't beating and you may not be conscious."[31] Again, since our normal experience is of stimuli coming into the brain from the outside, when one part of the brain abnormally generates these illusions, another part of the brain—quite possibly the left-hemisphere interpreter described by neuroscientist Michael Gazzaniga—interprets them as external events. Hence, the abnormal is interpreted as supernormal or paranormal.

In addition to localized neural networks, hallucinogenic drugs have been documented to trigger such preternatural experiences, such as the sense of floating and flying stimulated by atropine and other belladonna alkaloids. These can be found in mandrake and jimsonweed and were used by European witches and American Indian shamans, probably for this very purpose.[32] Dissociative anesthetics such as the ketamines are also known to induce out-of-body experiences. Ingestion of methylenedioxyamphetamine (MDA) may bring back long-forgotten memories and produce the feeling of age regression, while dimethyltryptamine (DMT)—also known as "the spirit molecule"—causes the dissociation of the mind from the body and is the hallucinogenic substance in *ayahuasca*, a drug taken by South American shamans. People who have taken DMT report "I no longer have a body," and "I am falling," "flying," or "lifting up."[33] Neuroscientist David Comings drew out the larger implications of such hallucinations for the relationship between our rational and spiritual brains:

> The psychedelic drugs like DMT often produce a sensation of "contact," of being in the presence of and interaction with a non-human being. Highly intelligent and sophisticated test subjects who knew these feelings were drug-induced nevertheless insisted the contact had really happened. The temporal lobe-limbic system's emotional tape recorder sometimes cannot distinguish between externally generated real events and internally generated non-real experience thus providing a system in which the rational brain and the spiritual brain are not necessarily in conflict.[34]

These studies, and countless others, continue to rain blows down upon the dualist head that brain and mind are separate. They are not. They are one and the same.[35] The brain, and the brain alone, is the source of our beliefs, and thus the template for our understanding of reality. The neural correlates of consciousness and subconsciousness elude us personally and can be gleaned only through careful scientific research using sophisticated tools such as brain scans and electrical stimulation of brain regions. As science marches onward it is inevitable that the paranormal and the supernatural either will be subsumed into the normal and the natural, or will simply disappear as a problem to be solved.

An Afterlife Interlude on *Larry King Live*

On Thursday, December 17, 2009, I filmed an episode of *Larry King Live*, which did not feature Larry King and was not live. No matter, it was a rockin' good time with a room full of guests, which Larry's show is wont to be.[36] Featured guests on this day included CNN's medical correspondent Dr. Sanjay Gupta (author of *Cheating Death: The Doctors and Medical Miracles That Are Saving Lives Against All Odds*), the New Age alt-med quantum guru Dr. Deepak Chopra (author of *Life After Death: The Burden of Proof*), the social commentator cum Christian apologist Dinesh D'Souza (who was touring for his new book *Life After Death: The Evidence*), a football referee named Bob Schriever who "died" on the playing field and saw the light, a reincarnation researcher who claims that birthmarks and bizarre dream images represent reincarnated dead people, and a young boy named James Leininger who believes he is the reincarnation of a World War II fighter pilot (accompanied by his parents there to promote their book, *Soul Survivor*). The guest host who artfully juggled all these guests was Jeff Probst, star of the television series *Survivor* (a title I thought ironically appropriate for the topic of the show). All the guests except me were in the New York CNN studio. I sat alone in the Hollywood CNN studio set staring into a camera with a video feed streaming in about three seconds ahead of the audio feed in my earpiece, which made me feel like I was being channeled from some other plane of existence. This was fitting because the subject of the show was life after death.[37]

Sanjay Gupta started us off with what turns out to be the first line of explanation for NDEs: the people who experience them are not actually dead! This is, in fact, why they're called *near*-death experiences. Gupta recalled that when he was in medical school the residents were taught to mark the time of death to the minute, as if one moment someone is alive and in the next moment . . . dead. "I mean, it just seemed so arbitrary even back then. And I think, in many ways, you know, that's been the hunt for me. That's what I've been searching for." What Gupta has discovered is that death can often take anywhere from a couple of minutes to a couple of hours to occur, depending on the conditions. As he demonstrates in his book (and CNN specials based on the book), people who have fallen into near-freezing lakes and rivers and "died" were actually not quite dead. Their core body temperatures were reduced so rapidly and dramatically that their vital brain and body tissues were preserved long enough for subsequent resuscitation. What appears to be something as miraculous as the resurrection of an actual dead person in fact has a nonmiraculous explanation in medical science.

So much of this debate on life after death turns on what is meant by *death*. People who believe in the afterlife and search for empirical evidence through NDEs, for example, will use such phrases as "he was dead and came back to life," or "she died and saw what was on the other side." When Probst introduced the football referee, for example, he said, "A man died on a football field seven years ago and came back to life." Gupta reinforced the point by explaining that Schriever "was dead for two minutes and forty seconds" (between collapse and revival). Schriever described what happened next: "It's very peaceful. It's very serene. And it's extremely, extremely bright. I mean, it is bright. And I was—I saw a place that I was supposed to go. I saw that halo, and something was saying, go toward the halo."

When I was asked for a scientific explanation for this apparent miracle, I gave the obvious answer that Gupta had earlier provided: "He wasn't dead. You started this hour off with Sanjay Gupta explaining we can't say somebody's dead at one given moment at a particular time on the clock. That's not how it works. It takes two, three, five, ten minutes to go through a dying process. The ref wasn't dead. He was in a near-death state." In fact, as the rest of the story revealed, the man had his heart restarted right there on the field by a portable automated external defibrillator available on the sidelines, and the entire event from collapse to revival was less than two minutes long. In this case, as in so many others, there is nothing miraculous to explain. The man was not brought back to life because he was never actually dead.

Whenever I appear on such shows I try to come up with a single message to leave viewers with, because in the chaos that is talk television a cacophony of voices often leads to confusion and obfuscation. For this show, the message I tried to convey based on what the other guests were saying is, in fact, a point that should be repeated like a mantra every time we encounter any mysteries: *the fact that we cannot fully explain a mystery with natural means does not mean it requires a supernatural explanation.*

Deepak Chopra made this error during the show when he responded to my argument that without the brain there is no mind because people who lose brain tissue due to injury, stroke, or surgery also lose the mind function associated with that brain tissue—no brain, no mind. Chopra challenged me with obviously intentional irony: "Well, I have to say of Michael that he is very superstitious. He's addicted to the superstition of materialism. The first thing he said about the brain, you know, that you destroy a certain part of the brain and that function will not come back—he hasn't kept up with the literature. There's a whole phenomenon called neural plasticity." Yes, indeed, I rejoined, and that makes my point even stronger: *it's the neural rewiring of the brain that saves the mind function.* Once again—no brain, no mind.

Chopra fired back that I had reversed the causal arrow: it is the ethereal nonphysical mind that causes the physical brain to rewire itself—no mind, no brain. In his book, Chopra defines *neuroplasticity* as "the notion that brain cells are open to change, flexibly responding to will and intention" and that "mind is the controller of the brain." Chopra is especially fond of quantum physics, and on such shows as this he loves to dazzle audiences with quantum pseudoscience, which is when you string together a series of terms and phrases from quantum physics and assume that explains something in the regular macro world in which we live. "The mind is like an electron cloud surrounding the nucleus of an atom," Chopra wrote in *Life After Death*. "Until an observer appears, electrons have no physical identity in the world; there is only the amorphous cloud. In the same way, imagine that there is a cloud of possibilities open to the brain at every moment (consisting of words, memories, ideas, and images I could choose from). When the mind gives a signal, one of these possibilities coalesces from the cloud and becomes a thought in the brain, just as an energy wave collapses into an electron."[38]

Baloney. The microscopic world of subatomic particles as described by the mathematics of quantum mechanics has no correspondence with the macroscopic world in which we live as described by the mathematics of Newtonian mechanics. These are two different physical systems at two different scales described by two different types of mathematics. The hydrogen atoms in the sun are not sitting around in a cloud of possibilities waiting for a cosmic mind to signal them to fuse into helium atoms and thereby throw off heat generated by nuclear fusion. By the laws of physics of this universe, a gravitationally collapsing cloud of hydrogen gas will, if large enough, reach a critical point of pressure to cause those hydrogen atoms to fuse into helium atoms and give off heat and light in the process, and it would do so even if there were not a single mind in the entire cosmos to observe it.

When we are dealing with such topics as the afterlife, there is the problem of fuzzy language in using words such as *mind, will, intention,* and *purpose.* Chopra writes, for example, "Neurologists have verified that a mere intention of purposeful act of will alters the brain. Stroke victims, for example, can force themselves, with the aid of a therapist, to use only their right hand if paralysis has occurred on that side of the body. Willing themselves day after day to favor the affected part, they can gradually cause the damaged sites in the brain to heal." Chopra also cites the work of UCLA neuroscientist Jeffrey Schwartz, an expert on OCD (obsessive-compulsive disorder), who has apparently had as much success controlling the obsessive thoughts and compulsive behaviors of patients using talk therapy as others have using Prozac, and that brain scans allegedly show that "the same impaired regions that become more normal with Prozac also become more normal with talk therapy."[39]

But what does it mean to "will" something, or to "intend" it, or to have "purpose"? Like *mind*, these are just words used to describe thoughts and behaviors, which are all driven by neural activity—every single one of them. There is not a behavior you perform or a thought you think that does not have a neural correlate to it. No neurons or neural activity, no thoughts or behaviors. Period. Calling a series of neural firings by a network of neurons "will" or "intention" or "purpose" does nothing to explain the process. You might as well say "he zlotted his leg to lift," or "she xekoned her hand to move." To describe neural activity as "zlotted" or "xekoned" is as meaningless as saying that it was "willed" or "intended." Saying that patients "talked" about their obsessions and compulsions and in the process improved does not explain how or why they improved. What we need to know is what neural activity involved in talking interacted with the neural activity associated with the obsessive thoughts or compulsive acts. Such terms are just linguistic placeholders for our ignorance, and only serve to push off the causal explanation to another day.

Most likely what we are observing in neuroplasticity is a neural network feedback loop whereby one cluster or series of neurons fires in a particular pattern that we describe as "will" or "intention" or "purpose," and these in turn interact with another cluster or series of neurons that are associated with the activity lost due to brain damage in that area. This signals dendrites to develop new synaptic connections, and the brain is therefore "rewired." We know from biofeedback research that talking or thinking about a particular problem sets up a feedback loop (either positively or negatively) that alters the neurophysiology of the brain. There is nothing mystical, paranormal, or woo-woo about any of this, but using such fuzzy language is unhelpful when we want to understand the underlying causal mechanisms of belief.

No one uses fuzzy language more adroitly than Deepak Chopra, who has an uncanny knack for stringing together words and phrases so that it actually sounds like something intelligible is being said. For example, what do you make of this explanation for near-death experiences? "There are traditions that say the in-body experience is a socially induced collective hallucination. We do not exist in the body. The body exists in us. We do not exist in the world. The world exists in us." Or this nugget on life and death: "Birth and death are space-time events in the continuum of life. So the opposite of life is not death. The opposite of death is birth.

And the opposite of birth is death. And life is the continuum of birth and death, which goes on and on." Uh? Read it again . . . and again . . . it doesn't become any clearer. When I asked what happened to little James Leininger's soul if his body was now occupied by the soul of a World War II fighter pilot, Chopra offered this jewel of Deepakese: "Imagine that you're looking at an ocean and you see lots of waves today. And tomorrow you see a fewer number of waves. It's not so turbulent. What you call a person actually is a pattern of behavior of a universal consciousness." He gestured toward our host. "There is no such thing as Jeff, because what we call Jeff is a constantly transforming consciousness that appears as a certain personality, a certain mind, a certain ego, a certain body. But, you know, we had a different Jeff when you were a teenager. We had a different Jeff when you were a baby. Which one of you is the real Jeff?" Jeff Probst looked as confused as I felt.

At one point in the show, when asked how he as a medical doctor and man of science deals with medical miracles that seem to border on religious and spiritual domains, Sanjay Gupta began by offering natural explanations, such as this one for the near-death experience: "The tunnel, for example, that potentially can be explained away by a lack of blood flow to the back of the eye. You start to lose your peripheral vision, see a tunnel. Bright lights, sort of the same thing. Even the seeing of deceased relatives, perhaps, that is a very cultural thing, for example, in Western cultures. In eastern Africa, people who are having near-death experiences tend to see things that they wish they had done in life. That tends to be their cultural thing they have." But then Gupta fell into the trap of the argument from ignorance ("if there isn't an explanation then there cannot be an explanation") when he said, "When I was researching this for a long time, I thought I was going to explain it all away physiologically. But things that I heard and validated and subsequently believed convinced me that there were things that I could not explain. There were things that were happening at that moment, that near-death experience moment, that simply could not be explained with existing scientific knowledge."

So what? Ignorance or incredulity simply means that we cannot explain every mystery we encounter. That's normal. No science can throw a comprehensive explanatory net over every mystery in the cosmos. The fact that we can "only" explain about 90 percent of all UFO sightings and crop circles does not mean that the other 10 percent represent actual visitations by extraterrestrial intelligences. The missing 10 percent—what is sometimes called the "residual problem" in science because for any given theory there will always be a residual of unexplained anomalies—just means that we can't explain *everything*. The fact that we cannot explain every cancerous tumor that has gone into remission does not mean that miraculous supernatural forces occasionally eliminate cancer. It just means that modern medicine has yet to catch up with the wonders and mysteries of the human body.

In the case of the afterlife, just because we do not have a 100 percent completely natural explanation for all of the experiences that people have near death does not mean that we will never understand death, or that there is some other mysterious force at work. It certainly does not mean that there is life after death. It just means that we don't know everything. Such uncertainty is at the very heart of science and is what makes it such a challenging enterprise.

Hoping and Knowing

I am, by temperament, a sanguine person, so I really hate to douse the flame of hope with the cold water of skepticism. But I care about what is *actually* true even more than what I *hope* is true, and these are the facts as I understand them.

I am occasionally accused of being skeptical of the wrong things, or of being too skeptical for my own good. Sometimes I'm even charged with denialism—I don't want X to be true, therefore I unfairly find reasons to reject X. That is undoubtedly sometimes the case. In fact, belief-dependent realism and the confirmation of beliefs after they are formed necessarily must apply to me as well as others.

On this particular issue of agenticity and its manifestation in dualism, mind, the supernatural, and the afterlife, however, I entertain no such denialist tendencies. In fact, I passively wish for their manifestation in reality. *The afterlife? I'm for it!* But the fact that I wish it were so does not make it so. And herein lies the problem of understanding the mind in order to know humanity: our belief systems are structured such that we will almost always find a way to support what we want to believe. Thus, the overwhelming desire to believe in something otherworldly—be it mind, spirit, or God—means that we should be especially vigilant in our skepticism of claims made in these arenas of belief.

Is scientific monism in conflict with religious dualism? Yes, it is. Either the soul survives death or it does not, and there is no scientific evidence that it does or ever will. Does science and skepticism extirpate all meaning in life? I think not; quite the opposite, in fact. If this is all there is, then how meaningful become our lives, our families, our friends, our communities—and how we treat others—when every day, every moment, every relationship, and every person counts, not as props in a temporary staging before an eternal tomorrow where ultimate purpose will be revealed to us but as valued essences in the here and now where we create provisional purpose.

Awareness of this reality elevates us all to a higher plane of humanity and humility, as we course through life together in this limited time and space—a momentary proscenium in the drama of the cosmos.

8

Belief in God

A<small>MONG THE MANY BINOMIAL DESIGNATIONS GRANTED OUR SPECIES</small>—
Homo sapiens, Homo ludens, Homo economicus—a strong case could be
made for *Homo religiosus.*

According to Oxford University Press's *World Christian Encyclopedia*, 84 percent of the world's population belongs to some form of organized religion, which at the end of 2009 equals 5.7 billion people. That's
a lot of souls. Christians dominate at around 2 billion adherents (with
Catholics accounting for half of these), Muslims come in at a little more
than a billion, Hindus at around 850 million, Buddhists at almost 400
million, and ethnoreligionists (animists and others in Asia and Africa
primarily) make up most of the remaining several hundred million
believers. Worldwide, there are about 10,000 distinct religions, each one
of which may be further subdivided and classified. Christians, for
example, may be apportioned among about 34,000 different denomina-
tions.[1]

Somewhat surprisingly—given that we are the most technologically
advanced and scientifically sophisticated nation in history—America is
among the most religious tribes of the species. A 2007 Pew Forum survey found the following percentages of belief:

God or a universal spirit	92 %
Heaven	74 %
Hell	59 %
Scripture is the word of God	63 %

Pray once a day	58 %
Miracles	79 %

Who or what God represents varies depending on religious faith. Is God a person with whom believers can have a relationship, or is he an impersonal force? According to the Pew survey, 91 percent of Mormons believe in a personal God, but only 82 percent of Jehovah's Witnesses, 79 percent of evangelicals, 62 percent of Protestants, and 60 percent of Catholics do. By contrast, 53 percent of Hindus, 50 percent of Jews, 45 percent of Buddhists, and 35 percent of unaffiliated believers believe in God as an impersonal force. Most striking to me and supporting one of the central themes of this book—agenticity—the dualistic belief that there must be something else out there is so pervasive that even 21 percent of those who identified themselves as atheists, and 55 percent who identify themselves as agnostics, expressed a belief in some sort of God or universal spirit.[2]

Why God Is Hardwired into Our Brains

Such statistics stagger the imagination. Any characteristic that is this common in a species cries out for an explanation. Why do so many people believe in God?

On one level, I have already answered this question in the chapters on patternicity and agenticity. God is the ultimate *pattern* that explains everything that happens, from the beginning of the universe to the end of time and everything in between, including and especially the fates of human lives. God is the ultimate intentional *agent* who gives the universe meaning and our lives purpose. As an ultimate amalgam, patternicity and agenticity form the cognitive basis of shamanism, paganism, animism, polytheism, monotheism, and all other forms of theisms and spiritualisms devised by humans.

Although there is much cultural variation among different religious faiths, all have in common the belief in supernatural agents in the form of a godhead or spirits who have intention and interact with us in the world. There are three lines of evidence pointing to the conclusion that such beliefs are hardwired into our brains and behaviorally expressed in consistent patterns throughout history and culture. These evidentiary lines come from evolutionary theory, behavior genetics, and comparative world religions, all of which support the larger thesis of this book that the belief comes first and the reasons for the belief follow. After reviewing this evidence, I will demonstrate why it is not possible to know for certain whether God exists, and why any scientific or rational attempt to prove God's existence can result only in our awareness of an intelligence greater than our own but considerably less than the omniscience traditionally associated with God.

Evolutionary Theory and God

In his 1871 book *The Descent of Man*, Charles Darwin noted that anthropologists conclude that "a belief in all-pervading spiritual agencies seems to be universal; and apparently follows from a considerable advance in the reasoning powers of man, and from a still greater advance in his faculties of imagination, curiosity and wonder."[3] What flummoxed Darwin about the universal nature of religious beliefs was how natural selection could account for them. On the one hand, he noted, "It is extremely doubtful whether the offspring of the more sympathetic and benevolent parents, or of those who were the most faithful to their comrades, would be reared in greater number than the children of selfish and treacherous parents of the same tribe. He who was ready to sacrifice his life, as many a savage has been, rather than betray his comrades, would often leave no offspring to inherit his noble nature."[4] On the other hand, although Darwin was a strident proponent of restricting the range and power of natural selection to operate strictly at the level of the individual organism, he conceded that selection might also operate at the group level when it came to religion and between-group competition: "There can be no doubt that a tribe including many members who, from possessing in a high degree the spirit of patriotism, fidelity, obedience, courage and sympathy, were always ready to aid one another, and to sacrifice themselves for the common good, would be victorious over most other tribes; and this would be natural selection [of the group]."[5]

Picking up where Darwin left off, in my book *How We Believe* I developed an evolutionary model of belief in God as one of a suite of mechanisms used by religion, which I define as *a social institution to create and promote myths, to encourage conformity and altruism, and to signal the level of commitment to cooperate and reciprocate among members of a community*. Around five thousand to seven thousand years ago, as bands and tribes began to coalesce into chiefdoms and states, government and religion co-evolved as social institutions to codify moral behaviors into ethical principles and legal rules, and God became the ultimate enforcer of the rules.[6] In the small populations of hunter-gatherer bands and tribes with a few dozen to a couple of hundred members, informal means of behavior control and social cohesion could be employed by capitalizing on the moral emotions, such as shaming someone through guilt for violating a social norm, or even excommunicating violators from the group. But when populations grew into the tens and hundreds of thousands, and eventually into millions of people, such informal means of enforcing the rules of society broke down because free riders and norm violators could more readily get away with cheating in large groups; something more formal was needed. This is one vital role that religion plays, such that even if violators think that they got away with a violation, believing that there is an invisible intentional agent who sees all and knows all and judges all can be a powerful deterrent of sin.

One line of evidence for this theory of religion can be found in human universals, or traits that are shared by all peoples. There are general universals, such as tool use, myths, sex roles, social groups, aggression, gestures, emotions, grammar, and phonemes, and there are specific universals, such as kinship classifications and specific facial expressions such as the smile, frown, or eyebrow flash. There are also specific universals directly related to religion and belief in God, including *anthropomorphizing animals and objects, general belief in the supernatural, specific supernatural beliefs and rituals about death, supernatural beliefs about fortune and misfortune, and especially divination, folklore, magic, myths, and rituals.*[7] Although such universals are not totally controlled by genes alone (almost nothing is), we can presume that there is a genetic predisposition for these traits to be expressed within their respective cultures, and that these cultures, despite their considerable diversity and variance, nurture these genetically predisposed natures in a consistent fashion.

A second line of evidence for the evolutionary origins of religion and belief in God can be found in anthropological studies of meat sharing practiced by all modern hunter-gatherer societies around the world. It turns out that these small communities—which can cautiously be used as a model for our own Paleolithic ancestors—are remarkably egalitarian. Using portable scales to measure precisely how much meat each family within the group received after a successful hunt, researchers found that the immediate families of successful hunters got no more meat than the rest of the families in the group, even when these results were averaged over several weeks of regular hunting excursions. Hunter-gatherers are egalitarian because individual selfish acts are effectively counterbalanced by the combined will of the rest of the group through the use of gossip to ridicule, shun, and even ostracize individuals whose competitive drives and selfish motives interfere with the overall needs of the group.[8] Thus, a human group is also a moral group in which "right" and "wrong" coincide with group welfare and self-serving acts, respectively.

Other hunter-gatherer groups employ supernatural beings and superstitious rituals to enforce fairness, such as the Chewong people of the Malaysian rain forest and the ritual *punen*, which is related to the calamities and misfortune that arise when you act too selfishly. In the Chewong world, the myth about Yinlugen Bud—a god who brought the Chewong out of a more primitive state by insisting that eating alone was improper human behavior—serves to ensure the sharing of food. When food is caught away from the village, it is promptly returned, publicly displayed, and equitably distributed among all households and even among all individuals within each home. Someone from the hunter's family touches the catch then proceeds to touch everyone present, repeating the word *punen*. Thus, both superstitious rituals and the belief in supernatural agents oversee the exchange process that reinforces group cohesiveness.

Your culture may dictate which god to believe in and which religion to adhere to, but the belief in a supernatural agent who operates in the world as an indispensable part of a social group is universal to all cultures because it is hardwired in the brain, a conclusion enhanced by studies on identical twins separated at birth and raised in different environments.

Behavior Genetics and God

Behavior geneticists attempt to tease apart the relative roles of heredity and environment on any given trait. Since there is variation in the expression of all traits, we are looking for a percentage of the variation accounted for by genes and environment, and one of the best natural experiments available for research are identical twins separated at birth and reared in different environments. In one study of fifty-three pairs of identical twins reared apart and thirty-one pairs of fraternal twins reared apart, Niels Waller, Thomas Bouchard, and their colleagues in the Minnesota twins project looked at five different measures of religiosity. They found that the correlations between identical twins were typically double those for fraternal twins, and subsequent analysis led them to conclude that genetic factors account for 41 to 47 percent of the observed variance in their measures of religious beliefs.[9]

Two much larger twin studies out of Australia (3,810 pairs of twins) and England (825 pairs of twins) found similar percentages of genetic influence on religious beliefs, comparing identical and fraternal twins on numerous measures of beliefs and social attitudes. They initially concluded that approximately 40 percent of the variance in religious attitudes was genetic.[10] These researchers also documented substantial correlations between the social attitudes of spouses. Because parents mate assortatively (like marries like because "birds of a feather flock together") for social attitudes, offspring tend to receive a double dose of whatever genetic propensities may underlie the expression of such attitudes. When these researchers included a variable for assortative mating in their behavioral genetics models, they found that approximately 55 percent of the variance in religious attitudes is genetic, approximately 39 percent can be attributed to the nonshared environment, approximately 5 percent is unassigned, and only about 3 percent is attributable to the shared family environment (and hence to cultural transmission via parents).[11] Based on these results, it would appear that people who grow up in religious families who themselves later become religious do so mostly because they have inherited a disposition, from one or both parents, to resonate positively with religious sentiments. Without such a genetic disposition, the religious teachings of parents appear to have few lasting effects.

Of course, genes do not determine whether one chooses Judaism, Catholicism, Islam, or any other religion. Rather, belief in supernatural agents (God, angels, and demons) and commitment to certain religious practices (church attendance, prayer, rituals) appear to reflect genetically based cognitive processes (inferring the existence of invisible agents) and personality traits (respect for authority, traditionalism). Why did we inherit this tendency?

One line of research that may help answer this question is related to dopamine, which as we saw in chapter 6 is directly connected to learning, motivation, and reward. There may be a genetic basis to how much dopamine each of our brains produces. The gene that codes for the production of dopamine receptors is called DRD4 (dopamine receptor D4) and is located on the short arm of the eleventh chromosome. When dopamine is released by certain neurons in the brain it is picked up by other neurons that are receptive to its chemical structure, thereby establishing dopamine pathways that stimulate organisms to become more active and reward certain behaviors that then get repeated. If you knock out dopamine from either a rat or a human, for example, they will become catatonic. If you overstimulate the production of dopamine, you get frenetic behavior in rats and schizophrenic behavior in humans.

The first people to associate the DRD4 gene with spirituality were medical researcher David Comings and his colleagues, when they went looking for genes associated with novelty seeking.[12] Their research was subsequently picked up and linked to risk-taking behavior by National Cancer Institute geneticist Dean Hamer. Most of us have four to seven copies of the DRD4 gene on chromosome eleven. Some people, however, have two or three copies, while others have eight to eleven copies. More copies of the DRD4 gene translate into lower levels of dopamine, which stimulates people to seek greater risks in order to artificially get their dopamine fix. Leaping off of buildings, antennae, spans, or earth (so-called BASE jumping) is one way to do it, although high-risk gambling in Las Vegas or Wall Street may also do the trick. As a test of this hypothesis, Hamer first had subjects take a survey that measures desire to seek novelty and thrills. (BASE jumpers score very high on this test.) He then took a sample of their DNA from chromosome eleven and discovered that people with high numbers on the risk-taking survey had more copies than normal of the DRD4 gene.[13]

From risk-taking behavior to religious belief, Hamer considered the possibility that dopamine might be implicated in faith, and he published his results in a controversial book entitled *The God Gene*. To his credit, Hamer disclaims the book's title (they are almost always determined by the sales and marketing departments of publishing companies), explaining that there is, of course, no single gene that could possibly represent something as complex and variegated as belief in God, much less the rich tapestry that is religious faith. But he does argue that some of us are born with genes that make us more or less "spiritual," which is a component in both belief in God and religious faith.[14] This time Hamer tagged another dopamine-related gene called *VMAT2* (vesicular monoamine transporter 2), which regulates the flow of serotonin, adrenaline, norepinephrine, and our friend dopamine. Starting with a database of siblings with cigarette addiction, Hamer wanted to know if there was a family genetic connection to an addictive personality, and so gave his subjects a battery of psychological questionnaires, one of which included the personality trait self-transcendence.

First identified by Washington University psychiatrist Robert Cloninger, people scoring high in self-transcendence tend toward "self-forgetfulness" (becoming totally absorbed in an activity), "transpersonal identification" (feeling connected to the larger world), and "mysticism" (a willingness to believe in things unprovable, such as ESP). Together, Cloninger believes, these measures add up to something like what we think of as spirituality. In twin studies conducted by Lindon Eaves and Nicholas Martin, self-transcendence was found to be heritable (as all personality characteristics are), so Hamer analyzed the DNA and the personality measures of more than a thousand people and found that those people in the study who scored high in self-transcendence had a dopamine-boosting version of the *VMAT2* gene. How does this gene lead to self-transcendence and spirituality?

VMAT2 is an integral membrane protein that acts to transport monoamines—an amine containing one amino group, such as the neurotransmitters dopamine, norepinephrine, and serotonin—from the fluid inside the neuron cell body to the synaptic vesicles at the ends of neuronal axons. The terminal buttons at the ends of these axons reach out to almost (but not quite) touch one another. Hamer thinks that the one variant of the *VMAT2* gene that is associated with increased self-transcendence leads to the production of more of these little transporters, and thus more neurotransmitter substances such as dopamine are delivered into those narrow synapses, thereby boosting such positive feelings as self-transcendence.

Hamer's studies have been strongly criticized by his fellow scientists—which is the norm in this profession—and admittedly identifying genes for this or that behavior or belief can be problematic. Nevertheless, the fact that dopamine is involved in this belief, as in so many beliefs, supports this book's thesis that there is a belief engine in the brain associated with specific areas that generate and evaluate beliefs across a wide variety of contexts. One role of this engine is to reward belief of all puta-tive claims, including and especially belief in God. In other words, it feels good and is rewarding to believe in God.

Comparative World Religions and God

The comparative study of why people believe in God and adhere to a religion has generated a wide variety of theories over the past century.[15] Although these theories vary considerably in their details about the ori-gins and purpose of religion, all have in common the belief in super-natural agents in the form of God, gods, or spirits as integral to religion, and it is this aspect of belief that we are exploring here. That is, I am less interested in why people believe in this or that god or join this or that religion, and more interested in why people believe in any gods or join any religion. To that end, I want to pull back and look at the bigger pic-ture of history. As a back-of-the-envelope calculation within an order-of-magnitude accuracy, we can safely say that over the past ten thousand years of history humans have created about ten thousand different reli-gions and about one thousand gods. What is the probability that Yahweh is the one true god, and Amon Ra, Aphrodite, Apollo, Baal, Brahma, Ganesha, Isis, Mithra, Osiris, Shiva, Thor, Vishnu, Wotan, Zeus, and the other 986 gods are false gods? As skeptics like to say, everyone is an athe-ist about these gods; some of us just go one god further.

There is, I believe, compelling evidence that humans created God and not vice versa. If you happened to be born in the United States in the twentieth century, for example, there is a very good chance that you are a Christian who believes that Yahweh is the all-powerful and all-knowing creator of the universe who manifested into flesh through Jesus of Naza-reth. If you happened to be born in India in the twentieth century, there is a very good chance that you are a Hindu who believes that Brahma is the unchanging, infinite, transcendent creator of all matter, energy, time, and space and who manifests into flesh through Ganesha, the blue elephant god who is the most worshipped divinity in India. To an anthro-pologist from Mars, all earthly religions would be indistinguishable at this level of analysis.

Even within the three great Abrahamic religions, who can say which one is right? Christians believe Jesus is the savior and that you must accept him to receive eternal life in heaven. Jews do not accept Jesus as the savior, and neither do Muslims. In fact, only roughly two billion of the world's

5.7 billion believers accept Jesus as their personal savior. Where Christians believe that the Bible is the inerrant gospel handed down from the deity, Muslims believe that the Koran is the perfect word of God. Christians believe that Christ was the latest prophet. Muslims believe that Muhammad is the latest prophet. Mormons believe that Joseph Smith is the latest prophet. And, stretching this track of thought just a bit, Scientologists believe that L. Ron Hubbard is the latest prophet. So many prophets, so little time.

Flood myths show similar cultural influence. Predating the biblical Noachian flood story by centuries, the Epic of Gilgamesh was written around 1800 BCE. Warned by the Babylonian Earth-god Ea that other gods were about to destroy all life by a flood, Utnapishtim was instructed to build an ark in the form of a cube that was 120 cubits (180 feet) in length, breadth, and depth, with seven floors, each divided into nine compartments, and to take aboard one pair of each living creature.

Virgin birth myths likewise spring up throughout time and geography. Among those alleged to have been conceived without the usual assistance from a male were Dionysus, Perseus, Buddha, Attis, Krishna, Horus, Mercury, Romulus, and, of course, Jesus. Consider the parallels between Dionysus, the ancient Greek god of wine, and Jesus of Nazareth. Both were said to have been born from a virgin mother, who was a mortal woman, but were fathered by the king of heaven; both allegedly returned from the dead, transformed water into wine, and introduced the idea of eating and drinking the flesh and blood of the creator, and both were said to have been liberator of mankind.

Resurrection myths are no less culturally constructed. Osiris is the Egyptian god of life, death, and fertility, and is one of the oldest gods for whom records have survived. Osiris first appears in the pyramid texts around 2400 BCE, by which time his following was already well established. Widely worshipped until the compulsory repression of pagan religions in the early Christian era, Osiris was not only the redeemer and merciful judge of the dead in the afterlife, he was also linked to fertility and, most notably (and appropriately for the geography), the flooding of the Nile and growth of crops. The kings of Egypt themselves were inextricably connected with Osiris in death. When Osiris rose from the dead, they would rise also in union with him. By the time of the New Kingdom, not only pharaohs but mortal men believed that they could be resurrected by and with Osiris at death if, of course, they practiced the correct religious rituals. Sound familiar? Osiris predates the Jesus messiah story by at least two and a half millennia.

Shortly after the crucifixion of Jesus there arose another messiah, Apollonius of Asia Minor. His followers claimed he was the son of God, that he was able to walk through closed doors, heal the sick, and cast out demons, and that he raised a dead girl back to life. He was accused of witchcraft, sent to Rome before the court, and was jailed but escaped. After he died his followers claimed he appeared to them and then ascended into heaven. Even as late as the 1890s, the Native American "ghost dance" centered on a Paiute Indian named Wovoka, who during a solar eclipse and fever-induced hallucination received a vision from God "with all the people who had died long ago engaged in their old-time sports and occupations, all happy and forever young. It was a pleasant land and full of game." Wovoka's followers believed that in order to resurrect their ancestors, bring back the buffalo, and drive the white man out of Indian territory, they needed to perform a ceremonial dance that went on for hours and days at a time. The ghost dance united the oppressed Indians but alarmed government agents, and this tension led to the massacre at Wounded Knee. This is what I call the "oppression-redemption" myth, a classic tale of cheating death, overcoming adversity, and throwing off the chains of bondage. You just can't keep a good story down. Why? Because the propensity to tell such stories is hardwired into our brains.

Does God Actually Exist?

Despite the overwhelming evidence that God is hardwired into our brains, believers could reasonably argue (1) that the question "Why do people believe in God?" is a separate question from "Does God exist?" and (2) that the deity hardwired himself into our brains so that we may know him. In other words, the biology of belief is a separate matter from the target of belief. Whether or not belief in God is hardwired into our brains, the question remains: does God actually exist?

What Is God?
Studies by religious scholars reveal that the vast majority of people in the industrial West who believe in God associate themselves with some form of monotheism, in which God is understood to be a being who is: *all pow-*

erful (omnipotent), all knowing (omniscient), and all good (omnibenevolent); who created out of nothing the universe and everything in it; who is uncreated and eternal, a noncorporeal spirit who created, loves, and can grant eternal life to humans. Synonyms include the Almighty, Supreme Being, Supreme Goodness, Most High, Divine Being, Deity, Divinity, God the Father, Divine Father, King of Kings, Lord of Lords, Creator, Author of All Things, Maker of Heaven and Earth, First Cause, Prime Mover, Light of the World, and Sovereign of the Universe.

Do you believe this God exists? Do you deny that this God exists? Or do you withhold judgment on this God's existence? These are the three questions the theologian Doug Geivett, a professor at the Talbot School of Theology at Biola University in Los Angeles, offers in our public debates on God's existence, demanding that I and audience members choose one. My response is twofold:

1. The burden of proof is on the believer to prove God's existence, not on the nonbeliever to disprove God's existence. Although we cannot prove a negative, I can just as easily argue that I cannot prove that there is no Isis, Zeus, Apollo, Brahma, Ganesha, Mithra, Allah, Yahweh, or even the Flying Spaghetti Monster. But the inability to disprove these gods in no way makes them legitimate objects of belief (let alone worship).
2. There is evidence that God and religion are human and social constructions based on research from psychology, anthropology, history, comparative mythology, and sociology.

Let's look at these two matters more closely.

Theist, Atheist, Agnostic, and the Burden of Proof

I once saw a bumper sticker that read "Militant Agnostic: I Don't Know and You Don't Either." This is my position on God's existence: I don't know and you don't either. But what does it mean to be an agnostic? Isn't that someone who is withholding judgment until more evidence is gathered? Earlier in the book I said that I don't believe in God, so doesn't that make me an atheist? It all depends on how these terms are defined, and for that we should turn to the Oxford English Dictionary, our finest source for the history of word usage: *Theism* is "belief in a deity, or deities" and "belief in one God as creator and supreme ruler of the universe."

Atheism is "Disbelief in, or denial of, the existence of a God." *Agnosticism* is "unknowing, unknown, unknowable."

Agnosticism was coined in 1869 by Thomas Henry Huxley—Darwin's friend and most enthusiastic public explainer of evolution—to describe his own beliefs: "When I reached intellectual maturity and began to ask myself whether I was an atheist, a theist, or a pantheist . . . I found that the more I learned and reflected, the less ready was the answer. They [believers] were quite sure they had attained a certain 'gnosis,'—had, more or less successfully, solved the problem of existence; while I was quite sure I had not, and had a pretty strong conviction that the problem was insoluble."[16] I, too, am convinced that the God question is insoluble.

Of course, no one is agnostic behaviorally. When we act in the world, we act as if there is a God or as if there is no God, so by default we must make a choice, if not intellectually then at least behaviorally. To this extent, I assume that there is no God and I live my life accordingly, which makes me an atheist. In other words, agnosticism is an intellectual position, a statement about the existence or nonexistence of the deity and our ability to know it with certainty, whereas atheism is a behavioral position, a statement about what assumptions we make about the world in which we behave.

Despite the fact that virtually everyone labels me an atheist, I prefer to call myself a skeptic. Why? Words matter and labels carry baggage. When most people employ the word *atheist*, they are thinking of *strong atheism* that asserts that God does not exist, which is not a tenable position (you cannot prove a negative). *Weak atheism* simply withholds belief in God for lack of evidence, which we all practice for nearly all the gods ever believed in history. As well, people tend to equate atheism with certain political, economic, and social ideologies, such as communism, socialism, extreme liberalism, moral relativism, and the like. Since I am a fiscally conservative civil libertarian, and most definitely not a moral relativist, this association does not fit. Yes, we can try redefining atheism in a more positive direction—which I do regularly—but since I publish a magazine called *Skeptic* and write a monthly column for *Scientific American* called "Skeptic," I prefer that as my label. A skeptic simply does not believe a knowledge claim until sufficient evidence is presented to reject the null hypothesis (that a knowledge claim is not true until proven otherwise). I do not know that there is no God, but I do not believe in God, and have good reasons to think that the concept of God is socially and psychologically constructed.

The problem we face with the God question is that certainty is not possible when we bump up against such ultimate questions as "What was there before time began?" or "If the big bang marked the beginning of all time, space, and matter, what triggered this first act of creation?" The fact that science presents us with a question mark on such questions doesn't faze scientists because theologians hit the same epistemological wall. You just have to push them one more step. In my debates and dialogues with theologians, theists, and believers, the exchange usually goes something like this for the question of what triggered the big bang, or the first act of creation:

God did it.

Who created God?

God is he who needs not be created.

Why can't the universe be "that which needs not be created"?

The universe is a thing or an event, whereas God is an agent or being, and things and events have to be created by something, but an agent or being does not.

Isn't God a thing if he is part of the universe?

God is not a thing. God is an agent or being.

Don't agents and beings have to be created as well? We're an agent, a being—a human being in fact. We agree that human beings need an explanation for our origin. So why does this causal reasoning not apply to God as agent and being?

God is outside of time, space, and matter, and thus needs no explanation.

If that is the case, then it is not possible for any of us to know if there is a God or not because, by definition, as finite beings operating exclusively within the world we can only know other natural and finite beings and objects. It is not possible for a natural finite being to know a supernatural infinite being.

At this point in the debate my erstwhile theological opponents typically turn to ancillary arguments for God's existence, such as personal revelation, which by definition is personal and thus cannot serve as evidence to others who have not shared that revelatory experience. Or, theists will invoke facts and miracles peculiar to their particular faith, such as Muslims as the fastest growing religion, or Judaism as the oldest religion that has survived millennia of attempts to eradicate it, or Christians who believe that the disciples would never have gone to their deaths defending their faith were such miracles as the resurrection not true. In all three cases the assumption is that millions of followers cannot be wrong.

Well, I counter, millions of Mormons believe that their sacred text was dictated in an ancient language onto gold plates by the angel Moroni and buried and subsequently dug up near Palmyra, New York, by Joseph Smith, who then translated them into English by burying his face in a hat containing magic stones. Millions of Scientologists believe that eons ago a galactic warlord named Xenu brought alien beings from another solar system to Earth, placed them in select volcanoes around the world, and then vaporized them with hydrogen bombs, scattering to the winds their thetans (souls), which attach themselves to people today, leading to drug and alcohol abuse, addiction, depression, and other psychological and social ailments that only Scientology can cure. Clearly the veracity of a proposition is independent of the number of people who believe it.

The burden of proof is on believers to prove God's existence—not on nonbelievers to disprove it—and to date theists have failed to prove God's existence, at least by the high evidentiary standards of science and reason. So we return again to the nature of belief and the origin of belief in God. I have built a strong case that belief in a supernatural agent with intention is hardwired in our brains, and that the agent as God was created by humans and not vice versa.

Shermer's Last Law and the Scientific Search for God

For most theists, God's existence is not a matter of blind faith, circumstantial geography, or cultural construction. They know that God is real, and they have as much confidence in that knowledge—and often much more—as they have in many other claims to knowledge. Atheists also affirm the belief that God's existence is knowable. By making the argument that there is insufficient evidence for God's existence, they are including God in the epistemological arena of the empirical sciences. If sufficient evidence did emerge that God is real, atheists should—at least in principle—assent to his existence. Would they? What evidence would be sufficient that both theists and atheists would agree to settle the issue once and for all? I contend that there is none. (This is another reason why I prefer to call myself an agnostic or a skeptic.) Here's why.

Most theists believe that God created the universe and everything in it, including stars, planets, and life. My question is this: how could we distinguish an omnipotent and omniscient God or Intelligent Designer (ID) from an extremely powerful and really smart extraterrestrial intelligence (ETI)? That is, if we go in search of such a being—as both theists and atheists claim to be doing—we encounter a problem that I call (pace Arthur C. Clarke[17]) Shermer's last law: *any sufficiently advanced extraterrestrial intelligence is indistinguishable from God.*[18]

My gambit (ET = ID = God) arises from an integration of evolutionary theory, intelligent design creationism, and the SETI (Search for Extraterrestrial Intelligence) program, and can be derived from the following observations and deductions.

Observation I. Biological evolution is glacially slow compared to technological evolution. The reason is that biological evolution is Darwinian and requires generations of differential reproductive success, whereas technological evolution is Lamarckian and can be implemented within a single generation.

Observation II. The cosmos is very big and space is very empty, so the probability of making contact with an ETI is remote. By example, the speed of our most distant spacecraft, *Voyager I*, relative to the sun is 17.246 kilometers per second, or 38,578 miles per hour. If *Voyager I* was heading toward the closest star system to us (which it isn't)—the Alpha Centauri system at 4.3 light-years away—it would take an almost unfathomable 74,912 years to get there.

Deduction I. The probability of making contact with an ETI who is only slightly more advanced than us is virtually nil. Any ETIs we would encounter will either be way behind us (in which case we could only encounter them by landing on their planet) or way ahead of us (in which case we would encounter them either through telecommunications or by their landing on our planet). How far ahead of us is an ETI likely to be?

Observation III. Science and technology have changed our world more in the past century than it changed in the previous hundred centuries—it took ten thousand years to get from the cart to the airplane, but only sixty-six years to get from powered flight to a lunar landing. Moore's law of computer power doubling every eighteen months continues unabated and is now down to about a year. Computer scientists calculate that there have been thirty-two doublings since World War II, and that as early as 2030 we may encounter the singularity—the point at which total computational power will rise to levels that are so far beyond anything we can imagine that they will appear nearly infinite and thus, relatively speaking, be indistinguishable from omniscience. When this happens the world will change more in a decade than it did in the previous thousand decades.[19]

Deduction II. Extrapolate these trend lines out tens of thousands, hundreds of thousands, or even millions of years—mere eye blinks on an evolutionary time scale—and we arrive at a realistic estimate of how far advanced an ETI will be. Consider something as relatively simple as DNA. We can already engineer genes after only fifty years of genetic science. An ETI that was fifty thousand years ahead of us would surely be able to construct entire genomes, cells, multicellular life, and complex ecosystems. (At the time of this writing the geneticist J. Craig Venter produced the first artificial genome and constructed synthetic bacteria that were chemically controlled by the artificial genome.[20]) The design of life is, after all, just a technical problem in molecular manipulation. To our not-so-distant descendants, or to an ETI we might encounter, the ability to create life will be simply a matter of technological skill.

Deduction III. If today we can engineer genes, clone mammals, and manipulate stem cells with science and technologies developed in only the past half century, think of what an ETI could do with fifty thousand years of equivalent powers of progress in science and technology. For an ETI who is a million years more advanced than we are, engineering the creation of planets and stars may be entirely possible.[21] And if universes are created out of collapsing black holes—which some cosmologists think is probable—it is not inconceivable that a sufficiently advanced ETI could even create a universe by triggering the collapse of a star into a black hole.[22]

What would we call an intelligent being capable of engineering life, planets, stars, and even universes? If we knew the underlying science and technology used to do the engineering, we would call it an extraterrestrial intelligence; if we did not know the underlying science and technology, we would call it God.

Einstein's God

Inevitably in discussions about science and God, the matter of Albert Einstein's religious beliefs arises, with theists and New Age spiritualists of various stripes clamoring to claim the great physicist as one of their own. With careful quote mining one can find support for Einstein as a believer of some sort. To wit: "God is cunning but He is not malicious," "God does not play dice," and "I want to know how God created the world. I am not interested in this or that phenomenon, in the spectrum of this or that element. I want to know His thoughts, the rest are details." In the final weeks of his life, when Einstein learned of the death of his old physicist friend Michele Besso, he wrote the Besso family: "He has departed from this strange world a little ahead of me. That means nothing. For us believing physicists, the distinction between past, present and future is only a stubborn illusion."

What did Einstein mean by "God" playing dice, or "us believing physicists"? Was he speaking literally or metaphorically about the deity? Did he mean belief in the models of theoretical physics that make no distinction between past, present, and future? Did he mean belief in some impersonal force that exists above such time constraints? Was he just being polite and consoling to Besso's family? Such is the enigma of the most well-known scientist in history whose fame was such that nearly everything he wrote or said was scrutinized for its meaning and import. It is easy to yank such quotes out of context and spin them in any direction one desires. Much has been written about Einstein, but until recently his literary executors protected his convoluted and controversial personal life so carefully that we knew only snippets of what was going on outside Einstein's scientific mind and social circle. Until now. Thanks to the Einstein Papers Project under the direction of Diana Kormos-Buchwald at the California Institute of Technology in Pasadena, California, the archival materials are now available to tell the full story, which Walter Isaacson did in his magisterial biography of Einstein.[23]

Einstein's Jewish identity was undeniably important to all aspects of his life, especially and including his politics. After declining the presidency of Israel, Einstein wrote: "My relationship to the Jewish people has become my strongest human tie."[24] The religiosity of his childhood still compelled him in midlife: "Try and penetrate with our limited means the secrets of nature and you will find that, behind all the discernible laws and connections, there remains something subtle, intangible and inexplicable. Veneration for this force beyond anything that we can comprehend is my religion. To that extent I am, in fact, religious."[25]

Being religious in some esoteric sense of the awe and wonder over the cosmos is one thing, but what about God, particularly Yahweh, the God of Abraham, Einstein's own patriarch? When he turned fifty, Einstein granted an interview in which he was asked point-blank, do you believe in God? "I am not an atheist," he began.

> The problem involved is too vast for our limited minds. We are in the position of a little child entering a huge library filled with books in many languages. The child knows someone must have written those books. It does not know how. It does not understand the languages in which they are written. The child dimly suspects a mysterious order in the arrangement of the books but doesn't know what it is. That, it seems to me, is the attitude of even the most intelligent human being toward God. We see the universe marvelously arranged and obeying certain laws but only dimly understand these laws.[26]

That almost sounds like Einstein is attributing the laws of the universe to a God of some sort. But what type of God, a personal deity or some amorphous force? To a Colorado banker who wrote and asked him the God question, Einstein responded:

> I cannot conceive of a personal God who would directly influence the actions of individuals or would sit in judgment on creatures of his own creation. My religiosity consists of a humble admiration of the infinitely superior spirit that reveals itself in the little that we can comprehend about the knowable world. That deeply emotional conviction of the presence of a superior reasoning power, which is revealed in the incomprehensible universe, forms my idea of God.[27]

The most famous Einstein pronouncement on God came in the form of a telegram, in which he was asked to answer the question in fifty words or less. He did it in thirty-two: "I believe in Spinoza's God, who reveals himself in the lawful harmony of all that exists, but not in a God who concerns himself with the fate and the doings of mankind."[28]

Finally, if any doubt remains, in a 1997 issue of *Skeptic* magazine we published an article by one of our editors, Michael Gilmore, who had recently met a World War II U.S. Navy veteran named Guy H. Raner, who corresponded with Einstein on this very question. We republished those letters in their entirety for the first time anywhere.[29] In the first letter, dated June 14, 1945, sent from the USS *Bougainville* in the Pacific Ocean, Raner recounts a conversation he had on the ship with a Jesuit-educated Catholic officer who claimed that Einstein converted from atheism to theism when he was confronted by a Jesuit priest with three irrefutable syllogisms. "The syllogisms were: A design demands a designer; the universe is a design; therefore there must have been a designer." Raner countered the Catholic by noting that cosmology and evolutionary theory adequately explain most apparent design in the world, "but even if there was a 'designer,' that would give only a re-arranger, not a creator; and again assuming a designer, you are back where you started by being forced to admit a designer of the designer etc. etc. Same as the account of the earth resting on an elephant's back—elephant standing on a giant turtle; turtle on turtle on turtle, etc."

At this point in his life Einstein was world famous and routinely received hundreds of such letters, many from prominent scholars and scientists, so for him to write a lowly ensign aboard a ship in the middle of the Pacific Ocean reveals how much this story got his goat. On July 2, 1945, Einstein fired back:

> I received your letter of June 10th. I have never talked to a Jesuit priest in my life and I am astonished by the audacity to tell such lies about me. From the viewpoint of a Jesuit priest I am, of course, and have always been an atheist. Your counter-arguments seem to me very correct and could hardly be better formulated. It is always misleading to use anthropomorphical concepts in dealing with things outside the human sphere—childish analogies. We have to admire in humility the beautiful harmony of the structure of this world—as far as we can grasp it. And that is all.

Four years later, in 1949, Raner wrote Einstein again, asking for clarification: "Some people might interpret (your letter) to mean that to a Jesuit priest, anyone not a Roman Catholic is an atheist, and that you are in fact an orthodox Jew, or a Deist, or something else. Did you mean to leave room for such an interpretation, or are you from the viewpoint of the dictionary an atheist; i.e., 'one who disbelieves in the existence of a God, or a Supreme Being'?" Einstein responded on September 28, 1949:

I have repeatedly said that in my opinion the idea of a personal God is a childlike one. You may call me an agnostic, but I do not share the crusading spirit of the professional atheist whose fervor is mostly due to a painful act of liberation from the fetters of religious indoctrination received in youth. I prefer an attitude of humility corresponding to the weakness of our intellectual understanding of nature and of our own being.

Has there ever been a prominent figure who was so clear about what he believes as Einstein, and yet so egregiously misunderstood? This is yet another example of belief blindness.

The Natural and the Supernatural

Science operates in the natural, not the supernatural. In fact, there is no such thing as the supernatural or the paranormal. There is just the natural, the normal, and mysteries we have yet to explain by natural causes. Invoking such words as *supernatural* and *paranormal* just provides a linguistic placeholder until we find natural and normal causes, or we do not find them and discontinue the search out of lack of interest. This is what usually happens in science. Mysteries once thought to be supernatural or paranormal happenings—such as astronomical or meteorological events—are incorporated into science once their causes are understood. For example, when cosmologists reference "dark energy" and "dark matter" to the so-called missing energy and mass needed to explain the structure and motion of galaxies and galaxy clusters, they do not intend these descriptors to be causal explanations. Dark energy and dark matter are merely cognitive conveniences until the actual sources of the energy and matter are discovered. When theists, creationists, and intelligent design theorists invoke miracles and acts of creation ex nihilo, that is the end of the search for them, whereas for scientists the identification of such mysteries is only the beginning. Science picks up where theology leaves off. When a theist says "and then a miracle happens," as wittily portrayed in my favorite Sydney Harris cartoon of the two mathematicians at the chalkboard with the invocation tucked in the middle of a string of equations, I quote from the cartoon's caption: "I think you need to be more explicit here in step two."

To our Bronze Age ancestors who created the great monotheistic religions, the ability to create the world and life was godlike. Once we know the technology of creation, however, the supernatural becomes the natural. Thus my gambit: the only God that science could discover would be a natural being, an entity that exists in space and time and is constrained by the laws of nature. A supernatural God who exists outside of space and time is not knowable to science because he is not part of the natural world, and therefore science cannot know God.

This was the argument I made in a Templeton Foundation–sponsored print debate with theist and Harvard professor of medicine Jerome Groopman, who in his comments argued that God is "without form, immeasurable," that he exists "in a dimension that cannot be quantitated or depicted by science," that "we are unable to grasp fully God's nature and dimensions," and that "God exists outside of time and cannot be bound by space." How then, I asked, do you know this God exists? As corporeal beings who form beliefs about the world based on percepts (from our senses) and concepts (from our minds), how can we possibly know a being who by definition lies outside of both our percepts and our concepts? At some point doesn't God need to step into our space-time to make himself known in some manner—say through prayer, providence, or miracles? And if so, why can't science measure such divine action? If there is some other way of knowing, say that of the mystics or the faithful through deep meditation or prayer, why couldn't neuroscience say something meaningful about that process of knowing? If we came to understand—as studies with meditating monks and praying priests have shown—that a part of the parietal lobe of the brain associated with the orientation of the body in space is quiescent during such meditative states (breaking down the normal distinction one feels between self and nonself and thus making one feel "at one" with the environment), wouldn't this imply that rather than being in touch with a being outside of space and time, it is actually just a change in neurochemistry?

In the end, in one of the most nakedly honest statements of belief that I have ever encountered, Groopman had to admit: "Why believe? I have no rational answer. The question seems to be in the domain of why do we love someone? You could reduce it to certain components, perhaps refer to neurotransmitters, but somehow the answer seems to transcend the truly knowable. This is the cognitive dissonance that people like me live with, and with which we often struggle."[30]

On one level I have no rebuttal to this belief statement because none is necessary. If no empirical claim is made, then there is little more that science can say on the matter. Life can be a painful struggle and filled with mysteries, so whatever one needs to do to get through the day to find happiness and to bring some resolution to those nagging mysteries . . . well . . . who am I to argue? As declared in Psalms 46:1: "God is our refuge and strength, a very present help in trouble." On another level, however, I can't help but think that had Groopman been born to Hindu parents in India rather than Jewish parents in the West, he would believe something entirely different about the ultimate nature of the universe that would be equally subject to justification through rational arguments.

What science offers for explaining the feelings we experience when believing in God or falling in love is complementary, not conflicting; additive, not detractive. I find it deeply interesting to know that when I fall in love with someone my initial lustful feelings are enhanced by dopamine, a neurohormone produced by the hypothalamus that triggers the release of testosterone, the hormone that drives sexual desire, and that my deeper feelings of attachment are reinforced by oxytocin, a hormone synthesized in the hypothalamus and secreted into the blood by the pituitary. Further, it is instructive to know that such hormone-induced neural pathways are exclusive to monogamous pair-bonded species as an evolutionary adaptation for the long-term care of helpless infants. We fall in love because our children need us! Does this in any way lessen the qualitative experience of falling in love and doting on one's children? Of course not, any more than unweaving a rainbow into its constituent parts reduces the aesthetic appreciation of the rainbow.

Religious faith and belief in God have equally adaptive evolutionary explanations. Religion is a social institution that evolved to reinforce group cohesion and moral behavior. It is an integral mechanism of human culture to encourage altruism, reciprocal altruism, and indirect altruism, and to reveal the level of commitment to cooperate and reciprocate among members of a social community. Believing in God provides an explanation for our universe, our world, and ourselves; it explains where we came from, why we are here, and where we are going. God is also the ultimate enforcer of the rules, the final arbiter of moral dilemmas, and the pinnacle object of commitment.

It is time to step out of our evolutionary heritage and our historical traditions and embrace science as the best tool ever devised for explaining how the world works. It is time to work together to create a social and political world that embraces moral principles and yet allows for natural human diversity to flourish. Religion cannot get us there because it has no systematic methods of explanation of the natural world, and no means of conflict resolution on moral issues when members of competing sects hold absolute beliefs that are mutually exclusive. Flawed as they may be, science and the secular Enlightenment values expressed in Western democracies are our best hope for survival.

9

Belief in Aliens

In the spring of 1999, I appeared on the southern California NPR affiliate radio station KPCC with Joe Firmage, author of the immodestly titled book *The Truth*. Firmage is a young man best known as the founder and original CEO of the Internet giant USWeb, a company then valued at around $3 billion. Unlike most CEO authors, however, Firmage was not on a book tour to tout his pearls of wisdom for constructing a Silicon Valley powerhouse; rather, Firmage wanted to talk about building a powerhouse of a different sort, one that can carry humans to the stars . . . and beyond.[1]

Where does a Silicon Valley Internet phenom get inspiration for such an undertaking? It began in the early morning hours of a fall day in 1997, when Firmage was awakened to see, in his words, "a remarkable being, clothed in brilliant white light hovering over my bed." This alien being spoke to Firmage, asking him "Why have you called me here?" Firmage replied, "I want to travel in space." The alien wondered why such a wish should be granted. "Because I'm willing to die for it," Firmage explained. Now *that's* a commitment any form of intelligence could understand. At this point, Firmage says, out of the alien being "emerged an electric blue sphere, just smaller than a basketball. . . . It left his body, floated down and entered me. Instantly I was overcome by the most unimaginable ecstasy I have ever experienced, a pleasure vastly beyond orgasm. . . . Something had been given to me."[2]

How powerful is such an experience to change the course of a person's life? Firmage promptly announced his resignation from his billion-dollar company and went out and founded the International Space Sciences Organization, which according to its Web page seeks "to advance human understanding of the fundamental nature and functions of matter and energy, yielding breakthroughs in propulsion, energy generation, and likely a deeper appreciation for the physical processes underlying consciousness."[3] Now *that* is a testament (literally) to the power of belief.

Firmage set fingers to keyboard and cranked out an ambitious 244-page manuscript. He titled it *The Truth*, since it includes his goal of convincing the "scientific establishment" of the reality of UFOs and such advanced technologies as zero-point energy from the vacuum of space, "propellantless propulsion" and "gravitational propulsion" for "greater-than-light" travel, "vacuum fluctuations" to alter "gravitational and inertial masses," and other forms of alternative space propulsion systems.[4] In fact, says Firmage, for thousands of years we humans have been "nudged" along our technological trajectory by periodic contact with advanced "teachers" willing to share their knowledge with us, the latest being in 1947 at Roswell, New Mexico. As he waxes poetic in his book:

Teachers have taught us
through the ages.
They are watching us now.
The Cosmos is their ocean
and they have been mindful
of our need to develop.[5]

To encourage further alien contact and technological development, Firmage invested $3 million into the founding of Project Kairos (Greek for "opportune moment") to prepare humanity for future contacts. "Imagine that one day a new city is constructed somewhere on Earth, a 'Universe City', where a spacetime port is established as a centerpoint of interaction among Earth-dwellers and visitors from elsewhere," Firmage fantasizes.[6]

"Why would a young, successful CEO risk his reputation on something this fantastic?" Firmage asked a reporter rhetorically. "Because I believe so much in this theory. And I am in a unique position to communicate an extremely important message. I have the money, credibility, scientific grounding and faith."[7]

Faith is the operative word here. Joe Firmage loves science, but it is his faith that powers his beliefs. In considering the nature of the cosmos and life, we see my thesis of belief-dependent realism that beliefs come first and reasons for belief come second once again borne out in Firmage's explanation that "there is one concept of which I am logically and totally convinced—which science has utterly failed to teach me directly—but which religion has long held and somewhat rationally explained in its internal structure: There can be no question that the Cosmos is the product of intent." Intention implies agency, and an agent is a being, in this case a being outside our world who gives us meaning and hope: "It is in this concept of intended creation, or being, that the emotive feeling of meaning has a place for discussion within the mechanical laws of physics. The physicality of intention allows the physicist in me to incorporate an understanding of emotion into the laws that govern the universe."[8]

The physicality of intention. This is the very embodiment of agenticity.

Interestingly, Firmage was raised Mormon, and one of the fundamental beliefs of the Mormon church is that its founder, Joseph Smith, was contacted by the angel Moroni, who directed him to the sacred golden tablets from which the Book of Mormon was written. In *The Truth*, Firmage explains that the revelation "was received by a man named Joseph Smith, whose descriptions of encounters with brilliant, white-clothed beings are almost indistinguishable from many modern-day accounts of first-hand encounters with 'visitors.'"[9] So, Joseph Smith had a close encounter of the third kind. And according to Firmage, Smith was by no means the first. Eighteen centuries earlier, St. John the Divine received his "revelation" from which the last book in the Bible was written, and shortly before that a Jewish carpenter from Nazareth encountered an intentional agent of the highest order. Before Jesus there was Moses and the burning bush, who spoke to him as "I am who I am." From Moses to Jesus to St. John the Divine to Joseph Smith to Joseph Firmage—an unbroken lineage of mortal humans touched by alien agents.

Alien Agenticity

Over the years I have appeared on numerous television shows with alien abductees. I have little doubt that most of them are genuine in their recounting of the emotional trauma of the experience of being abducted. One of these abductees was Whitley Strieber, author of the spectacularly best-selling account of his abduction, *Communion*, which has become the bible of the alien abduction community. I met Strieber in the green room at Bill Maher's television series *Politically Incorrect*. While we were chatting before the taping of the show began, I asked him what he did when he wasn't writing about being abducted by aliens. He told me that he writes science fiction, fantasy, and horror novels. "Of course!" I thought to myself. "He either made it all up or fantasized it in his creative imagination."

The key word here is *imagination*. People often seem incredulous that anyone could concoct such fantastic stories of alien encounters, implying that they must therefore have some measure of verisimilitude. In fact, people make up such encounters every day. They're called science fiction and fantasy writers. Consider the alternative worlds of Harry Potter, *The Lord of the Rings*, *Star Wars*, *Star Trek*, *Avatar*, and the rest. We have the fantastic ability to project ourselves into other worlds of make-believe, and the line between conscious fiction and subconscious imagining is a fine one. Reality and fantasy may blur in the recesses of the mind and come to the forefront under certain conditions, such as hypnosis and sleep.

Hypnosis. Many of these abduction experiences are "remembered" years or decades after the fact through a technique called *hypnotic regression*, in which a subject is hypnotized and asked to imagine regressing back in time to retrieve a memory from the past, and then play it back on the imaginary screen of the mind, as if there's a diminutive homunculus sitting inside a little theater in the head reporting to the brain's director what he is seeing. This is not at all how memory works. The metaphor of memory as a videotape-playback system is completely wrong. There is no recording device in the brain. Memories are formed as part of the association learning system of making connections between things and events in the environment, and repetitive associations between them generate new dendritic and synaptic connections between neurons, which are then strengthened through additional repetition or weakened through disuse. Use it or lose it.

Do you remember your tenth birthday, or do you remember your mother's memory of your tenth birthday that she recalled for you when you were fifteen, or is it the photographs of your tenth birthday that you reviewed when you were twenty? It is likely all of the above, and much more. So, when an alien abductee is "recovering" a memory of an abduction experience, what is actually being recovered? Analysis of hypnotic regression tapes used by abduction "therapists" who employ hypnosis shows that they ask leading questions and construct imaginary scenarios through which their subjects may concoct an entirely artificial event of something that never happened.[10] In fact, memory contamination through suggestive questioning by the hypnotist and by the imagination of the hypnotized person is what happened in the disastrous "recovered memory movement" of the 1990s that resulted in dozens of fathers being convicted of child molestation based on nothing more than adult women's "recovered memories" planted by therapists.

Sleep anomalies. Abduction experiences that are not generated through hypnotic regression typically occur late at night or early in the morning during sleep cycles that strongly resemble *hypnagogic* (just after falling asleep) and *hypnopompic* (just before waking up) hallucinations, and appear to be related to lucid dreams and *sleep paralysis*, which have been well documented among subjects in experiments and patients in sleep labs and contain most of the components of the abduction experience. Hypnagogic and hypnopompic hallucinations occur in the fuzzy borderlands between wakefulness and sleep, when our conscious brain slips into unconsciousness as we fall asleep, or transitions into wakefulness from sleep. Reality and fantasy blur. Multiple sensory modalities may be involved, including and especially seeing and hearing things that are not actually there, such as speckles, lines, geometrical patterns, or representational images. Such hallucinatory images may be in black and white or in color, still or moving, flat or 3-D, and sometimes even include the spiraling tunnels reported by people who have out-of-body and near-death experiences.

Auditory components are also sometimes part of the hallucinatory experience, such as hearing your own name called out, the sound of a doorbell or rapping sound on the door, and even fragments of speech from others imagined to be in the room. A *lucid dream* is stronger still. This is a dream in which the sleeping person is aware that he or she is asleep and dreaming, but can participate in and alter the dream itself. *Sleep paralysis* is a type of lucid dream in which the dreamer, aware of the dream, also senses paralysis, pressure on chest, presence of a being in the room, floating, flying, falling, or leaving one's body, with an emotional component that includes an element of terror, but sometimes also excitement, exhilaration, rapture, or ecstasy. Psychologist J. Allan Cheyne has documented thousands of cases of sleep paralysis and believes that they are associated with the temporal lobes as well as the parietal lobes, which are associated with how the brain orients the body in space.[11]

Several centuries ago, the English referred to nighttime sensations of chest pressure from witches or other supernatural beings as the "mare," from Anglo-Saxon *merran*, or "to crush." So a *nightmare* was believed to represent a crusher who comes in the night. Since they lived in a demon-haunted world, they called these crushers *demons*. Since we live in an alien-haunted world, we call them *aliens*. Your culture dictates what labels to assign these anomalous brain experiences.

The power of these beliefs is unmistakable and the experience can lead to a condition similar to post-traumatic stress disorder (PTSD), a fact demonstrated by Harvard University psychologists Richard J. McNally and Susan A. Clancy in a 2004 paper entitled "Psychophysiological Responding During Script-Driven Imagery in People Reporting Abduction by Space Aliens." McNally, Clancy, and their colleagues measured heart rate, skin conductance, and brain wave activity of people who claimed to have been abducted by aliens, as they relived their experiences through script-driven imagery. "Relative to control participants," the authors concluded, "abductees exhibited greater psychophysiological reactivity to abduction and stressful scripts than to positive and neutral scripts."[12] That is, some fantasies are indistinguishable from reality and they can be just as traumatic. McNally noted in his 2003 book, *Remembering Trauma*, "The fact that people who believe they have been abducted by space aliens respond like PTSD patients to audiotaped scripts describing their alleged abductions, underscores the power of belief to drive a physiology consistent with actual traumatic experience."[13] In addition, McNally found that abductees "were much more prone to exhibit false recall and false recognition in the laboratory than were control subjects" and they scored significantly higher than normal on a questionnaire measuring "absorption," a trait related to fantasy proneness that also predicts false recall.

The vividness of a traumatic memory cannot be taken as evidence of its authenticity, an effect subsequently documented by Susan Clancy in her follow-up 2005 book-length study of the phenomenon, *Abducted*, noting that abduction beliefs provide "the same things that millions of people the world over derive from their religions: meaning, reassurance, mystical revelation, spirituality, transformation."[14] Respectfully disagreeing with Carl Sagan, who argued that belief in pseudoscience was directly proportional to misunderstanding of science, Clancy concluded her study by noting:

> The abductees taught me that people go through life trying on belief systems for size. Some of these belief systems speak to powerful emotional needs that have little to do with science—the need to feel less alone in the world, the desire to have special powers or abilities, the longing to know that there is something out there, something more important than you that's watching over you. Belief in alien abduction is not just bad science. It's not just an explanation for misfortune and a way to avoid taking responsibility for personal problems. For many people, belief in alien abduction gratifies spiritual hungers. It reassures them about their place in the universe and their own significance.[15]

I have often recounted my own alien abduction experience that happened in the 1983 bicycle Race Across America while I was traversing Nebraska. I had decided that I had slept too much in the 1982 race and I was curious to see how far I could ride in the 1983 event without stopping for sleep. I made it 1,259 miles in eighty-three hours, to the outskirts of a tiny town called Haigler. I was sleepily weaving down the road when my support motor home flashed its brights and pulled alongside while my crew entreated me to take a sleep break. At that moment, a distant memory of the 1960s television series *The Invaders* was inculcated into my waking dream. In that TV series, alien beings were taking over the earth by replicating actual people but, inexplicably, they each retained a stiff little finger. Suddenly my support team was transmogrified into aliens. I stared intensely at their fingers, grilled my mechanic on bike technology, and interrogated my girlfriend on intimacies that aliens could not possibly know (could they?). There, on the side of the road in the middle of the night, in full cycling regalia with my bike firmly between my legs for a quick getaway, I argued with the aliens, trying to avoid being abducted into the mother craft hovering nearby. I finally relented and went inside, only to discover that the interior of the UFO looked remarkably like a GMC motor home, so I lay down for the proverbial examination probe. Ninety minutes later, after a refreshing sleep break (and thankfully no probes), I was back on the bike cruising down the highway mildly amused by what had just happened. When the sun came up I had a good laugh about it with my support crew, and that evening I recounted the hallucination to the ABC *Wide World of Sports* camera crew, which can be viewed on YouTube.[16]

The bottom line is this: stories of UFOs and alien abductions are far more likely to be due to known psychological effects of terrestrial beings rather than to the unknown physical characteristics of extraterrestrial beings.[17]

Are We Alone in the Universe?

Are we alone in the universe? It is a legitimate question irrespective of how belief systems operate, and at this point science offers us an unambiguously ambiguous answer: we do not know. The answer still eludes us because no contact has yet been made. Why not? Whole books have been written to answer the question,[18] and there are at least fifty answers to what is known as Fermi's paradox—assuming the Copernican principle that we are not special, there should be lots of ETIs out there, and if so then at least some of them would have figured out self-replicating robotic spacecraft and/or practical interstellar space travel themselves, and assuming that at least some of those would be millions of years ahead of us on an evolutionary time scale, their technologies would be advanced enough to have found us by now, but they haven't, so . . . where are they?[19] Here is my Twitter-sized answer (140 characters): *ETIs are probably out there but they have not been here because of the vast interstellar distances and their extreme rarity. Keep searching!*

The Search for Extraterrestrial Intelligence (SETI) is a problem in patternicity and trying to discern a meaningful pattern of a communication signal from the background noise of space. SETI scientists have worked out systematic algorithms and rigorous standards for what would constitute a legitimate signal, a process that was simplified effectively by Carl Sagan in *Contact*, where the ETIs reasoned that sending a sequence of prime numbers would be distinguishable from, say, the signals produced by rotating neutron stars. To date, no such signal has been detected, and

SETI scientists continue to improve the technologies to broaden the spectrum of electromagnetic energy from which they can search the skies, along with the number of possible star systems that can be scanned at any one time. It is truly a needle-in-a-haystack problem, with a couple of hundred billion stars in our galaxy alone boggling the technological minds that do the searching.

Will ET Look Anything Like Us?

One aspect of alien agenticity that has always bothered me is the depiction of ET as a bipedal primate with very humanlike characteristics. What are the chances of that happening on some other planet? Of the hundreds of millions (perhaps billions) of species that evolved here on our planet, only one lineage evolved into bipedal primates, and only one subspecies of that lineage has survived to this day. If we do encounter extraterrestrial intelligences, what are the odds that they will be anything remotely like us, much less what are typically portrayed by alien abductees as bipedal primates with bulbous heads, large almond-shaped eyes, and some gnarly stuff on their foreheads speaking broken English with a peculiar accent? The odds are not high—not even low, I contend.

Nevertheless, I could be wrong, and no less an evolutionary theorist than Richard Dawkins has challenged me on this very point after the director of his foundation produced a short YouTube video of me in alien garb explaining why I think that the chances are close to zero that intelligent and technically advanced aliens would evolve to be anything like the ones we see in films and hear about in abductee accounts.[20] Dawkins wrote:

> I would agree with [Shermer] in betting against aliens being bipedal primates and I think the point is worth making, but I think he greatly overestimates the odds against. [University of Cambridge paleontologist] Simon Conway-Morris, whose authority is not to be dismissed, thinks it positively likely that aliens would be, in effect, bipedal primates. [Harvard University evolutionary biologist] Ed Wilson gave at least some time to the speculation that, if it had not been for the end-Cretaceous catastrophe, dinosaurs might have produced something like the attached.

Figure 8. A Bipedal Dinosaur as Alien ET
In a rerun of the history of life on earth, if dinosaurs had survived might some of them become bipedal tool users? Paleontologist Dale A. Russell speculated as much in a projection of how a bipedal dinosaur might have evolved into a reptilian humanoid, rendered here by Matt Collins, after Russell's original illustration in D. A. RUSSELL AND R. SEGUIN, *RECONSTRUCTIONS OF THE SMALL CRETACEOUS THEROPOD* STENONYCHOSAURUS INEQUALIS *AND A HYPOTHETICAL DINOSAUROID,* NATIONAL MUSEUMS OF CANADA, NATIONAL MUSEUM OF NATURAL SCIENCES, 1982.

I replied to Dawkins along the lines above—that if something like a smart, technological, bipedal hominoid has a certain level of inevitability because of how evolution unfolds, then it should have happened more than once here. Dawkins's rejoinder to me is enlightening:

> But you are leaping from one extreme to the other. In the film vignette, you implied a quite staggering rarity, so rare that you

don't expect two humanoid life forms in the entire universe. Now you are talking about "a certain inevitability," and pointing out, correctly, that a certain inevitability would predict that humanoids should have evolved more than once on Earth! So yes, we can say that humanoids are *fairly* improbable, but not necessarily all *that* improbable! Anything approaching "a certain inevitability" would mean millions or even billions of humanoid life forms in the universe, simply because the number of available planets is so huge. Now, my guess is intermediate between your two extremes. I agree with you that humanoids are rare; that is indeed suggested by the fact that they have only evolved once on Earth. But I suspect that humanoids are not so very rare as to justify the statistical superlatives that you permitted yourself in the vignette.[21]

Good point. But the problem for both Dawkins and myself is our chauvinism. As Carl Sagan liked to say, we are carbon chauvinists. But we are also oxygen chauvinists, temperature chauvinists, vertebrate chauvinists, mammal chauvinists, primate chauvinists, and many others. The chauvinism that ETIs will communicate via radio signals, that their intelligence will take a form similar to ours, and especially that they are social beings who live in civilizations, are anthropomorphisms that have no basis whatsoever in reality. We cannot even communicate with terrestrial intelligences such as apes and dolphins, so what hubris of us to think that we will be able to decode the communiqués of an ETI millions of years our superior.

Here I strongly suspect that we are blinded by what I call *Protagoras's bias*—"Man is the measure of all things"—when we project ourselves into the alien Other. Consider Neanderthals by comparison. If primate intelligence is so vaunted, why did they not survive?

Neanderthals as ETs

Neanderthals split off from the common ancestor shared with us between 690,000 and 550,000 years ago, and they arrived in Europe at least 242,000 (and perhaps 300,000) years ago, giving them free rein there for a quarter of a million years. They had a cranial capacity just as large as ours (ranging from 1,245 to 1,740 cc, with an average of 1,520 cc compared to our average of 1,560 cc), were physically more robust than us with barrel chests and heavy muscles, and they sported a reasonably complex toolkit of about sixty different tools. On paper it certainly seems reasonable to argue that Neanderthals had a good shot at "becoming us," in the sense of a technologically advanced intelligent species capable of space travel and interstellar communication.

But if we dig deeper we see that there is almost no evidence that Neanderthals would have ever "advanced" beyond where they were when they disappeared 30,000 years ago. Even though paleoanthropologists disagree about a great many things, there is near total agreement in the literature that Neanderthals were not on their way to becoming "us." They were perfectly well-adapted organisms for their environments.[22]

Paleoanthropologist Richard Klein, in his authoritative work *The Human Career*, concluded that "the archeological record shows that in virtually every detectable aspect—artifacts, site modification, ability to adapt to extreme environments, subsistence, and so forth—the Neanderthals were behaviorally inferior to their modern successors, and to judge from their distinctive morphology, this behavioral inferiority may have been rooted in their biological makeup."[23] Neanderthals had Europe to themselves for at least 250,000 years unrestrained by the presence of other hominids, yet their tools and culture are not only simpler than those of *Homo sapiens*; they show almost no sign of change at all, let alone progress toward social globalization. Paleoanthropologist Richard Leakey noted that Neanderthal tools "remained unchanged for more than 200,000 years—a technological stasis that seems to deny the workings of the fully human mind. Only when the Upper Paleolithic cultures burst onto the scene 35,000 years ago did innovation and arbitrary order become pervasive."[24]

Likewise, Neanderthal art objects are comparatively crude, and there is much controversy over whether many of them were the product of natural causes instead of artificial manipulation.[25] The most striking exception to this is the famous Neanderthal bone flute dated from between 40,000 to 80,000 years ago, which some archaeologists speculate means that the maker was musical. Yet even biologist Christopher Wills, a rare dissenting voice who rejects the inferiority of the Neanderthals, admitted that it is entirely possible that the holes were naturally created by an animal gnawing on the bone, not by some Paleolithic Ian Anderson. And even though Wills argued "Recent important discoveries suggest that toward the end of their career, the Neanderthals might have progressed considerably in their technology," he had to confess that "it is not yet clear whether this happened because of contact with the Cro-Magnons and other more advanced peoples or whether they accomplished these advances without outside help."[26]

Probably the most dramatic claim for the Neanderthals' "humanity" is the burial of their dead, which often included flowers strewn over carefully laid-out bodies in fetal positions. I used this example in my book *How We Believe*, on the origins of religion,[27] but new research is challenging this interpretation. Klein noted that graves "may have been dug simply to remove corpses from habitation areas" and that in sixteen of twenty of the best documented burial sites "the bodies were tightly flexed (in near fetal position), which could imply a burial ritual or simply a desire to dig the smallest possible burial trench."[28] Paleoanthropologist Ian Tattersall agreed: "Even the occasional Neanderthal practice of burying the dead may have been simply a way of discouraging hyena incursions into their living spaces, or have a similar mundane explanation, for Neanderthal burials lack the 'grave goods' that would attest to ritual and belief in an afterlife."[29]

Much has been made about the possibility of Neanderthal language—that quintessential component of modern intelligence. This is inferential science at best, since soft brain tissue and vocal box structures do not fossilize. Inferences can be drawn from the hyoid bone, which is part of the vocal box structure, as well as the shape of the basicranium, or the bottom of the skull. But the discovery of part of an apparent Neanderthal hyoid bone is inconclusive, said Tattersall: "However the hyoid argument works out, however, when you put the skull-base evidence together with what the archaeological record suggests about the capacities of the Neanderthals and their precursors, it's hard to avoid the conclusion that articulate language, as we recognize it today, is the sole province of fully modern humans."[30]

As for the cranial structure, in mammals the bottom of the cranium is flat but in humans it is arched (related to how high up in the throat the larynx is located). In ancestral hominids the basicranium shows no arching in australopithecines, some in *Homo erectus*, and even more in archaic *Homo sapiens*. In Neanderthals, however, the arching largely disappears, evidence that does not bode well for theories about Neanderthal language, as Leakey concluded: "Judging by their basicrania, the Neanderthals had poorer verbal skills than other archaic sapiens that lived several hundred thousands years earlier. Basicranial flexion in Neanderthals was less advanced even than in *Homo erectus*."[31]

Leakey then speculated, counterfactually, what might have happened had even earlier hominid ancestors survived: "I conjecture that if, by some freak of nature, populations of *Homo habilis* and *Homo erectus* still existed, we would see in them gradations of referential language. The gap between us and the rest of nature would therefore be closed, by our own ancestors."[32] That "freak of nature" is the contingency in our time line that allowed us to survive while no other hominids did, and thus Leakey concluded, "*Homo sapiens* did eventually evolve as a descendant of the first humans, but there was nothing inevitable about it."[33] Ian Tattersall also reasoned in the contingent mode: "If you'd been around at any earlier stage of human evolution, with some knowledge of the past, you might have been able to predict with reasonable accuracy what might be coming up next. *Homo sapiens*, however, is emphatically not an organism that does what its predecessors did, only a little better; it's something very—and potentially very dangerously—different. Something extraordinary, if totally fortuitous, happened with the birth of our species."[34]

Had Neanderthals won and we lost, there is every reason to believe that they would still be living in a Stone Age culture of hunting, fishing, and gathering, roaming the hinterlands of Europe in small bands of a couple of dozen individuals, surviving in a world without towns and cities, without music and art, without science and technology . . . a world so different from our own that it is almost inconceivable.

As for the great apes or monkeys succeeding had humans, Neanderthals, and the rest of our hominid ancestors gone extinct, apes have never shown any inclination toward progressive cultural evolution, now or in the fossil record, and monkeys proliferated throughout Asia and the New World for tens of millions of years without any interference from hominids, yet they didn't take step one toward developing a complex culture.

The fossil record, while still fragmented and desultory, is complete enough now to show us that over the past thirty million years we can conservatively estimate that hundreds of primate species have lived out their lives in the nooks and crannies of rain forests around the world; over the past ten million years dozens of great ape species have forged specialized niches on the planet; and over the last six million years—since the hominid split from the common ancestor of gorillas, chimps, and orangutans—dozens of bipedal, tool-using hominid species have struggled for survival. If these hominids were so inevitable by the laws of evolutionary progress, why is it that only a handful of those myriad pongids and hominids survived? If braininess is such a predictable product of the unfolding powers of nature, then why has only one hominid species managed to survive long enough to ask the question? What happened to those bipedal, tool-using *Australopithecines: anamensis, afarensis, africanus, aethiopicus, robustus, boisei*, and *garhi*? What happened to those big-brained culture-generating *Homos: habilis, rudolfensis, ergaster, erectus, heidelbergensis*, and *neanderthalensis*? If big brains are so great, why did all but one of their owners go extinct?

Historical experiment after experiment reveals the same answer: we are a fluke of nature, a quirk of evolution, a glorious contingency. It is tempting to fall into the oldest trap of all pattern-seeking, storytelling animals: writing yourself into the story as the central pattern in order to find purpose and meaning in this gloriously contingent cosmos. But skeptical alarms should toll whenever anyone claims that science has discovered that our deepest desires and oldest myths are true after all. If there is an inevitability in this story, it is that a purpose-seeking animal will find itself as the purpose of nature. That is what lies at the very core of alien agenticity.

Aliens and Gods

Aliens as intentional agents links the belief to religion and equates aliens with gods. This connection is well documented by the technology historian George Basalla's intriguing book *Civilized Life in the Universe*. Basalla observes, "The idea of the superiority of celestial beings is neither new nor scientific. It is a widespread and old belief in religious thought. Aristotle divided his universe into two distinct regions, the superior celestial realm and the inferior terrestrial realm." The incorporation of Aristotle into Christian theology carried this belief into the Middle Ages. "Christians populated the celestial regions with God, the saints, angelic beings of varying ranks, and the souls of the dead. These immortal celestial beings were superior to mortals, who inhabited the inferior terrestrial realm." Even though the Copernican revolution overturned Aristotelian cosmology, "the belief that creatures living on a distant planet were superior to the human species" hung on into the modern age, and "religious elements continue to adhere to the perception of extra-terrestrial life even as we study it in the twenty-first century."[35]

In 2001, I conducted a study on the pioneers of SETI, most of whom were once religious but became either atheists or agnostics as adults.[36] Radio astronomer Frank Drake—creator of the canonical "Drake equation"—was raised "Very strong Baptist. Sunday school every Sunday," and made this observation: "A strong influence on me, and I think on a lot of SETI people, was the extensive exposure to fundamentalist religion. You find when you talk to people who have been active in SETI that there seems to be that thread. They were either exposed or bombarded with fundamentalist religion. So to some extent it is a reaction to firm religious upbringing."[37] In his 1992 book on the subject, *Is Anyone Out There?*, Drake even suggested that "immortality may be quite common among extraterrestrials."[38] Contact with ETIs would amount to a type of second coming for many people. SETI pioneer Melvin Calvin noted: "It would have a marked effect. It's such a broad, major subject of concern to everyone, no matter where they are, that I think people would listen. It's like introducing a new religion, I suppose, and having it picked up by a lot of people."

Many other scientists and science fiction visionaries agree. The scientist and science fiction writer David Brin suggested that SETI combines "serious and far-reaching science with a kind of gosh-wow zeal that seems (at times) to border on the mystical—perhaps as much religious as a product of science or science fiction. Indeed, to some, contact with advanced alien civilizations may carry much the same transcendental or hopeful significance as any more traditional notion of 'salvation from above.'"[39] In a 2003 speech at Caltech, the science fiction writer extraordinaire Michael Crichton opined that "SETI is unquestionably a religion," noting: "Faith is defined as the firm belief in something for which there is no proof. The belief that there are other life forms in the universe is a matter of faith. There is not a single shred of evidence for any other life forms, and in forty years of searching, none has been discovered. There is absolutely no evidentiary reason to maintain this belief."[40]

"What I am more concerned with is the extent to which the modern search for aliens is, at rock-bottom, part of an ancient religious quest," the astrobiologist (and SETI consultant) Paul Davies wrote in his 1995 book, *Are We Alone?*[41] Fifteen years later, with the skies still quiet, Davies noted in *The Eerie Silence* that "a project with the scope and profundity of SETI cannot be divorced from this wider cultural context, for it too offers us the vision of a world transformed, and holds the compelling promise that this could happen any day soon."[42] Even Carl Sagan, the scientist more equated with aliens than anyone before or since, and who was equally notorious for his religious skepticism, nevertheless said of SETI's importance: "It touches deeply into myth, folklore, religion, mythology; and every human culture in some way or another has wondered about that type of question."[43] He even seemingly wrote the deity back into the cosmos through the extraterrestrial intelligences in *Contact*, when his heroine Ellie discovers that pi—the ratio of the circumference of a circle to its diameter—is numerically encoded in the cosmos, providing proof that a superintelligence designed the universe:

> The universe was made on purpose, the circle said. In whatever galaxy you happen to find yourself, you take the circumference of a circle, divide it by its diameter, measure closely enough, and uncover a miracle—another circle, drawn kilometers downstream of the decimal point. In the fabric of space and in the nature of matter, as in a great work of art, there is, written small, the artist's signature. Standing over humans, gods, and demons, subsuming Caretakers and Tunnel builders, there is an intelligence that antedates the universe.[44]

Why should so many people—theists and atheists, theologians and scientists—believe in the existence of superior celestial beings? Basalla cited the work of psychologist Robert Plank, who suggests that humans have an emotional need to believe in imaginary beings.[45] "Despite all their scientific trappings," Basalla wrote, "the extraterrestrials discussed by scientists are as imaginary as the spirits and gods of religion or myth."[46] In his magisterial two-volume history of the conception of extraterrestrial intelligences, *Plurality of Worlds* and *The Biological Universe*, science historian Steven Dick posited that when the Newtonian mechanical universe displaced the spiritual world of the Middle Ages, it left a vast and lifeless void, which was filled by modern science with ETIs.[47] Susan Clancy concluded her study of alien abductees somewhat wistfully, wishing she could believe in such transcendent beings:

> Alien-abduction beliefs can be considered a type of religious creed, based on faith, not facts. Indeed, a vast body of scientific

data indicates that the believers are psychologically benefiting: they're happier, healthier, and more optimistic about their lives than people who lack such beliefs. We live in an age when science and technology prevail and traditional religions are under fire. Doesn't it make sense to wrap our angels and gods in space suits and repackage them as aliens?[48]

ETIs are secular gods—deities for atheists.

The indefatigable ETI searcher Jill Tarter, who brooks no sloppiness or sentimentality in her rigorous research program, in response to my initial suggestion in a *Science* review essay that ETIs are secular gods,[49] expressed her contempt at such a characterization. She correctly noted that "physics, not faith, dictates that any successful SETI detection will be with a long-lived technology (and perhaps the technologists who invented it)," and that "we work on the search because we want to know the answer to a very old question, popularly phrased as 'are we alone?'" That's true. Why does Jill Tarter search the skies for a sign?

> I search because I'm curious, not to find some deity, secular or otherwise! I do not know the answer to this old question, but I am as excited about using whatever tools are available to try to find the answer as I am excited about the possibility of using other tools to understand the nature of dark matter, or the state of dark energy, or whether giant planets form by aggregation or runaway gravitational instability. All are perfectly valid scientific questions to ask about the universe in which we find ourselves. Nevertheless, Basalla and you sling your accusations of special religious motivations at me and my colleagues and let the cosmologists (and their publishers), who pepper their book titles with the "God" word, off the hook.[50]

Fair enough. And let me add that I do not in any way equate SETI scientists with alien abductees and flying saucer searchers. SETI is science; UFOlogy is pseudoscience. SETI is elitist; UFOlogy is populist. SETI is dominated by Ph.D. astronomers, physicists, and mathematicians; UFOlogy is predominantly the domain of noncredentialed amateurs. SETI assumes the null hypothesis that aliens do not exist until contact is made; UFOlogy rejects the null hypothesis outright by starting with the assumption that contact has already been made.

What I am after is the deeper motivation for the search, the psychology behind the belief that somewhere out there in the vast cosmos filled with trillions of stars and planets, there exist other intentional and intelligent beings who are vastly superior to us. I contend here that the belief comes first, the search for evidence of the target of belief follows. There is nothing wrong with this; it is how most of science operates. Darwin and Wallace believed that there was a natural force at work creating new species (as opposed to a supernatural creator), and they found it in the form of natural selection. Einstein and Hubble believed that the large-scale structure of the universe could be understood through the operation of natural laws instead of supernatural interventions, and they found it in the principles of relativity and gravity. We search for such ultimate explanations because we are pattern-seeking agent-postulating primates whose brains are wired to find patterns and agents, even if the patterns are purely natural and the agents are just the laws of nature or other corporeal beings. Of course, we must search. It is what we do. We are explorers. So in the spirit of scientific inquiry, the search must go on.

10

---◆◆---

Belief in Conspiracies

AGENTICITY NEED NOT BE SO EPHEMERAL AS GHOSTS, GODS, ANGELS, and demons. Agents may be flesh and blood, even while retaining an element of near-invisibility, cloaked from our normal senses, secretive in their actions, and inferred by their effects. This form of agenticity is more familiarly known as a *conspiracy*, and the inference is a *conspiracy theory*.

Pattern of Conspiracy

Conspiracy theories are a different breed of animal than conspiracies themselves. Whether there was or was not a conspiracy behind the assassination of JFK (I contend that there was not), theories of JFK conspiracies abound, as they do for the assassinations of RFK, MLK Jr., and Malcolm X; the disappearance of Jimmy Hoffa; and the deaths of Princess Diana and assorted rock stars, not to mention conspiracy theories behind the fluoridation of water supplies, jet contrails depositing chemical and biological agents in the atmosphere (chemtrails), the spread of AIDS and other infectious diseases, the dispersal of cocaine and guns to inner cities, peak oil and related oil company suppression of alternative energy technologies, the moon landing that never happened, UFO landings that did happen, and the nefarious goings-on of the Federal Reserve, the New World Order, the Trilateral Commission, the Council on Foreign Relations, the Committee of 300, Skull and Bones, the Knights Templar, the Freemasons, the Illuminati, the Bilderberg Group, the Rothschilds, the Rockefellers, the Learned Elders of Zion and the Zionist Occupation Government, satanists and satanic ritual cults, and the like. The list is seemingly endless.

The term *conspiracy theory* is often used derisively to indicate that someone's explanation for an event is highly improbable or even on the lunatic fringe, and that those who proffer such theories are most probably crackpots. Since conspiracies do happen, however, we cannot just automatically dismiss any and all conspiracy theorists a priori. So what should we believe when we encounter a conspiracy theory? What are some of the characteristics of a conspiracy theory that indicate that it is likely untrue?

1. There is an obvious pattern of connected dots that may or may not be connected in a causal way. When the Watergate conspirators confessed to the burglary, or Osama bin Laden boasts about the triumph of 9/11, we can be confident that the pattern is real. But when there is no forthcoming evidence to support a causal connection between the dots in the pattern, or when the evidence is equally well explained through some other causal chain—or through randomness—the conspiracy theory is likely false.

2. The agents behind the pattern of the conspiracy are elevated to near superhuman power to pull it off. We must always remember how flawed human behavior is, and the natural tendency we all have to make mistakes. Most of the time in most circumstances most people are not nearly as powerful as we think they are.

3. The more complex the conspiracy, and the more elements involved for it to unfold successfully, the less likely it is to be true.

4. The more people involved in the conspiracy, the less likely they will all be able to keep silent about their secret goings-on.

5. The grander and more worldly the conspiracy is believed to be— the control of an entire nation, economy, or political system, especially if it suggests world domination—the less likely it is to be true.

6. The more the conspiracy theory ratchets up from small events that might be true into much larger events that have much lower

probabilities of being true, the less likely it is to be grounded in reality.

7. The more the conspiracy theory assigns portentous and sinister meanings and interpretations to what are most likely innocuous or insignificant events, the less likely it is to be true.

8. The more facts and speculation are commingled without distinguishing between the two and without assigning degrees of probability of factuality, the less likely the conspiracy theory represents reality.

9. Extreme hostility about and strong suspicions of any and all government agencies or private organizations in an indiscriminate manner indicates that the conspiracy theorist is unable to differentiate between true and false conspiracies.

10. If the conspiracy theorist defends the conspiracy theory tenaciously to the point of refusing to consider alternative explanations for the events in question, rejecting all disconfirming evidence for his theory and blatantly seeking only confirmatory evidence to support what he has already determined is the truth, he is likely wrong and the conspiracy is probably a figment of his imagination.

Why People Believe Conspiracies

Why do people believe in highly improbable conspiracies? I contend that it is because their pattern-detection filters are wide open, thereby letting in any and all patterns as real, with little to no screening of potential false patterns. Conspiracy theorists connect the dots of random events into meaningful patterns, and then infuse those patterns with intentional agency. Add to those propensities the *confirmation bias* and the *hindsight bias* (in which we tailor after-the-fact explanations to what we already know happened), and we have the foundation for conspiratorial cognition.

Examples of these processes can be found in Arthur Goldwag's 2009 book, *Cults, Conspiracies, and Secret Societies*, which covers everything from the Freemasons, the Illuminati, and the Bilderberg Group to black helicopters and the New World Order. "When something momentous happens, everything leading up to and away from the event seems momentous too. Even the most trivial detail seems to glow with signifi-cance," Goldwag explained, noting the JFK assassination as a prime example.

> Knowing what we know now . . . film footage of Dealey Plaza from November 22, 1963, seems pregnant with enigmas and ironies—from the oddly expectant expressions on the faces of the onlookers on the grassy knoll in the instants before the shots were fired (*What were they thinking?*), to the play of shadows in the background (*Could that flash up there on the overpass have been a gun barrel gleaming in the sun?*). Each odd excrescence, every ran-dom lump in the visual texture seems suspicious.[1]

Add to these factors how compellingly a good narrative story can tie it all together—think Oliver Stone's *JFK* or Dan Brown's *Angels and Demons*, both equally fictional—and you've got a formula for conspira-torial agenticity.

I experienced this effect firsthand when I visited Dealey Plaza, where on any given day conspiracy theorists are at the ready (for a modest tip) to give you a tour of where the shooters were hiding on that fateful day. In the photo (Figure 9), my guide reveals that one shooter was hiding in a sewer pipe; he then showed me where another shooter was behind the fence atop the grassy knoll. For more than an hour this conspiracist connected the dots into meaningful patterns that he infused with intentional agency.

Why do people believe in conspiracies? A useful distinction here is between transcendentalists and empiricists. *Transcendentalists* tend to believe that everything is interconnected and all events happen for a reason. *Empiricists* tend to think that randomness and coincidence interact with the causal net of our world and that belief should depend on evidence for each individual claim. The problem for skepticism is that transcendentalism is intuitive; empiricism is not. Our propensity for pat-ternicity and agenticity leads us naturally into the transcendental camp of seeing events in the world as unfolding according to a preplanned logic, whereas the empirical method of being skeptical until a claim is proven otherwise requires concerted effort that most of us do not make. Thus, the psychology of belief first and evidence second is once again borne out. Or as Buffalo Springfield once intoned: *Paranoia strikes deep. Into your life it will creep. . . .*

Figure 9. Dealey Plaza and JFK Conspiracy Theorists
On any given day in Dealey Plaza, conspiracy theorists will give you a tour of where the shooters were hiding. Here my guide reveals that one shooter was hiding in a sewer pipe. AUTHOR COLLECTION, PHOTOGRAPH BY REGINA HUGHES.

How to Test a Conspiracy Theory:
The Truth About the 9/11 Truthers

My experience with the 9/11 truthers will serve as a case study in how to test the validity of a conspiracy theory. It began after a public lecture in 2005, when I was buttonholed by a documentary filmmaker with Michael Mooreish ambitions of exposing the conspiracy behind 9/11.

"You mean the conspiracy by Osama bin Laden and al-Qaeda to attack the United States?" I asked rhetorically, knowing what was to come.

"That's what they want you to believe," he said.

"Who is *they*?" I queried.

"The government," he whispered, as if "they" might be listening at that very moment.

"But didn't Osama and some members of al-Qaeda not only say they did it," I reminded him, "they gloated about what a glorious triumph it was?"

"Oh, you're talking about that video of Osama," he rejoined knowingly. "That was faked by the CIA and leaked to the American press to mislead us. There has been a disinformation campaign going on ever since 9/11."

"How do you know?" I inquired.

"Because of all the unexplained anomalies surrounding 9/11," he answered.

"Such as?"

"Such as the fact that steel melts at a temperature of 2,777 degrees Fahrenheit, but jet fuel burns at only 1,517 degrees Fahrenheit. No melted steel, no collapsed towers."

At this point I ended the conversation and declined to be interviewed, knowing precisely where the dialogue was going next—if I cannot explain every single minutia about the events of that fateful eleventh day of September 2001, that lack of knowledge equates to direct proof that 9/11 was orchestrated by Bush, Cheney, Rumsfeld, and the CIA in order to implement their plan for global domination and a New World Order, to be financed by GOD (gold, oil, drugs) and launched by a Pearl Harbor–like attack on the World Trade Center and the Pentagon, thereby providing the justification for war. The evidence is in the details, he explained, handing me a faux dollar bill ("9-11" replacing the "1" and Bush supplanting Washington) chockablock with Web sites. Where had I heard all this before?

In the early 1990s I launched a full-scale investigation of the Holocaust deniers, initially as the cover story for *Skeptic* magazine and subsequently expanded into a book-length treatment, *Denying History*.[2] The deniers employ this tactic of anomalies-as-proof to great effect. David Irving, for example, claims that there are no holes in the roof of the gas chamber at Krema 2 at Auschwitz-Birkenau. So what? So plenty, he says. No holes in the roof of the gas chamber at Krema 2 means that the eyewitness account of SS guards climbing up on the roof and pouring Zyklon-B gas pellets through the holes and into the gas chamber below is wrong, which means that no one was gassed in Krema 2, which means that no one was gassed at Auschwitz-Birkenau, which means that no one was gassed at any prison camp, which means that no Jews anywhere were systematically exterminated by the Nazis. In short, "no holes, no Holocaust," says David Irving. The slogan was emblazoned on the T-shirts of his supporters at his London trial in which he sued a historian for calling him a Holocaust denier.

No holes, no Holocaust. No melted steel, no al-Qaeda attack. The parallels are equal, and equally flawed. And just as I never imagined that Holocaust denial would wend its way into the mainstream press (Irving's trial was front-page news for months), after my above conversation with the filmmaker I never imagined that 9/11 denial would get media legs. But now it has legs for days, and so *Skeptic* magazine published a full rebuttal of all the 9/11 truthers' claims.[3]

The belief that a handful of unexplained anomalies can undermine a well-established theory lies at the heart of all conspiratorial thinking. It is easily refuted by noting that beliefs and theories are not built on single facts alone, but on a convergence of evidence from multiple lines of inquiry. All of the "evidence" for a 9/11 conspiracy falls under the rubric of this fallacy. I could apply this principle to any number of conspiracy theories, but I shall focus on 9/11 because it is so recent and topical.

Let's begin with this issue of the melting temperature of steel. According to 911research.wtc7.net, steel melts at a temperature of 2,777 degrees Fahrenheit (other sources put it at 2,750), but jet fuel burns at only 1,517 degrees Fahrenheit. No melted steel, no collapsed towers.[4] Wrong. In an article in the *Journal of the Minerals, Metals, and Materials Society*, MIT engineering professor Dr. Thomas Eager explains why: steel loses 50 percent of its strength at 1,200 degrees Fahrenheit; the 90,000 liters of jet fuel ignited other combustible materials such as rugs, curtains, furniture, and paper, which continued burning after the jet fuel was exhausted, raising temperatures above 1,400 degrees Fahrenheit and spreading the fire throughout the building; temperature differentials of hundreds of degrees across single steel horizontal trusses caused them to sag, straining and then breaking the angle clips that held them to the vertical columns; once one truss failed, others failed, and when one floor collapsed (along with the ten stories above it) onto the next floor below, that floor then gave way, creating a pancaking effect that triggered the 500,000-ton building to collapse.

Conspiracists also argue that if the buildings had collapsed due to the impact of the planes, they should have fallen over on their sides. This is also wrong. With 95 percent of each building consisting of empty space (these were office buildings after all), they could only have collapsed straight down—there simply isn't enough structural support integrity to take an entire building down in one piece.

The truthers also claim—in direct contradiction of the above claim—that the buildings fell straight down into their own footprint, which, they say, could have happened only if they had been deliberately brought down by explosive charges carefully and deliberately set ahead of time. Not true. The buildings did not fall down perfectly straight. Their collapse began on the side where the planes impacted, and so were tilted slightly toward that weakened collapse point, which you can clearly see in the numerous videos of the collapsing buildings.

Another conspiracy claim is that the buildings fell from the top down, precisely in the manner that controlled demolition buildings collapse. False. Controlled demolitions are done from the bottom up, not the top down. If you search "building demolition" on YouTube you will find hundreds of video clips of buildings collapsing by controlled demolition. I could not find one that collapsed from the top down, as did the World Trade Center buildings. Instead, you see what demolition experts tell us is how it is done: the charges are set to explode from the bottom up.

For our special 9/11 issue of *Skeptic* we consulted a demolition expert named Brent Blanchard, who is director of field operations for Protec Documentation Services, a company that documents the work of building demolition contractors. Since the rise in popularity of 9/11 conspiracy theories, he, too, has been inundated with requests to explain why the buildings appeared to have "collapsed as if by a controlled demolition."[5] Blanchard and his team of experts at Protec have worked with all major American demolition companies and many foreign ones to study the controlled demolition of more than one thousand of the largest and tallest buildings around the world. Their duties include engineering studies, structural analysis, vibration/air overpressure monitoring, and photographic services. On September 11, 2001, Protec had portable field seismic monitoring systems operating at other sites in Manhattan and Brooklyn. Demolition specialists were hired to clean up Ground Zero and remove the remaining damaged structures, and these experts called on Blanchard's company to document both the deconstruction and the debris removal. Here are nine of the best arguments made by 9/11 conspiracy theorists and their rebuttal by Protec:

Claim #1: The collapse of the towers looked exactly like controlled demolitions.

Protec: No they did not. The key to any demolition investigation is in finding out the "where"—the actual point at which the building failed. All photographic evidence shows World Trade Center buildings 1 and 2 failed at the point of impact. Actual implosion demolitions always start with the bottom floors. Photo evidence shows the lower floors of WTC 1 and 2 were intact until destroyed from above.

Claim #2: But they fell right down into their own footprints.

Protec: They did not. They followed the path of least resistance, and there was a lot of resistance. Buildings of twenty stories or more do not topple over like trees or reinforced towers or smokestacks. Imploding demolitions fall into a footprint because lower stories are removed first. WTC debris was forced out, away from the building, as the falling mass encountered intact floors.

Claim #3: Explosive charges are seen shooting from several floors just prior to collapse.

Protec: No, air and debris can be seen being violently ejected from the building—a natural and predictable effect of rapid structure collapse.

Claim #4: Witnesses heard explosions.

Protec: All seismic evidence from many independent sources on 9/11 showed none of the sudden vibration spikes that result from explosive detonations.

Claim #5: A heat-generating explosive (perhaps thermite) melted steel at Ground Zero.

Protec: To a man, demolition workers do not report encountering molten steel, cut beams, or any evidence of explosions. Claims of detected traces of thermite are at this time inconclusive.

Claim #6: Ground Zero debris—particularly the large steel columns from WTC 1 and 2—were quickly shipped overseas to prevent scrutiny.

Protec: Not according to those who handled the steel. The chain of procession is clearly documented, first at Ground Zero by Protec and later at the Fresh Kills site by Yannuzzi Demolition. The time frame (months) before it was shipped to China was normal.

Claim #7: WTC 7 was intentionally "pulled down" with explosives. The building owner himself was quoted as saying he decided to "pull it."

Figure 10. World Trade Center Buildings Collapse

a. The circled area of one of the World Trade Center buildings shows a volume of smoke being compressed out the windows below from the compressing floors above. The 9/11 conspiracy theorists claim that these are explosive "squibs" setting off charges to bring the buildings down through explosive devices. PHOTO COURTESY OF FEMA: www.fema.gov/pdf/library/fema403_ch2.pdf.

b. Contrary to what 9/11 conspiracy theorists claim, the World Trade Center buildings did not fall evenly straight from the top down, but instead began their collapse and tilted on the side of the building of the plane's strike. PHOTO COURTESY OF FEMA: www.fema.gov/pdf/library/fema403_ch2.pdf.

c. The image of WTC 7 commonly presented by 9/11 conspiracy theorists as showing what appears to be only minimal damage to the building. PHOTO COURTESY OF FEMA: http://www.fema.gov/pdf/library/fema.403_ch5.pdf.

d. WTC 7 seen from the southwest side, showing the true extent of fire and structural damage. PHOTO COURTESY OF FEMA: http://www.fema.gov/pdf/library/fema403_ch5.pdf.

Protec: Building owners do not have authority over emergency personnel at a disaster scene. We have never heard "pull it" used to refer to an explosive demolition. Demolition explosive experts anticipated the collapse of WTC 7 and witnessed it from a few hundred feet away, and no one heard detonations.

Claim #8: Steel-frame buildings do not collapse due to fire.

Protec: Many steel-framed buildings have collapsed due to fire.

Claim #9: Anyone who denies that explosives were used is ignoring evidence.

Protec: Most of our comments apply to the differences between what people actually saw on 9/11 and what they should have seen had explosives been present. The hundreds of men and women who worked to remove debris from Ground Zero were some of the country's most experienced and respected demolition veterans. They of all people possessed the experience and expertise to recognize evidence of a controlled demolition if it existed. None of these people has come forward with suspicions that explosives were used.

The collapse of World Trade Center building 7, in fact, has grown in importance to conspiracy theorists, especially since standard nonconspiracy explanations for the demise of WTC buildings 1 and 2 became accepted. Since WTC 7 was not struck by a plane, and it did not collapse until 5:20 p.m. on 9/11, the cause of its collapse must be different from that of WTC 1 and 2. According to wtc7.net, "fires were observed in Building 7 prior to its collapse, but they were isolated in small parts of the building, and were puny by comparison to other building fires"; furthermore, any damage from falling debris from WTC 1 and WTC 2 would have needed to be symmetrical to trigger the pancaking collapse of WTC 7.

In point of fact, the fires burning in WTC 7 were extensive, not isolated. Conspiracy theorists tend to only show the north side of WTC 7, which does not look nearly as damaged as the other side. (Compare the photographs in figure 10.)

As the building burned all day, emergency response workers realized that collapse was imminent, and at 3 p.m. they began evacuation of all emergency personnel. When the building did collapse, the south side of the building—which sustained the most extensive damage from the falling debris of WTC 1 and 2—went first. As for the claim that WTC 7 leaseholder Larry Silverstein gave the order to "pull it," here is the actual quote from a September 2002 PBS special called *America Rebuilds*: "I remember getting a call from the, er, fire department commander, telling me that they were not sure they were gonna be able to contain the fire, and I said, 'We've had such terrible loss of life, maybe the smartest thing to do is pull it.' And they made that decision to pull and we watched the building collapse."

Here is Silverstein's own explanation for this quote, issued through a spokesperson on September 9, 2005:

> In the afternoon of September 11, Mr. Silverstein spoke to the Fire Department Commander on site at Seven World Trade Center. The Commander told Mr. Silverstein that there were several fire-fighters in the building working to contain the fires. Mr. Silverstein expressed his view that the most important thing was to protect the safety of those firefighters, including, if necessary, to have them withdraw from the building.
>
> Later in the day, the Fire Commander ordered his firefighters out of the building and at 5:20 p.m. the building collapsed. No lives were lost at Seven World Trade Center on September 11, 2001.
>
> As noted above, when Mr. Silverstein was recounting these events for a television documentary he stated, "I said, you know, we've had such terrible loss of life. Maybe the smartest thing to do is to pull it." Mr. McQuillan has stated that by "it," Mr. Silverstein meant the contingent of firefighters remaining in the building.

Silverstein's explanation is supported by eyewitness accounts of that day, including that of one rescue worker who noted that there were "tremendous, tremendous fires going on. Finally they *pulled* us out." Note the verb.

For my money, the oddest of all the 9/11 conspiracy theories is one involving the Pentagon. The idea, first floated in Thierry Meyssan's book *9/11: The Big Lie*, was that the Pentagon was struck by a missile because the damage was too narrow and limited to be the result of an impact from a Boeing 757. In the 9/11 conspiracy film *Loose Change*, dramatic reenactments are presented, showing that the hole in the Pentagon was too small to have been made by American Airlines Flight 77. There is nothing like selective visuals. Yet structural engineer Allyn E. Kilsheimer, who arrived on the scene shortly after the impact, reported: "I saw the marks of the plane wing on the face of the building. I picked up parts of the plane with the airline markings on them. I held in my hand the tail section of the plane, and I found the black box." Kilsheimer's eyewitness account is backed up by photos of plane wreckage inside and outside the building. Kilsheimer adds: "I held parts of uniforms from crew members in my hands, including body parts. Okay?"

Okay for me, but not for conspiracy theorists hell-bent on conforming the facts to fit the theory.

All of the 9/11 conspiracy claims are easily refuted. On the Pentagon "missile strike," for example, I queried my documentary antagonist about what happened to Flight 77, which disappeared at the same time as the Pentagon was struck. "The plane was destroyed and the passengers were murdered by Bush operatives," he solemnly revealed. "Do you mean to tell me that not *one* of the thousands of conspirators needed to pull all this off," I retorted, "is a whistle-blower who would go on TV or write a tell-all book?"

Think about all the examples of disgruntled government bureaucrats and ex-politicians who can't wait to go public with their insider information that we taxpayers will presumably want to know about. Not one of these 9/11 insiders, witness to what is arguably the greatest conspiracy and cover-up in the history of Western civilization, wants to go on *Larry King Live* or *60 Minutes* or *Dateline* to reveal his or her secret? Not one of them wants to cash in on what could very well be one of the best-selling books of the year, if not the decade? Not one of them, after a couple of drinks and a twinge or two of guilt, has leaked to a friend (or a friend of a friend) his or her deep secret? Not one? My rejoinder was met with the same grim response I get from UFOlogists when I ask them for concrete evidence: men in black silence witnesses and dead men tell no tales.

Was 9/11 a Conspiracy?

Was 9/11 a conspiracy? Yes, it was. By definition, a conspiracy is a secret plan by two or more people to commit an illegal, immoral, or subversive action against another without their knowledge or agreement. So, nineteen members of al-Qaeda plotting to fly planes into buildings without telling us constitutes a conspiracy. The ultimate failure of the 9/11 conspiracy theorists is their inability to explain away the overwhelming evidence of the real conspiracy by Osama bin Laden and al-Qaeda. For example, how do they explain these facts?

- The 1983 attack on the U.S. Marine barracks in Lebanon by a radical Hezbollah faction.
- The 1993 truck bomb attack on the World Trade Center.
- The 1995 attempt to blow up twelve planes heading from the Philippines to the United States.
- The 1995 bombings of U.S. embassy buildings in Kenya and Tanzania that killed twelve Americans and two hundred Kenyans and Tanzanians.
- The 1996 attack on Khobar Towers in Saudi Arabia that killed nineteen U.S. military personnel.
- The 1999 attempt to attack Los Angeles International airport by Ahmed Ressam.
- The 2000 suicide boat attack on the USS *Cole* that killed seventeen sailors and injured thirty-nine others.
- The well-documented evidence that Osama bin Laden is a major financier for and the leader of al-Qaeda.
- The 1996 fatwa by bin Laden that officially declared a jihad against the United States.
- Bin Laden's 1998 fatwa that said "to kill the Americans and their allies—civilian and military—is an individual duty for any Muslim who can do it in any country in which it is possible to do it."

Given this background, since Osama bin Laden and al-Qaeda have officially claimed responsibility for the attacks of 9/11, we should take them at their word that they did it.

Conspiracy Mongering

One rebuttal I often hear from conspiracy theorists is that I am spreading negative information as a means of distracting the public from "the truth." This is neither the first nor the last time that I've been accused of being a governmental agent of disinformation. UFOlogists suspected as much when I pooh-poohed their contention that the government is hiding alien spacecraft and bodies in Area 51. Holocaust deniers think that

I'm Jewish (I'm not) and that I'm being paid off by the Zionist lobby (whoever they are). Most recently the 9/11 truthers have fingered me as a patsy of the inside jobbers. This accusation came after I wrote one of my monthly columns in *Scientific American* about the 9/11 conspiracy theory and why it is wrong. To date, after ten years of writing monthly columns for the magazine, I have never received so many angry and hostile letters. I reprint a few excerpts here as a window into the conspiratorial mind-set:

> It is obvious that the name "Shermer" will go down in history as meaning "one who lies" or "a shill" or a "stooge." Example: "That guy was lying." "Yeah, he's nothing but a shermer." Or, "What a shermer he is!" and everyone will know what THAT means. I may start using this "word" immediately in my daily conversations. It certainly applies regarding the so called "article" Shermer wrote about 9/11.

One correspondent identified who he thinks is behind the conspiracy:

> The broadcast and print media are almost totally controlled by the Zionist criminals who are behind the evil undertakings of our government. They operate through blackmail, and bribery, and have taken complete control of this government and foreign policy to further their expansion in the middle east.

Sadly, he wasn't the only one to identify Zionists as conspirators:

> Please, accept my cancellation for *Scientific American* as your 9-11 report is neither scientific nor American but religious and Zionist. SHAME, SHAME, SHAME—another quisling to the Israeli overlords—START THINKING and STOP PROSTITUTING TO YOUR HIGHER POWER.

And this one as well:

> Your whitewash on the 9-11 does not work. Your Zionist front guys are treating your readership as fools. I have been a life long subscriber of your magazine and I have all the issues since 1971. I

will cancel my subscription due to your treasonous servility to the foreign power (Israel).

Another correspondent fingered me and the magazine as part of the conspiracy:

> I'm deeply shocked *Scientific American* could so obviously discredit its reputation, with such nonsense. Why not run stories about little green men on the moon? I mean, you've gone this low, why not go further? Don't be suprised if the scientific community starts laughing at you, and sales dry up. You can't publish crap like this AND keep your reputation. Mere pawns for the military-industrial complex—thats what you are.

Here is one equating America with Nazi Germany:

> Its so sad to see all our institutions being forced to lie about 9/11. And now you too! Sirs shame on you. Do you not realize this is EXACTLY what happened in Germany in the 1930s. Surely you do.

My 9/11 correspondence died down for a while, until I publicly commented on the Muslim wannabe terrorist Umar Farouk Abdulmutallab, who lit his underwear on fire on a Northwest Airlines flight on Christmas Day, 2009. If all these acts of terrorism are really an "inside job" by the Bush administration, I wrote, why did al-Qaeda issue this statement?: "Be prepared to suffer because the killing is coming and we prepared you men who love death just as you love life and by God's permission, we will come to you with more things that you have never seen before. Because, as you kill, you will be killed and tomorrow is coming soon. The martyrdom brother was able to reach his objective with the grace of God but due to a technical fault, the full explosion did not take place." Are we to believe that Abdulmutallab worked for the U.S. government? His own father ratted him out after he was radicalized by Muslim extremists—was that all part of the "inside job" as well? What was that sewn up in his underwear, the same superthermite that Bush operatives used to bring down the World Trade Center buildings with planted explosive devices?

Undaunted, and powered by conspiratorial agenticity, the 9/11 truthers fired back.[6] One told me to

Wipe the smile off your smug gob right now Michael Shermer. Whatever may have happened on Xmas day still doesn't change the fact that two of the highest buildings in the world could not have fallen in free fall time through the line of most resistance through gravity alone as NIST suggests.

Another growled:

Your glee about this retarded patsy trying to light his underwear on fire shows your bias. You so want the mainstream media conspiracy theory to be true that you can almost taste it. This story reminds me of the "Let's Roll" story and the Jessica Lynch story and the Pat Tillman story and the WMD stories and the official 911 conspiracy theory about a bunch of guys with box cutters defeating the most sophisticated air defense system in the world, hitting 3 of their 4 targets, including the most defended building in the world. Explain WTC 7 to me Mr. Shermer. It is still the 47 story elephant in the living room.

But the crème de la crème of conspiracy mongering was this explanation for the underwear bomber:

This guy was let through on purpose. He was a known terror risk. He was handed to the CIA on a plate by his own father! Remember all those Cheney/neocon warnings? They desperately want to blot Obamas copybook. Obama still has nests of neocon vipers in the CIA/Blackwater nexus and the justice department that for some unexplained reason he has been unable to eradicate. As with the 911 horror the Al Qaeda operators were tracked all the way. They were compromised and coordinated by black ops agents working on behalf of the PNAC conspirators. As a sceptic Mr Shermer should be less prepared to swallow the guff dished up to him by the neocon operators.[7]

How Conspiracies Actually Work

As acknowledged, conspiracies do happen, so I do not automatically dismiss them straight out of the gate. Abraham Lincoln was the victim of an assassination conspiracy, as was the Austrian archduke Franz Ferdinand, gunned down by a Serbian secret society on the eve of World War I. The attack on Pearl Harbor was a Japanese conspiracy (although some conspiracists think Franklin D. Roosevelt was in on it), and Watergate was a conspiracy (that Richard Nixon *was* in on). How can we tell the difference between the pattern of a real conspiracy and the pattern of conspiracy mongering? As Kurt Cobain, the rock star of Nirvana, once snarled in his grunge lyrics shortly before his death from a self-inflicted (or was it?) gunshot to the head, "Just because you're paranoid don't mean they're not after you."

But as G. Gordon Liddy once told me, the problem with government conspiracies is that bureaucrats are incompetent and people can't keep their mouths shut. Liddy should know, as he was an aide to President Nixon and one of the masterminds behind the break-in of the Democratic National Committee offices at the Watergate Hotel. Complex conspiracies are difficult to pull off—in this case even something as simple as a hotel burglary was foiled by a security guard, and under the pressure of congressional hearings and journalistic investigations many of the conspiracists cracked and talked. So many people want their quarter hour of fame that even the men in black couldn't squelch the squealers from spilling the beans. Once again, there's a good chance that the more elaborate a conspiracy theory is, and the more people that would need to be involved to pull it off, the less likely it is true.

As an example of how conspiracies actually operate in the highly random and massively contingent real world (as opposed to the hypothetical perfect world of conspiracy theorists), let's examine in detail the assassination of the Austrian archduke Franz Ferdinand and his wife, Sophie, who were together in Sarajevo on June 28, 1914. This is one of the most important and consequential assassinations in history, for it promptly triggered a military buildup over the summer that led to the guns of August and the outbreak of the First World War. This was unquestionably a conspiracy organized by a secret radical organization called Black Hand, whose political objective was to unite all of the territories containing Serb populations annexed by Austro-Hungary. The assassins were backed by an underground railroad of Serbian civilians and military officers who provided them with weapons, maps, and training to pull off the conspiracy.

Archduke Franz Ferdinand, the heir to the Austro-Hungarian throne, was in Sarajevo to observe military maneuvers and to open a new state museum. He arrived at the train station in the morning, and he and his entourage were driven to the first stop in six automobiles. Franz Ferdinand and Sophie were in the third vehicle, a convertible, and he instructed the drivers to proceed at a leisurely pace so that he could take in the local sights of beautiful downtown Sarajevo as the procession wended its way down the historic boulevard Appel Quay. There, the conspiracy ringleader Danilo Ilic had arranged his six assassins at strategic locations, arming them at the last moment.

As the motorcade entered the kill zone, the first two assassins, Muhamed Mehmedbasic, armed with a hand grenade, and Vaso Cubrilovic, equipped with a pistol and hand grenade, failed to act, either out of fear or an inability to get a clean line on the targets. Next in line was Nedeljko Cabrinovic, who hurled his hand grenade directly at the target third vehicle. It bounced off the rolled-down roof behind Franz Ferdinand and Sophie, skirted across the back of the car, and landed under the following vehicle where it then detonated, wounding the passengers and a number of police and bystanders in the crowd.

In a panic, Cabrinovic swallowed the cyanide pill given to him in the event of capture, and jumped into the nearby Miljacka River. But the river was too shallow at that time of year for drowning, and the cyanide resulted only in violent vomiting, so Cabrinovic was captured, beaten by the crowd, and hauled off to the police station. The vehicles sped off to safety while the other three assassins—Cvjetko Popovic, Trifun Grabez, and Gavrilo Princip—slunk away in defeat, the assassination conspiracy foiled by incompetence and bad luck.

Even the best-planned conspiracies hardly ever go according to plan, and this one was not yet over. Remarkably, Franz Ferdinand decided to complete his appointed rounds, and so continued on to the town hall reception for him, where he upbraided Sarajevo's elected leader: "Mr. Mayor, I came here on a visit and I get bombs thrown at me. It is outrageous." The archduke then delivered his speech, read from blood-soaked sheets of paper that were retrieved from car number four, acknowledging what he thought he saw in the faces of his audience, "an expression of their joy at the failure of the attempt at assassination." He spoke too soon, as conspiracies often turn on the quirkiest of events. In this case, Franz Ferdinand decided to visit the hospital where his wounded comrades from car four were being treated. Sophie canceled her plans and thought it best to join her husband.

Meanwhile, dejected by the failed conspiracy, Gavrilo Princip mean-dered over to a delicatessen on the corner of Appel Quay and Franz Joseph Street for a sandwich and private consolation. Finishing his meal, he emerged from Schiller's café and lo and behold what appeared before his startled eyes was the convertible vehicle making its way from the town hall to the hospital back along Appel Quay, with Franz Ferdinand and Sophie sticking out of the back like sitting ducks. Princip instantly saw this as his glorious moment of good fortune and took it, moving to the right of the car and firing his pistol, hitting the archduke in the jugu-lar vein of the neck and hitting Sophie in the torso. Both bled out and died shortly after.

This is how conspiracies really work—as messy events that unfold according to real-time contingencies. They often turn on the minutiae of chance and the reality of human error. Our propensity to think other-wise—to believe that conspiracies are well-oiled machines of Machiavel-lian manipulations—is to fall into the trap of conspiratorial patternicity and agenticity, where the patterns are too well delineated and the agents superhuman in knowledge and power.

PART IV

BELIEF IN THINGS SEEN

When people thought the earth was flat, they were wrong. When people thought the earth was spherical, they were wrong. But if you think that thinking the earth is spherical is just as wrong as thinking the earth is flat, then your view is wronger than both of them put together.

—ISAAC ASIMOV, THE RELATIVITY OF WRONG, 1989

11

Politics of Belief

ARE YOU A POLITICAL LIBERAL OR A CONSERVATIVE? IF YOU ARE A liberal, I predict that you read the *New York Times*, listen to progressive talk radio, watch CNN, hate George W. Bush and loathe Sarah Palin, adore Al Gore and revere Barack Obama, are pro-choice, anti-gun, adhere to the separation of church and state, are in favor of universal health care, vote for measures to redistribute wealth and tax the rich in order to level the playing field, and believe that global warming is real, human caused, and potentially disastrous for civilization if the government doesn't do something dramatic and soon. If you are a conservative, I predict that you read the *Wall Street Journal*, listen to conservative talk radio, watch FOX News, love George W. Bush and venerate Sarah Palin, despise Al Gore and abhor Barack Obama, are pro-life, anti–gun control, believe that America is a Christian nation that should meld church and state, are against universal health care, vote against measures to redistribute wealth and tax the rich, and are skeptical of global warming and/or government schemes to dramatically alter our economy in order to save civilization.

Although this cluster of specific predictions may not be a perfect match for any one person's positions, the fact that most Americans do fall into one of these two sets of attitudes indicates that even political, economic, and social beliefs form distinct patterns that we can identify and assess. In this chapter on our journey into the believing brain, I want to pull back for a grander visage of belief systems and how they operate in the realm of politics, economics, and ideologies of various types.

The Power of Political Beliefs, or Why People Divide Themselves into Liberals and Conservatives

In 2003, Stanford University social psychologist John Jost and his colleagues published a paper in the prestigious journal *Psychological Bulletin* entitled "Political Conservatism as Motivated Social Cognition," which was a synthesis of fifty years of findings published in eighty-eight papers encompassing 22,818 subjects that led the researchers to conclude that conservatives suffer from "uncertainty avoidance" and "terror management," and have a "need for order, structure," and "closure" along with "dogmatism" and "intolerance of ambiguity," all of which leads to "resistance to change" and "endorsement of inequality" in their beliefs and practices.

"Understanding the psychological underpinnings of conservatism has for centuries posed a challenge for historians, philosophers, and social scientists," the authors concluded.

> We regard political conservatism as an ideological belief system that is significantly (but not completely) related to motivational concerns having to do with the psychological management of uncertainty and fear. Specifically, the avoidance of uncertainty (and the striving for certainty) may be particularly tied to one core dimension of conservative thought, resistance to change. Similarly, concerns with fear and threat may be linked to the second core dimension of conservatism, endorsement of inequality.[1]

The paper was picked up by the news dailies and the story broke that scientists had at long last discovered what makes conservatives tick. One commentator for *Psychology Today* asked "Is Political Conservatism a Mild Form of Insanity?"[2] The British paper the *Guardian* reported: "A study funded by the US government has concluded that conservatism can be explained psychologically as a set of neuroses rooted in 'fear and aggression, dogmatism and the intolerance of ambiguity.'" If this was not enough to get the blood boiling of conservatives everywhere, the report's authors linked Ronald Reagan and the right-wing talk show host Rush Limbaugh to Hitler and Mussolini, arguing they all suffered from the same affliction.[3] Needless to say, conservatives were none too keen on having their political beliefs biopsied like so many cancerous tumors.

Why are people conservative? Why do people vote Republican? The questions are typically posed without even a whiff of awareness of the inherent bias in asking it in this manner—that because Democrats are so indisputably right and Republicans so unquestionably wrong, conservatism must be a mental disease, a flaw in the brain, a personality disorder that leads to cognitive malfunctioning. Much as medical scientists study cancer in order to cure the disease, liberal political scientists study political attitudes and voting behavior in order to cure people of the cancer of conservatism. This liberal belief bias in academia is so deeply entrenched that it becomes the political water through which the liberal fish swim— they don't even notice it.

University of Virginia psychologist Jonathan Haidt noticed the bias and called attention to it in a widely read and commented upon essay on Edge.org, "What Makes People Vote Republican?" The standard liberal line—as reflected in the Jost study—is that people vote Republican because they are "cognitively inflexible, fond of hierarchy, and inordinately afraid of uncertainty, change, and death." Haidt inveigled his fellow academics to move beyond such "diagnoses" and remember "the second rule of moral psychology is that morality is not just about how we treat each other (as most liberals think); it is also about binding groups together, supporting essential institutions, and living in a sanctified and noble way. When Republicans say that Democrats 'just don't get it,' this is the 'it' to which they refer."[4]

Why do liberals characterize conservatives in such a slanted manner? To answer the question let's begin by reversing the process and characterize Democrats and liberals as suffering from a host of equally defective mental states: a lack of moral compass that leads to an inability to make clear ethical choices, an inordinate lack of certainty about social issues, a pathological fear of clarity that leads to indecisiveness, a naive belief that all people are equally talented, and a blind adherence in the teeth of contradictory evidence that culture and environment alone determine one's lot in society and therefore it is up to the government to remedy all social injustices. Once you set up the adjectives in the form of operationally defined personality traits and cognitive styles, it is easy to collect the data to support them. The flaw is in the characterization process itself.

Two popular book-length examples that fall into the same belief bias trap are the 2008 book *The Political Mind* by University of California–Berkeley cognitive scientist George Lakoff and the 2007 book *The*

Political Brain by Emory University psychologist Drew Westen. The tropes are familiar: liberals are generous to a fault ("bleeding hearts"), rational, intelligent, optimistic, and appeal to voters' reason through cogent arguments; conservatives are stingy ("heartless"), dour, and dim-witted authoritarians who appeal to voters' emotions through threat and fearmongering. But conservatives win most elections because of their Machiavellian manipulation of voters' emotional brains, and therefore liberal politicians need to ramp up their campaigns with an appeal to voters' hearts instead of their heads.

Not only is the characterization driven entirely by a liberal belief bias, but the very premise that conservatives are winning the battle for voters' hearts is erroneous. In congressional races Democrats have seized the day: in the Senate, Democrats edged out Republicans 3,395 to 3,323 in contesting 6,832 seats from 1855 to 2006, and in the House, Democrats trounced Republicans 15,363 to 12,994 in the 27,906 seats contested from 1855 to 2006.

As for the personality traits and temperament of conservatives versus liberals, and the supposedly dour nature of the former, according to the National Opinion Research Center's *General Social Surveys, 1972–2004*, 44 percent of people who reported being "conservative" or "very conservative" said they were "very happy" versus only 25 percent of people who reported being "liberal" or "very liberal." A 2007 Gallup Poll found that 58 percent of Republicans versus only 38 percent of Democrats said that their mental heath is "excellent." One reason may be that conservatives are so much more generous than liberals, giving 30 percent more money (even when controlled for income), donating more blood, and logging more volunteer hours. And it isn't because conservatives have more expendable income. The working poor give a substantially higher percentage of their incomes to charity than any other income group, and three times more than those on public assistance of comparable income. In other words, poverty is not a barrier to charity, but welfare is.[5] One explanation for these findings is that conservatives believe charity should be private (through nonprofit organizations) whereas liberals believe charity should be public (through government). Here we see a pattern of political party preferences grounded in different moral foundations, which we will explore below.

One reason that liberals characterize conservatives in this manner may be the liberal bias of academic social scientists. To wit, a 2005 study

by George Mason University economist Daniel Klein using voter regis-
trations found that Democrats outnumbered Republicans by a staggering ratio of 10 to 1 among the faculty at the University of California–Berkeley
and by 7.6 to 1 among the faculty at Stanford University. In the humanities and social sciences, the ratio was 16 to 1 at both campuses (30 to 1
among assistant and associate professors). In some departments, such as
anthropology and journalism, there wasn't a single Republican to be
found. The ratio for all departments in all colleges and universities
throughout the United States, said Klein, is 8 to 1 Democrats over
Republicans.[6]

Smith College political scientist Stanley Rothman and his colleagues
found a similar bias in a 2005 national study: only 15 percent of professors describe themselves as conservative, compared to 72 percent who
said they were liberal (80 percent in humanities and social sciences).[7] A
more nuanced nationwide study conducted in 2001 by UCLA's Higher
Education Research Institute found that 5.3 percent of faculty members
were far left, 42.3 percent were liberal, 34.3 percent were middle of the
road, 17.7 percent were conservative, and 0.3 percent were far right.
Comparing the extremes in this sample, there are seventeen times more
far left liberals than far right conservatives. The bias appears even in law
schools, where one would hope for a more balanced education in our
future lawmakers. In 2005, Northwestern law professor John McGinnis
surveyed the faculties of the top twenty-one law schools rated by *U.S.
News & World Report* and found that politically active professors overwhelmingly tend to be Democrat, with 81 percent contributing "wholly
or predominantly" to Democratic campaigns while just 15 percent did
the same for Republicans.[8]

The liberal slant also appears to dominate many forms of the media.
A 2005 study by UCLA political scientist Tim Groseclose and University
of Missouri economist Jeffrey Milyo measured media bias by counting
the times that a particular media outlet cited various think tanks and
policy groups, and then compared this with the number of times that
members of Congress cited the same groups. "Our results show a strong
liberal bias: all of the news outlets we examine, except *Fox News' Special
Report* and the *Washington Times*, received scores to the left of the average member of Congress." Predictably, the *CBS Evening News* and the
New York Times "received scores far to the left of center." The three
most politically neutral media outlets were PBS's *NewsHour*, CNN's

NewsNight, and ABC's Good Morning America. Interestingly, the most politically centrist of all news sources was USA Today.[9]

Of course, liberals do not have a monopoly on political bias. Whenever I listen to conservative talk radio I find it distressingly easy to predict what the hosts are going to say about X, even before they open their mouths to speak, and this is the case for whatever X happens to be: health care, the war in Iraq, abortion, gun control, gay marriage, global warming, and most other issues. I don't even bother to listen to Rush Limbaugh anymore because I already know what he is going to say. Ditto Bill O'Reilly, Sean Hannity, and Glenn Beck, who are as predictable as death and taxes, neither one of which they believe in.

The political commentators who are more difficult to predict are the ones who do not just toe the party line but seem willing to break the ideological pattern in response to new data or a better theory. An example is Dennis Praeger, perhaps owing to his extensive training in the rabbinical style of thought in which each moral issue is to be weighed carefully, debated extensively, and thought through deeply. Of course, this more nuanced style may not appeal to as many listeners, and Praeger's show does lag behind the more black-and-white conservative talk shows in the ratings. Andrew Sullivan and Christopher Hitchens are also difficult to predict, but I attribute this to the fact that both are closer to being libertarian—socially liberal and economically conservative. Not placing yourself squarely in the middle of an ideological pattern makes it easier to break out of that pattern (and thus be more unpredictable). On the openly libertarian front, John Stossel is very predictable, but since he echoes many of my own ideological beliefs I tend not to notice the bias.

And that's the point. It's not that any of these social commentators (or many others—the specific examples are not important) are not original thinkers in and of themselves, or that they are not intelligent, educated, and live by the courage of their convictions (they are all of these things and more); it is that when you strap on an ideological belief you slot yourself into a set pattern of specific positions within that belief and parrot those back to your social group—the audience, in the case of public intellectuals—who listen mostly in order to have their own ideological beliefs bolstered.

Partisan Hearts and Political Minds

In their book *Partisan Hearts and Minds*, political scientists Donald Green, Bradley Palmquist, and Eric Schickler demonstrated that most people do not select a political party because it reflects their views; instead, they first *identify* with a political position, usually inherited from their parents, peer groups, or upbringing. Once they have made a commitment to that political position they choose the appropriate party and then follow the dictates of it.[10] This is the power of political belief, and it shows in the very tribal nature of modern politics and the stereotypes of each tribe.

Anyone who follows political commentary on a regular basis through the standard channels of talk radio and television, newspaper and magazine editorials, popular books, blogs, vlogs, tweets, and the like, knows the standard stereotype of what liberals think of conservatives:

> *Conservatives are a bunch of Hummer-driving, meat-eating, gun-toting, small-government-promoting, tax-decreasing, hard-drinking, Bible-thumping, black-and-white-thinking, fist-pounding, shoe-stomping, morally dogmatic blowhards.*

And what conservatives think of liberals:

> *Liberals are a bunch of hybrid-driving, tofu-eating, tree-hugging, whale-saving, sandal-wearing, big-government-promoting, tax-increasing, bottled-water-drinking, flip-flopping, wishy-washy, namby-pamby bedwetters.*

Such stereotypes are so ingrained into our culture that everyone understands them and comedians and commentators exploit them. Like many stereotypes, they both have an element of truth to them that reflects an emphasis on differing moral values, especially those we derive intuitively. In fact, research now overwhelmingly demonstrates that most of our moral decisions are grounded in automatic moral feelings rather than deliberatively rational calculations. We do not reason our way to a moral decision by carefully weighing the evidence for and against; instead, we make intuitive leaps to moral decisions and then rationalize the snap decision after the fact with rational reasons. Our moral intuitions—reflected in such conservative-liberal stereotypes—are more emotional than rational. As with most of our beliefs about most things in life, our moral beliefs come first; the rationalization of those moral beliefs comes second.

According to Jonathan Haidt, in fact, such stereotypes can be better understood in the context of moral intuition theory,[11] which explains why we have a natural aversion to certain behaviors such as incest, even if we cannot articulate those reasons. For example, read the following scenario and consider if you think the actions of the characters are morally acceptable or wrong:

> Julie and Mark are brother and sister. They are traveling together in France on summer vacation from college. One night they are staying alone in a cabin near the beach. They decide that it would be interesting and fun if they tried making love. At very least it would be a new experience for each of them. Julie was already taking birth control pills, but Mark uses a condom too, just to be safe. They both enjoy making love, but they decide not to do it again. They keep that night as a special secret, which makes them feel even closer to each other. What do you think about that, was it OK for them to make love?

Almost everyone who reads this vignette, constructed by Haidt to test people's moral intuitions, says that it is morally wrong. When asked why, they give answers such as Julie might get pregnant (but she can't) or that it will hurt their sibling relationship (but it didn't), or that others will find out (but they won't). Eventually people give up reasoning and just blurt out something like, "I don't know. I can't explain it. I just know it's wrong."[12]

Haidt concludes from this and similar research findings that we have moral emotions that evolved to help us survive and reproduce. In the Paleolithic environment of our ancestors, incest led to the very real problem of genetic mutations from close inbreeding. Of course, no one before our generation understood the underlying genetic reasons for the incest taboo, but evolution endowed us with moral emotions for avoiding close sexual relations with our kin and kind through the natural selection against those who practiced it extensively. Haidt proposes that the foundations of our sense of right and wrong rest within five innate and universally available psychological systems.[13]

1. *Harm/care*, related to our long evolution as mammals with attachment systems and an ability to feel (and dislike) the pain of others. We have evolved a deep sense of empathy and sympathy for others as we imagine ourselves in their position and what a situation would feel like if it were to happen to us. This foundation underlies such moral virtues as kindness, gentleness, and nurturance.

2. *Fairness/reciprocity*, related to the evolutionary process of reciprocal altruism, in which "I'll scratch your back if you'll scratch mine." This eventually evolved into genuine feelings of right and wrong over fair and unfair exchanges—a foundation that leads to such political ideals of justice, rights, and autonomy for individuals.

3. *In-group/loyalty*, related to our long history as a tribal species able to form shifting coalitions. We evolved the propensity to form within-group amity for our fellow tribesmen and between-group enmity for anyone in another group. This foundation creates within a tribe a "band-of-brothers" effect and underlies such virtues as patriotism and self-sacrifice for the group.

4. *Authority/respect*, shaped by our long primate history of hierarchical social interactions. We evolved a natural tendency to defer to authority, show deference to leaders and experts, and follow the rules and dictates given by those above us in social rank. This foundation underlies such virtues as leadership and followership, including esteem for legitimate authority and respect for traditions.

5. *Purity/sanctity*, shaped by the psychology of disgust and contamination. We evolved emotions to direct us toward the clean and away from the dirty. This foundation underlies religious notions of striving to live in a less carnal and more elevated and noble way, and it emphasizes the belief that the body is a temple that can be desecrated by immoral activities and contaminants.

Over the years Haidt and his University of Virginia colleague Jesse Graham have surveyed the moral opinions of more than 118,000 people from over a dozen different countries and regions around the world, and they have found this consistent difference between liberals and conservatives: Liberals are higher than conservatives on 1 and 2 (*harm/care* and *fairness/reciprocity*), but lower than conservatives on 3, 4, and 5 (*in-group/loyalty*, *authority/respect*, and *purity/sanctity*). Conservatives are roughly equal on all five dimensions: lower than liberals on 1 and 2 but

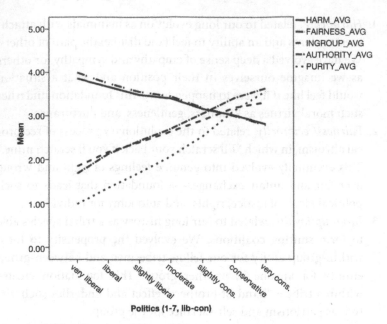

Figure 11. The Five Moral Foundations

Based on surveys of the moral opinions of 118,240 people from more than a dozen countries conducted by Jonathan Haidt and Jesse Graham of the University of Virginia, there is a consistent difference between liberals and conservatives in which liberals score higher than conservatives on moral foundations numbers 1 and 2 (*harm/care* and *fairness/reciprocity*), but score lower than conservatives on moral foundations numbers 3, 4, and 5 (*in-group/loyalty, authority/respect,* and *purity/sanctity*). Conservatives are roughly equal on all five dimensions, lower than liberals on 1 and 2 but higher on 3, 4, and 5. The graph is of responses to five subscales of the Moral Foundations Questionnaire. GRAPH COURTESY OF JONATHAN HAIDT. SURVEY AVAILABLE AT www.yourmorals.org.

higher on 3, 4, and 5. (Take the survey yourself at http://www.yourmorals .org.) The breakdown can be seen in figure 11.

In other words, liberals question authority, celebrate diversity, and often flaunt faith and tradition in order to care for the weak and oppressed. They want change and justice even at the risk of political and economic chaos. By contrast, conservatives emphasize institutions and traditions, faith and family, and nation and creed. They want order even at the cost of those at the bottom falling through the cracks. Of course, there are exceptions to such generalizations, but the point here is that instead of viewing the Left and the Right as either right or wrong (depending on which one you are), a more reflective approach is to recognize that liberals and conservatives emphasize different moral values and tend to sort themselves into these two clusters.

Consider just one study among many on the relationship between generosity and the rule of law. In a 2002 experiment by the economists Ernst Fehr and Simon Gachter on "moralistic punishment," subjects were given the opportunity to punish others who refuse to cooperate in a group activity that calls for altruistic giving. The study employed a cooperation game in which the subjects could give money into a shared commons. In the experimental condition in which there was no punishment for "free riding" (people could receive the benefits of being in the group without giving anything into the commons), the experimenters discovered that cooperation among the subjects quickly decayed within the first six rounds of the game. In the seventh round Fehr and Gachter introduced a new condition in which subjects were allowed to punish free riders by taking money from them. This they did with impunity, which immediately triggered a rise in the levels of cooperation and giving by the former free riders.[14] Conclusion: in order for there to be social harmony society needs to have in place a system that both encourages generosity and punishes free riding.

There are two such systems in the modern world—religion and government—and both arose about five thousand to seven thousand years ago to meet the needs of social control and political harmony when small bands and tribes of hunter-gatherers, fishermen, and herders coalesced into much larger chiefdoms and states of agriculturalists, craftsmen, and tradesmen. When populations became too large for informal means of social control (such as gossip and shunning), religion and government evolved as social watchdogs and enforcers of the rules.[15] Both conservatives and liberals agree that society needs rules, but for most behaviors conservatives prefer more private regulation through religion, community, and family while liberals favor more public regulation through government (except for sexual mores, when the opposite is the case). The problem with both institutions is that our moral minds also evolved to unite us into teams, divide us against other teams, and convince ourselves that we are right and other groups are wrong. This fact has had dire consequences, from December 7, 1941, to September 11, 2001.

My favorite example of the tension created by these differences comes from the 1992 film *A Few Good Men*, which I think well illustrates the conservative-liberal differences in moral foundations. In the courtroom ending, Jack Nicholson's conservative marine colonel Nathan R. Jessup is being cross-examined by Tom Cruise's liberal navy lieutenant Daniel Kaffee, who is defending two marines accused of accidentally killing a fellow soldier. Kaffee thinks Jessup ordered a "code red"—an off-the-books command to rough up a disloyal marine trainee named Santiago who was in need of discipline—and that matters got tragically out of hand. Kaffee wants individual justice for his clients even at the cost of group unity in the military. Jessup wants freedom and security for the nation even at the cost of individual liberty. Kaffee thinks that he is "entitled" to "the truth," but Jessup suspects that Kaffee "can't handle the truth." Why? Because, Jessup explains,

> Son, we live in a world that has walls. And those walls have to be guarded by men with guns. Who's going to do it? You? . . . You don't want the truth because deep down in places you don't talk about at parties, you want me on that wall. You need me on that wall. We use words like *honor, code, loyalty.* We use these words as the backbone to a life spent defending something. You use 'em as a punch line. I have neither the time nor the inclination to explain myself to a man who rises and sleeps under the blanket of the very freedom that I provide, and then questions the manner in which I provide it. I would rather you just said 'thank you' and went on your way. Otherwise I suggest you pick up a weapon and stand a post. Either way, I don't give a damn what you think you are entitled to.

Personally, I am conflicted, and that conflict reflects the fact that there are times when moral beliefs are irreconcilable, as is the case here. On the one hand, I lean toward the liberal emphasis on individual fairness, justice, and liberty, and I worry that overemphasis on group loyalty will trigger our inner tribalism and its corresponding xenophobia.[16] On the other hand, the evidence from history, anthropology, and evolutionary psychology reveals just how deep our tribal instincts run. Good fences make good neighbors because evil people really are part of the moral landscape. I am a civil libertarian who holds the value of individual liberty and autonomy above almost all other values, but ever since 9/11, 7/7, 12/25, and the countless other assaults on our liberties by other tribes, I am especially grateful to all the brave soldiers on those walls who have allowed us to sleep under a blanket of freedom.

Tragic, Utopian, and Realistic Visions of Human Nature

Identifying the moral values that make up the beliefs of liberals and conservatives perhaps helps attenuate our natural propensity to demonize those in the other party as evil. Through understanding comes tolerance. At least that's what the idealized liberal circuits in my brain tell me. In reality, I strongly suspect that the two-party system evolved as it did over the centuries because of the natural tendency to emphasize these equally important but often irreconcilable moral values.

Recall from chapter 8 the research by behavior geneticists on identical twins separated at birth and raised in different environments that found that about 40 percent of the variance in their religious attitudes was accounted for by their genes. These same studies also showed that about 40 percent of the variance in their political attitudes is due to inheritance.[17] Of course, just like genes do not code for particular religious faiths, we don't inherit political party affiliation directly. Instead, genes code for temperament and people tend to sort themselves into the left and right clusters of moral values based on their personality preferences, with liberals emphasizing the *harm/care* and *fairness/reciprocity* values and conservatives underscoring the *in-group/loyalty, authority/ respect*, and *purity/sanctity* values. This would explain why people are so predictable in their beliefs on such a wide range of issues that are seemingly unconnected—why someone who believes that the government should stay out of the private bedroom nevertheless believes that the government should be deeply involved in private business; why someone who believes that taxes should be lowered nevertheless wants to spend heavily on military, police, and the judicial system.

In his book *A Conflict of Visions*, economist Thomas Sowell argued that these two clusters of moral values are intimately linked to the vision one holds about human nature, either as constrained (conservative) or unconstrained (liberal). He called these the *constrained vision* and the *unconstrained vision.* Sowell showed that controversies over a number of seemingly unrelated social issues such as taxes, welfare, Social Security, health care, criminal justice, and war repeatedly reveal a consistent ideological dividing line along these two conflicting visions. "If human options are not inherently constrained, then the presence of such repugnant and disastrous phenomena virtually cries out for explanation—and for solutions. But if the limitations and passions of man himself are at the heart of these painful phenomena, then what requires explanation are the ways in which they have been avoided or minimized."

Which of these natures you believe is true will largely shape which solutions to social ills you think will be most effective. "In the unconstrained vision, there are no intractable reasons for social evils and therefore no reason why they cannot be solved, with sufficient moral commitment. But in the constrained vision, whatever artifices or strategies restrain or ameliorate inherent human evils will themselves have costs, some in the form of other social ills created by these civilizing institutions, so that all that is possible is a prudent trade-off."

It's not that conservatives think we're evil and liberals believe we're good. "Implicit in the unconstrained vision is the notion that the potential is very different from the actual, and that means exist to improve human nature toward its potential, or that such means can be evolved or discovered, so that man will do the right thing for the right reason, rather than for ulterior psychic or economic rewards," Sowell elaborated. "Man is, in short, 'perfectible'—meaning continually improvable rather than capable of actually reaching absolute perfection."[18]

In his masterpiece analysis of human nature, *The Blank Slate*, Harvard psychologist Steven Pinker relabeled these two visions the *Tragic Vision* and the *Utopian Vision*, and reconfigures them slightly:

> The Utopian Vision seeks to articulate social goals and devise policies that target them directly: economic inequality is attacked in a war on poverty, pollution by environmental regulations, racial imbalances by preferences, carcinogens by bans on food additives. The Tragic Vision points to the self-interested motives of the people who would implement these policies—namely, the expansion of their bureaucratic fiefdoms—and to their ineptitude at anticipating the myriad consequences, especially when the social goals are pitted against millions of people pursuing their own interests.

The distinct left-right divide consistently cleaves the (respectively) Utopian Vision and Tragic Vision along numerous specific contests, such as the size of the government (big versus small), the amount of taxation (high versus low), trade (fair versus free), health care (universal versus individual), environment (protect it versus leave it alone), crime (caused by social injustice versus caused by criminal minds), the constitution (judicial activism for social justice versus strict constructionism for original intent), and many others.[19]

Personally I agree with Sowell and Pinker that the unconstrained vision is utopian, which in its original Greek means "no place." An unconstrained utopian vision of human nature largely accepts the blank-slate model and believes that custom, law, and traditional institutions are sources of inequality and injustice and should therefore be heavily regulated and constantly modified from the top down; it holds that society can be engineered through government programs to release the natural unselfishness and altruism within people; it deems physical and intellectual differences largely to be the result of unjust and unfair social systems that can be reengineered through social planning, and therefore people can be shuffled across socioeconomic classes that were artificially created through unfair and unjust political, economic, and social systems inherited from history. I believe that this version of human nature exists in literally *no place*.

Although some liberals embrace just such a vision of human nature, I strongly suspect that when pushed on specific issues most liberals realize that human behavior is constrained to a certain degree—especially those educated in the biological and evolutionary sciences who are aware of the research in behavior genetics. Therefore, the debate turns on *degrees* of constraint. Rather than there being two distinct and unambiguous categories of constrained and unconstrained (or tragic and utopian) visions of human nature, I think there is just one vision with a sliding scale. Let's call this the *Realistic Vision*.

If you believe that human nature is partly constrained in all respects—morally, physically, and intellectually—then you hold a Realistic Vision of human nature. In keeping with the research from behavioral genetics and evolutionary psychology, let's put a number on that constraint at 40 to 50 percent. In the Realistic Vision, human nature is relatively constrained by our biology and evolutionary history, and therefore social and political systems must be structured around these realities, accentuating the positive and attenuating the negative aspects of our natures. A Realistic Vision rejects the blank-slate model that people are so malleable and responsive to social programs that governments can engineer their lives into a great society of its design, and instead believes that family, custom, law, and traditional institutions are the best sources for social harmony. The Realistic Vision recognizes the need for strict moral education through parents, family, friends, and community members because people have a dual nature of being selfish and selfless, competitive and cooperative, greedy and generous, and so we need rules and guidelines and encouragement to do the right thing. The Realistic Vision acknowledges that people vary widely both physically and intellectually—in large part because of natural inherited differences—and therefore will rise (or fall) to their natural levels. Therefore governmental redistribution programs are not only unfair to those from whom the wealth is confiscated and redistributed, but the allocation of the wealth to those who did not earn it cannot and will not work to equalize these natural inequalities.

I think most moderates on both the Left and the Right embrace a Realistic Vision of human nature. They should, as should the extremists on both ends, because the evidence from psychology, anthropology, economics, and especially evolutionary theory and its application to all three of these sciences supports it. There are at least a dozen lines of evidence that converge to this conclusion:[20]

1. Clear and quantitative physical differences among people in size, strength, speed, agility, coordination, and other physical attributes translate into some being more successful than others; at least half of these differences are inherited.

2. Clear and quantitative intellectual differences among people in memory, problem-solving ability, cognitive speed, mathematical talent, spatial reasoning, verbal skills, emotional intelligence, and other mental attributes translate into some being more successful than others; at least half of these differences are inherited.

3. Evidence from behavioral genetics and twin studies indicate that 40 to 50 percent of the variance among people in temperament, personality, and many political, economic, and social preferences are accounted for by genetics.

4. Failed communist and socialist experiments around the world throughout the twentieth century revealed that top-down draconian controls over economic and political systems do not work.

5. Failed communes and utopian community experiments tried at various places throughout the world over the past 150 years demonstrated that people by nature do not adhere to the Marxian principle "from each according to his ability, to each according to his need."

6. Family ties are powerful and the connectedness between blood relatives is deep. Communities who have tried to break up the family and have children raised by others provide counterevidence to the claim that "it takes a village" to raise a child. The continued practice of nepotism further reinforces the practice that "blood is thicker than water."

7. The principle of reciprocal altruism—I'll scratch your back if you'll scratch mine—is universal; people do not by nature give generously unless they receive something in return, even if what they receive is only social status.

8. The principle of moralistic punishment—I'll punish you if you do not scratch my back after I have scratched yours—is universal; people do not long tolerate free riders who continually take but almost never give.

9. Hierarchical social structures are almost universal. Egalitarianism works (barely) only among tiny bands of hunter-gatherers in resource-poor environments where there is next to no private property. When a precious game animal is hunted, extensive rituals and religious ceremonies are required to ensure equal sharing of the food.

10. Aggression, violence, and dominance are almost universal, particularly among young males seeking resources, women, and especially status. Status seeking in particular explains many heretofore unexplained phenomena, such as high risk taking, costly gifts, excessive generosity beyond one's means, and especially attention seeking.

11. Within-group amity and between-group enmity are almost universal. The rule of thumb is to trust in-group members until they prove to be distrustful, and to distrust out-group members until they prove to be trustful.

12. The desire of people to trade with one another is almost universal—not for the selfless benefit of others or the society, but for the selfish benefit of one's own kin and kind; it is an unintended consequence that trade establishes trust between strangers and lowers between-group enmity, as well as produces greater wealth for both trading partners and groups.

The founders of our republic established our system of government as they did based on this Realistic Vision of human nature. The tension between individual liberty and social cohesiveness can never be resolved to everyone's satisfaction, and so the moral pendulum swings left and right, and politics is played mostly between the two forty-yard lines of the political playing field. This tension between freedom and security, in fact, would explain why third parties have such a difficult time finding a toehold on the political rock face of America, and typically crater after an election or cower in the shadows of two behemoths that have come to define the left-right system. In Europe, where third, fourth, and even fifth parties receive substantial support at the polls, they are, in fact, barely distinguishable from the parties on either side of them, and political scientists find that they can easily classify them as largely emphasizing either liberal or conservative values. Haidt's data on the differing foundational values of American liberals and conservatives, in fact, extend to all countries that have been tested, and the chart lines from country to country are virtually indistinguishable from one another.

I believe that the Realistic Vision of human nature is what James Madison was thinking of when he penned his famous dictum in "Federalist Paper Number 51": "If men were angels, no government would be necessary. If angels were to govern men, neither external nor internal controls on government would be necessary."[21] Abraham Lincoln also had something like the Realistic Vision in mind when he wrote in his first inaugural address in March 1861, on the eve of the bloodiest conflict in our nation's history: "Though passion may have strained, it must not break our bonds of affection. The mystic chords of memory, stretching from every battlefield and patriot grave to every living heart and hearthstone all over this broad land, will yet swell the chorus of the Union, when again touched, as surely they will be, by the better angels of our nature."[22]

Left, Right, and Off the Charts

In my Realpolitik mode, I do not see this left-right system changing any time soon because it is so deeply grounded in our evolved human natures as exhibited in the five moral foundations and as evidenced by the twelve lines of evidence for the Realistic Vision. In my Idealpolitik[23] mode, however, I have found a political position beyond the traditional left-right spectrum that well suits my beliefs and temperament, and this is called *libertarian*. Libertarian? I know what you're thinking:

> *Libertarians are a bunch of electric-car driving, fusion-food eat-*
> *ing, pot-smoking, porn-watching, prostitution-supporting, gold-*
> *hoarding, gun-stashing, Constitution-waving, secession-mongering,*
> *tax-revolting, anti-government anarchists.*

Yes, like the other two stereotypes, there is some element of truth in this one as well. But, basically, libertarians are for freedom and liberty for individuals, and yet we recognize that in order to be free we must also be protected. Your freedom to swing your arms ends at my nose. As John Stuart Mill explained in his 1859 book *On Liberty*, "The sole end for which mankind are warranted, individually or collectively, in interfering with the liberty of action of any of their number, is self-protection. That the only purpose for which power can be rightfully exercised over any member of a civilized community, against his will, is to prevent harm to others."[24] The development of democracy was an important step to defeating the *tyranny of the magistrate* that reigned for centuries in European monarchies, but as Mill noted, the problem with democracy is that it can lead to the *tyranny of the majority*: "There needs protection also against the tyranny of the prevailing opinion and feeling, against the tendency of society to impose, by other means than civil penalties, its own ideas and practices as rules of conduct on those who dissent from them; to fetter the development and, if possible, prevent the forma-tion of any individuality not in harmony with its ways, and compel all characters to fashion themselves upon the model of its own."[25] This is, in fact, why our country's founders produced the Bill of Rights. These are rights that cannot be taken away no matter how big the majority in a democratic election.

Libertarianism is grounded in the *Principle of Freedom*: *all people are free to think, believe, and act as they choose, so long as they do not infringe on the equal freedom of others*. Of course, the devil is in the details of what constitutes "infringement," but there are at least a dozen essentials to liberty and freedom that need shielding from encroachment:

1. The rule of law.
2. Property rights.
3. Economic stability through a secure and trustworthy banking and monetary system.
4. A reliable infrastructure and the freedom to move about the country.
5. Freedom of speech and the press.
6. Freedom of association.
7. Mass education.
8. Protection of civil liberties.
9. A robust military for protection of our liberties from attacks by other states.
10. A potent police force for protection of our freedoms from attacks by other people within the state.
11. A viable legislative system for establishing fair and just laws.
12. An effective judicial system for the equitable enforcement of those fair and just laws.

These essentials incorporate the moral values embraced by both liberals and conservatives, and as such form the foundation for a bridge between the Left and the Right. Will the Libertarian Party ever grow large enough to challenge the two dominant political parties and form a viable three-party system? I doubt it, for the very reason that libertarians tend to dislike large and powerful political parties. Organizing libertarians is like herding cats. Nevertheless, in the context of the pattern of political parties and the moral values on which they are based, the libertarian position is just a reshuffling of the foundations of the other two. Nothing new needs to be invented or introduced into the system. These are values deeply ingrained in our nature and thus will likely remain a relatively permanent part of future political patterns.

Belief and Truth

Belief statements in politics are not always the same as belief statements in science. When I say, "I believe in evolution" or "I believe in the big bang," this is something different than when I say, "I believe in a flat tax" or "I believe in liberal democracy." Either evolution and the big bang happened or they did not, and the overwhelming evidence is that they did. The matter of the origin of species and the origin of the universe are, in principle, puzzles that can be solved with more data and better theory. But the matter of the right form of taxation or governmental structure depends on the overall goals to be accomplished, and for that more data and better theory can help us only once the goal has been established. The determination of that overarching political goal, however, depends on the very subjective process of political debate in which both sides build a case for what they think is the better way to live. I happen to think that a flat tax is a much fairer system than a progressive tax, because I don't think that people should be punished with higher taxes just because they earn more income through hard work and creativity. But my liberal friends argue that a progressive tax is fairer because people lower on the income scale are hit harder by the same tax rate than people higher on the income scale.

Although science may not be able to adjudicate such issues of fairness to everyone's satisfaction, a case can and should be made for science informing political beliefs—sometimes belief statements in politics are not dissimilar from belief statements in science. I have crossed this boundary myself many times, most notably in *The Science of Good and Evil* and *The Mind of the Market*. I reject in practice the *naturalistic fallacy* (sometimes called the *is-ought fallacy*), which holds that the *is* should not determine the *ought*; that *the way things are* is not necessarily *how they should be*, or that just because something is *natural* does not make it *right*. Sometimes that is the case, but sometimes it is not the case. I firmly believe that how we structure society should be informed by and even based on a Realistic Vision of human nature and the twelve lines of evidence I presented for it; the failed communist and socialist experiments demonstrate what happens when you ignore the way things are naturally—people die by the hundreds of millions.

Another example of crossing the is-ought divide can be found in Timothy Ferris's book *The Science of Liberty*, in which he weds democracy and science.[26] Ferris argues, for example, that the political belief of John Locke that people should be treated equally under the law—which factored heavily in the construction of the U.S. Constitution—was an untested theory in the seventeenth century. It could have been falsified. We could have given women and blacks the vote and discovered that democracy doesn't work unless it is practiced by white males only, which it was at the time of Locke. But that is not what happened. We ran the experiment and the results were unequivocally positive.

"Liberalism and science are methods, not ideologies," Ferris explained to me when I initially doubted his thesis by suggesting *all* political beliefs are ideologies. "Both incorporate feedback loops through which actions (e.g., laws) can be evaluated to see whether they continue to meet with general approval. Neither science nor liberalism makes any doctrinaire claims beyond the efficacy of their respective methods—that is, that science obtains knowledge and that liberalism produces social orders generally acceptable to free peoples." But, I rejoined, aren't all political claims types of *beliefs*? No, Ferris responded: "To put it another way, (classical) liberalism is not a belief. It was a proposed method, which could easily have been found wanting in practice. As it has instead succeeded, it deserves support. Belief is not required at any step along the way—except in the sense, say, that John Locke 'believed' (or rather reasonably thought) that he was on to something promising."[27]

Unfortunately, not everyone agrees that the overall goal of a society should be greater equality, liberty, freedom, wealth, and prosperity for more people in more places more of the time, as commentators such as myself, Timothy Ferris, and most other Western observers believe. Some societies—extreme Islamic theocracies, for example—believe that too much equality, liberty, freedom, wealth, and prosperity leads to decadence, licentiousness, promiscuity, pornography, prostitution, teen pregnancy, suicides, abortions, STDs, and sex, drugs, and rock 'n' roll. Ed Husain recalled in *The Islamist*, his book about Islamic extremism and his indoctrination into the Muslim brotherhood in Britain, that their motto was "The Quran Is Our Constitution; *Jihad* Is Our Way; Martyrdom Is Our Desire." One cell member told him: "Democracy is *haram*! Forbidden in Islam. Don't you know that? Democracy is a Greek concept, rooted in *demos* and *kratos*—people's rule. In Islam, we don't rule; Allah rules. . . . The world today suffers from the malignant cancers of freedom and democracy."[28]

Some Islamists hold as a higher goal obedience to God and his holy book, which leads them to believe in a rigid and hierarchical social structure in which, for example, women should obey men, should be punished for adultery by the death penalty, and should be treated under the law as property little different from chattel or cattle. In the words of the Pakistani journalist and pro-Islamic ideologue Abul Ala Mawdudi: "Islam wants the whole earth and does not content itself with only a part thereof. It wants and requires the entire inhabited world. . . . It is not satisfied by a piece of land but demands the whole universe [and] does not hesitate to utilize the means of war to implement its goal."[29]

While science and liberty go hand in hand, what do you say to someone who does not believe in either? "*Try winning an election*," is what Timothy Ferris would tell them, although this would likely fall on deaf ears since such people are almost never able to do so in a free and fair democratic election. Nevertheless, Ferris told me that he is optimistic about the future of democracy: "In practice there is more consensus around the world than is generally realized—at least within those parts of the world that have reasonably free media so that people can make fact-based decisions. It is not the case, for instance, that Muslim countries 'believe' that wealth and freedom are undesirable. That position, taken by radical Islamists, appeals to but a small minority. Polls repeatedly show that the majority of Muslims who do not already live in democratic countries prefer liberal democracy to other systems of government."[30] In fact, most Muslims in Indonesia, Egypt, Pakistan, Morocco, and other Islamic nations oppose Islamism and extremism of any kind. It isn't hard to understand why when you outline the problem, as David Frum and Richard Perle did so clearly and succinctly in their book *An End to Evil*, from which we may derive a scientific solution:

> Take a vast area of the earth's surface, inhabited by people who remember a great history. Enrich them enough that they can afford satellite television and Internet connections, so that they can see what life is like across the Mediterranean or across the Atlantic. Then sentence them to live in choking, miserable, polluted cities ruled by corrupt, incompetent officials. Entangle them in regulations and controls so that nobody can ever make much of a living except by paying off some crooked official. Subordinate them to elites who have suddenly become incalculably wealthy

from shady dealings involving petroleum resources that supposedly belong to all. Tax them for the benefit of governments that provide nothing in return except military establishments that lose every war they fight: not roads, not clinics, not clean water, not street lighting. Reduce their living standards year after year for two decades. Deny them any forum or institution—not a parliament, not even a city council—where they may freely discuss their grievances. Kill, jail, corrupt, or drive into exile every political figure, artist, or intellectual who could articulate a modern alternative to bureaucratic tyranny. Neglect, close, or simply fail to create an effective school system—so that the minds of the next generation are formed entirely by clerics whose own minds contain nothing but medieval theology and a smattering of third world nationalist self-pity. Combine all this, and what else would one expect to create but an enraged populace.[31]

Slipping back into my Idealpolitik mode, the scientific solution to the political problem of oppressive governments is the tried-and-true method of spreading liberal democracy and market capitalism through the free and open exchange of information, products, and services across porous economic borders. Liberal democracy is not just the least bad political system compared to all others (pace Winston Churchill); it is the best system yet devised for giving people a chance to be heard, an opportunity to participate, and a voice to speak truth to power. Market capitalism is the greatest generator of wealth in the history of the world and it has worked everywhere that it has been tried. Combine the two and Idealpolitik may become Realpolitik.

⚬

A final note on belief and truth: To many of my liberal and atheist friends and colleagues, an explanation for religious beliefs such as what I have presented in this book is tantamount to discounting both its internal validity and its external reality. Many of my conservative and theist friends and colleagues take it this way as well and therefore bristle at the thought that explaining a belief explains it away. This is not necessarily so. Explaining why someone believes in democracy does not explain away democracy; explaining why someone holds liberal or conservative values within a democracy does not explain away those values. In principle, the formation and reinforcement of political, economic, or social beliefs is no different from religious beliefs.

Explaining that people are conservative because their parents voted Republican, that they were raised in or now live in a red state, that their religion leans conservative instead of liberal, or that by temperament they prefer ordered social hierarchies and strict rules, does not automatically discount the validity of the conservative principles and values, any more than explaining that people are liberal because their parents voted Democratic, that they were raised in or now live in a blue state, that their religion leans liberal instead of conservative, or that by temperament they prefer the leveling of social hierarchies and more flexible rules, automatically discounts the validity of the liberal position.

Nevertheless, the fact that our beliefs are so heavily laden with emotional baggage should give one pause to at least consider the position of others and to be skeptical of one's own beliefs. The fact that we tend not to do so is a result of some very powerful cognitive biases that work to ensure that we are always right. I will examine them in the next chapter.

12

Confirmations of Belief

HAVE YOU EVER GONE TO THE PHONE TO CALL A FRIEND, ONLY TO have the phone ring first and find your friend on the other end of the line? What are the odds of that? Not high, and your patternicity intuition probably signaled to you that there was something special about this event. Was there? Probably not. Here's why: the sum of all probabilities equals one. Given enough opportunities, outlier anomalies will inevitably happen. The question is not *What is the probability that a friend would phone while being thought about?*—which is very low—but *In the total population of all people making phone calls and thinking about friends, what is the probability that at least one phone call will overlap with at least one simultaneous thought?* which is very high. Analogously, the chance of any one person winning the lottery is extremely low, but in the lottery system as a whole, someone will win.

In his insightful book *The Drunkard's Walk*, the mathematician and science writer Leonard Mlodinow computed the odds of a mutual fund manager named Bill Miller beating Standard & Poor's 500 index fifteen years in a row.[1] For this feat Miller was hailed as "the greatest money manager of the 1990s" and CNN computed the odds of him doing so at 372,529 to 1. Those are long odds indeed. And Mlodinow notes that if you had picked Bill Miller at the start of the streak in 1991 and computed the odds of him beating the S&P 500 every year for the next fifteen years, they would indeed be very slim. But this principle would apply to any mutual fund manager that you happened to pick. "You would have had the same odds against you if you had flipped a coin once a year for fifteen years with the goal of having it land heads up each time," Mlodinow notes. But, in fact, there are more than six thousand mutual fund managers, "so the relevant question is, if thousands of people are tossing coins once a year and have been doing so for decades, what are the chances that one of them, for some period of fifteen years or longer, will toss all heads?" That probability is much higher. In fact, Mlodinow demonstrates that over the past forty years of active mutual fund trading, the odds that at least one mutual fund manager would beat the market every year for fifteen years in a row turn out to be about three out of four, or 75 percent!

I have applied this principle of probability thinking to miracles. Let us define a miracle as an event with million-to-one odds of occurring (intuitively that seems rare enough to earn the moniker). Let us also assign a number of one bit per second of data that flows into our senses as we go about our day, and assume that we are awake for twelve hours a day. That nets us 43,200 bits of data per day, or 1,296,000 per month. Even assuming that 99.999 percent of these bits are totally meaningless (and so we filter them out or forget them entirely), that still leaves 1.3 "miracles" per month, or 15.5 miracles per year. Thanks to selective memory and the *confirmation bias*, we will remember only those few astonishing coincidences and forget the vast sea of meaningless data.

We can employ a similar back-of-the-envelope calculation to explain death premonition dreams. The average person has about five dreams per night, or 1,825 dreams per year. If we remember only a tenth of our dreams, then we recall 182.5 dreams per year. There are about 300 million Americans, who thus produce 54.7 billion remembered dreams per year. Sociologists tell us that each of us knows about 150 people fairly well, thus producing a network social grid of 45 billion personal relationship connections. With an annual death rate of 2.4 million Americans per year (all ages, all causes), it is inevitable that some of those 54.7 billion remembered dreams will be about some of these 2.4 million deaths among the 300 million Americans and their 45 billion relationship connections. In fact, it would be a *miracle* if some death premonition dreams did not come true! Here is a television talk show episode you will never see: "Next, we have a very special guest who has experienced a number of vivid dreams about the deaths of prominent people, not one of which has come true. But stay tuned because you never know when the next one will be confirmed." Instead, of course, television talk shows focus on the million-to-one events and ignore the rest of the noise.

These examples show the power of what I call *folk numeracy,* a form of patternicity. Folk numeracy is our natural tendency to misperceive probabilities, to think anecdotally instead of statistically, and to focus on and remember short-term trends and small-number runs. We notice a short stretch of cool days and ignore the long-term global warming trend. We note with consternation a downturn in the housing and stock markets, forgetting the half century upward-pointing trend lines. Sawtooth data trend lines, in fact, are exemplary of folk numeracy, where our senses are geared to focus on each tooth's up or down angle while the overall direction of the blade is nearly invisible to us. Folk numeracy is just one of many cognitive biases that influence and often distort the way that we process information, and together these biases reinforce our intuitively derived belief systems.

How Our Brains Convince Us That We Are Always Right

Once we form beliefs and make commitments to them, we maintain and reinforce them through a number of powerful *cognitive heuristics* that guarantee they are correct. A *heuristic* is a mental method of solving a problem through intuition, trial and error, or informal methods when there is no formal means or formula for solving it (and often even when there is). These heuristics are sometimes called *rules of thumb,* although they are better known as *cognitive biases* because they almost always distort percepts to fit preconceived concepts. Beliefs configure perceptions. No matter what belief system is in place—religious, political, economic, or social—these cognitive biases shape how we interpret information that comes through our senses and mold it to fit the way we want the world to be and not necessarily how it really is; once again, the basis of belief-dependent realism.

I call this general process *belief confirmation.* There are a number of specific cognitive heuristics that operate to confirm our beliefs as true. When integrated into the processes of patternicity and agenticity, these heuristics support my thesis that beliefs are formed for a variety of subjective, emotional, psychological, and social reasons, and then are reinforced, justified, and explained with rational reasons.

The Confirmation Bias: The Mother of All Cognitive Biases

Throughout this book I have referenced the confirmation bias in various contexts. Here I would like to examine it in detail, as it is the mother of all the cognitive biases, giving birth in one form or another to most of the other heuristics. Example: as a fiscal conservative and social liberal I can find common ground whether I am talking to a Republican or a Democrat. In fact, I have close friends in both camps, and over the years I have observed the following: no matter what the issue is under discussion, both sides are equally convinced that the evidence overwhelmingly supports their position. I'm sure it does because of the confirmation bias, or *the tendency to seek and find confirmatory evidence in support of already existing beliefs and ignore or reinterpret disconfirming evidence.* The confirmation bias is best captured in the biblical wisdom *Seek and ye shall find.*

Experimental examples abound.[2] In 1981, psychologist Mark Snyder tasked subjects to assess the personality of someone whom they were about to meet, but only after they reviewed a profile of the person. Subjects in one group were given a profile of an introvert (shy, timid, quiet), while subjects in another group were given a profile of an extrovert (sociable, talkative, outgoing). When asked to make a personality assessment, those subjects who were told that the person would be an extrovert tended to ask questions that would lead to that conclusion; the introvert group did the same in the opposite direction.[3] In a 1983 study, psychologists John Darley and Paget Gross showed subjects a video of a child taking a test. One group was told that the child was from a high socioeconomic class while the other group was told that the child was from a low socioeconomic class. The subjects were then asked to evaluate the academic abilities of the child based on the results of the test. Even though both groups of subjects were evaluating the exact same set of numbers, those who were told that the child was from a high socioeconomic class rated the child's abilities as above grade level, and those who thought that the child was from a low socioeconomic class rated the child as below grade level in ability.[4] This is a striking indictment of human reason but a testimony to the power of belief expectations.

The power of expectation was displayed in a 1989 study by psychologists Bonnie Sherman and Ziva Kunda, who presented a group of subjects with evidence that contradicted a belief they held deeply, and with evidence that supported those same beliefs. The results showed that the subjects recognized the validity of the confirming evidence but were skeptical of the value of the disconfirming evidence.[5] In another 1989 study, by psychologist Deanna Kuhn, when children and young adults were exposed to evidence inconsistent with a theory they preferred, they failed to notice the contradictory evidence, or if they did acknowledge its existence, they tended to reinterpret it to favor their preconceived beliefs.[6] In a related study, Kuhn exposed subjects to an audio recording of an actual murder trial and discovered that instead of evaluating the evidence first and then coming to a conclusion, most subjects concocted a narrative in their mind about what happened, made a decision of guilt or innocence, then riffled through the evidence and picked out what most closely fit the story.[7]

The confirmation bias is particularly potent in political beliefs, most notably the manner in which our belief filters allow in information that confirms our ideological convictions and filters out information that disconfirms those same convictions. This is why it is so easy to predict which media outlets liberals and conservatives choose to monitor. We now even have an idea of where in the brain the confirmation bias is processed thanks to an fMRI study conducted at Emory University by Drew Westen.[8]

During the run-up to the 2004 presidential election, while undergoing a brain scan, thirty men—half self-described "strong" Republicans and half "strong" Democrats—were tasked with assessing statements by both George W. Bush and John Kerry in which the candidates clearly contradicted themselves. Not surprisingly, in their assessments of the candidates, Republican subjects were as critical of Kerry as Democratic subjects were of Bush, yet both let their own preferred candidate off the evaluative hook. Of course. But what was especially revealing were the neuroimaging results: the part of the brain most associated with reasoning—the *dorsolateral prefrontal cortex*—was quiescent. Most active were the *orbital frontal cortex*, which is involved in the processing of emotions, and the *anterior cingulate cortex*—our old friend the ACC, which is so active in patternicity processing and conflict resolution. Interestingly, once subjects had arrived at a conclusion that made them emotionally comfortable, their *ventral striatum*—a part of the brain associated with reward—became active.

In other words, instead of rationally evaluating a candidate's positions on this or that issue, or analyzing the planks of each candidate's platform, we have an emotional reaction to conflicting data. We rationalize away the parts that do not fit our preconceived beliefs about a candidate, then receive a reward in the form of a neurochemical hit, probably dopamine. Westen concluded:

> We did not see any increased activation of the parts of the brain normally engaged during reasoning. What we saw instead was a network of emotion circuits lighting up, including circuits hypothesized to be involved in regulating emotion, and circuits known to be involved in resolving conflicts. Essentially, it appears as if partisans twirl the cognitive kaleidoscope until they get the conclusions they want, and then they get massively reinforced for it, with the elimination of negative emotional states and activation of positive ones.

Hindsight Bias

In a type of time-reversal confirmation bias, the *hindsight bias* is *the tendency to reconstruct the past to fit with present knowledge*. Once an event has occurred, we look back and reconstruct how it happened, why it had to happen that way and not some other way, and why we should have seen it coming all along.[9] Such "Monday-morning quarterbacking" is literally evident on the Monday mornings following a weekend filled with football games. We all know what plays should have been called . . . after the outcome. Ditto the stock market and the endless parade of financial experts whose prognostications are quickly forgotten as they shift to post hoc analysis after the market closes. It's easy to "buy low, sell high" once you have perfect information, which is available only after the fact when it is too late.

The hindsight bias is on prominent display after a major disaster, when everyone thinks that they know how and why it happened, and why our experts and leaders should have seen it coming. NASA engineers should have known that the O-ring on the space shuttle *Challenger*'s solid rocket booster joints would fail in freezing temperatures leading to a massive explosion, or that a small foam strike on the leading edge of the wing of the space shuttle *Columbia* would result in its destruction upon reentry. Such highly improbable and unpredictable events become not only probable but practically certain *after they happen*. The hand-wringing and finger-pointing by the members of NASA's investigative commissions tasked with determining the causes of the two space shuttle disasters were case studies in the hindsight bias. Had such certainty really existed before the fact, then of course different actions would have been taken.

The hindsight bias is equally evident in times of war. Almost immediately following the Japanese attack on Pearl Harbor on December 7, 1941, for example, conspiracy theorists went to work to prove that President Roosevelt must have known it was coming because of the so-called bomb plot message that U.S. intelligence intercepted in October 1941: a Japanese agent in Hawaii had been instructed by his superiors in Japan to monitor warship movements in and around the naval base at Pearl. That sounds fairly damning and, in fact, there were eight such messages dealing with Hawaii as a possible target that were intercepted and decrypted by U.S. intelligence before December 7. How could our leaders not have seen it coming? They must have, and therefore they let it happen for nefarious and Machiavellian reasons. So say the conspiracy theorists with their hindsight bias dialed up to full.

Between May and December of that year, however, there were no less than fifty-eight messages intercepted regarding Japanese ship movements indicating an attack on the Philippines, twenty-one messages involving Panama, seven messages affiliated with attacks in Southeast Asia and the Netherlands East Indies, and even seven messages connected to the United States' West Coast. There were so many intercepted messages, in fact, that army intelligence stopped sending memos to the White House out of concern that there might be a breach in security leading the Japanese to realize that we had broken their codes and were reading their mail.[10]

President George W. Bush was subject to the same type of conspiratorial hindsight bias after 9/11, when a memo surfaced dated August 6, 2001, entitled "Bin Laden Determined to Strike in U.S." Reading the memo in hindsight is eerie, with references to hijacked planes, bombing the World Trade Center, and attacks on Washington, D.C., and the Los Angeles International Airport. But if you read it in a pre-9/11 mind-set, and in the context of the hundreds of intel memos tracking the various comings and goings and potential targets of al-Qaeda—an international organization operating in dozens of countries and targeting numerous American embassies, military bases, navy ships, and the like—it is not at all clear when, where, or if such attacks might happen. Think about the hindsight bias in today's context in which we know with near certainty that al-Qaeda will strike again, but we lack the information to know where and when and how they will attack. This leads us to defend against the *last* attack.

Self-Justification Bias

This heuristic is related to the hindsight bias. The *self-justification bias is the tendency to rationalize decisions after the fact to convince ourselves that what we did was the best thing we could have done.* Once we make a decision about something in our lives we carefully screen subsequent data and filter out all contradictory information related to that decision, leaving only evidence in support of the choice we made. This bias applies to everything from career and job choices to mundane purchases. One of the practical benefits of self-justification is that no matter what decision we make—to take this or that job, to marry this or that person, to purchase this or that product—we will almost always be satisfied with the decision, even when the objective evidence is to the contrary.

This process of cherry-picking the data happens at even the highest levels of expert assessment. Political scientist Philip Tetlock, for example, in his book *Expert Political Judgment*, reviewed the evidence for the ability of professional experts in politics and economics to make accurate predictions and assessments. He found that even though all of them claimed to have data in support of their positions, when analyzed after the fact such expert opinions and predictions turned out to be no better than those of nonexperts—or even chance. Yet, as the self-justification heuristic would predict, experts are significantly less likely to admit that they are wrong than nonexperts.[11] Or as I like to say, *smart people believe weird things because they are better at rationalizing their beliefs that they hold for nonsmart reasons.*

As we saw in the previous chapter, politics is filled with self-justifying rationalizations. Democrats see the world through liberal-tinted glasses, while Republicans filter it through conservative-shaded lenses. When you listen to both "conservative talk radio" and "progressive talk radio" you will hear current events interpreted in ways that are 180 degrees out of phase. So incongruent are the interpretations of even the simplest goings-on in the daily news that you wonder if they can possibly be talking about the same event. Social psychologist Geoffrey Cohen quantified this effect in a study in which he discovered that Democrats are more accepting of a welfare program if they believe it was proposed by a fellow

Democrat, even if the proposal came from a Republican and is quite restrictive. Predictably, Cohen found the same effect for Republicans, who were far more likely to approve of a generous welfare program if they thought it was proposed by a fellow Republican.[12] In other words, even when examining the exact same data people from both parties arrive at radically different conclusions.

A very disturbing real-world example of the self-justification heuristic can be seen in the criminal justice system. According to Northwestern University law professor Rob Warden,

> You get in the system and you become very cynical. People are lying to you all over the place. Then you develop a theory of the crime, and it leads to what we call tunnel vision. Years later overwhelming evidence comes out that the guy was innocent. And you're sitting there thinking, "Wait a minute. Either this overwhelming evidence is wrong or I was wrong—and I couldn't have been wrong because I'm a good guy." That's a psychological phenomenon I have seen over and over.[13]

Attribution Bias

Our beliefs are very much grounded in how we attribute the causal explanations for them, and this leads to a fundamental *attribution bias*, or *the tendency to attribute different causes for our own beliefs and actions than that of others.* There are several types of attribution bias.[14] There is a *situational attribution bias*, in which we identify the cause of someone's belief or behavior in the environment ("her success is a result of luck, circumstance, and having connections") and a *dispositional attribution bias*, in which we identify the cause of someone's belief or behavior in the person as an enduring personal trait ("her success is due to her intelligence, creativity, and hard work"). And, thanks to the self-serving bias, we naturally attribute our own success to a positive disposition ("I am hardworking, intelligent, and creative") and we attribute others' success to a lucky situation ("he is successful because of circumstance and family connections").[15] The attribution bias is a form of personal spin-doctoring.

My colleague Frank Sulloway and I discovered another form of attribution bias in a research project that we conducted several years ago. Frank and I wanted to know why people believe in God, so we polled ten thousand random Americans. In addition to exploring various demographic and sociological variables, we also directly asked subjects in an essay question why they believed in God and why they thought others believe in God. The top two reasons that people gave for why they believed in God were "the good design of the universe" and "the experience of God in everyday life." Interestingly, and tellingly, when subjects were asked why they thought other people believed in God, these two answers dropped to sixth and third place, respectively, and the two most common reasons given were that belief is "comforting" and "fear of death."[16] These answers revealed a sharp distinction between an *intellectual attribution bias*, in which people consider their own beliefs as being rationally motivated, and an *emotional attribution bias*, in which people see the beliefs of others as being emotionally driven.

You can see this attribution bias in political as well as religious beliefs. For example, on the issue of gun control, you will hear someone attribute their own position to reasoned intellectual choice ("I am for gun control because statistics show that crime decreases when gun ownership decreases" or "I'm against gun control because studies show that more guns means less crime"), and attribute the other person's opinion on the same subject to emotional need ("He is for gun control because he is a bleeding-heart liberal who needs to identify with the victim" or "He is against gun control because he's a heartless conservative who needs to feel emboldened by a weapon").[17] This was, in fact, what political scientists Lisa Farwell and Bernard Weiner discovered in their study on the attribution bias in political attitudes, with conservatives justifying their beliefs with rational arguments but accusing political liberals of being "bleeding hearts"; liberals, in turn, offered intellectual justifications for their positions, while accusing conservatives of being "heartless."[18]

The attribution bias of perceiving intellectual reasons for belief as superior to emotional reasons appears to be a manifestation of a broader form of self-serving bias through which people slant their perceptions of the world, especially the social world, in their favor.

Sunk-Cost Bias

Leo Tolstoy, one of the deepest thinkers on the human condition in the history of literature, made this observation on the power of deeply held and complexly entwined beliefs: "I know that most men, including those at ease with problems of the greatest complexity, can seldom accept even the simplest and most obvious truth if it be such as would oblige them to admit the falsity of conclusions which they have delighted in explaining to colleagues, which they have proudly taught to others, and which they have woven, thread by thread, into the fabric of their lives." Upton Sinclair said it more succinctly: "It is difficult to get a man to understand something when his job depends on not understanding it."

These observations are examples of the *sunk-cost bias*, or *the tendency to believe in something because of the cost sunk into that belief.* We hang on to losing stocks, unprofitable investments, failing businesses, and unsuccessful relationships. With the attribution bias throttled up we concoct rational reasons to justify those beliefs and behaviors in which we have made sizable investments. The bias leads to a basic fallacy: that past investment should influence future decisions. If we were rational we would just compute the odds of succeeding from this point forward and then decide if additional investment warrants the potential payoff. But we are not rational, not in business, certainly not in love, and most especially not in war. Consider the cost we've sunk into the wars in Iraq and Afghanistan. These wars are costing us $4.16 billion a year in military expenditures alone, not to mention the billions of dollars spent in nonmilitary expenditures, along with the 5,342 Americans killed (at the time of writing, a figure that grows by the day). No wonder most members of Congress from both parties, along with presidents Obama, Bush, Clinton, and Bush have all stated that we've got to "stay the course" and not just "cut and run." President George W. Bush explained in a July 4, 2006, speech at Fort Bragg, North Carolina: "I'm not going to allow the sacrifice of 2,527 troops who have died in Iraq to be in vain by pulling out before the job is done."[19] This is the very embodiment of the sunk-cost bias.

Status Quo Bias

Are you an organ donor? I am, but in my state (California) I had to punch out a little tab and stick it on my driver's license to indicate my preference, and this little requirement means that far fewer people in my state are organ donors compared to states where the default position is that you are an organ donor unless you punch out a little tab indicating that you do not wish to participate. This is an opt-in versus opt-out choice architecture design dilemma, and it is an example of the *status quo bias*, or *the tendency to opt for whatever it is we are used to, that is, the status quo.* We tend to prefer existing social, economic, and political arrangements over proposed alternatives, even sometimes at the expense of individual and collective self-interest. Other examples abound.

Economists William Samuelson and Richard Zeckhauser discovered that when people are offered a choice among four different financial investments with varying degrees of risk, they select one based upon how risk averse they are, and their choices range widely. But when people are told that an investment tool has been selected for them and that they then have the opportunity to switch to one of the other investments, 47 percent stayed with what they already had compared to the 32 percent who chose those particular investment opportunities when none were presented first as a default option.[20] In the early 1990s, citizens in New Jersey and Pennsylvania were offered two options for their automobile insurance: a high-priced option that granted them the right to sue and a cheaper option that restricted their rights to sue. Corresponding options in each state were roughly equivalent. In New Jersey the default option was the more expensive one, that is, if you did nothing you were automatically given that choice, and so 75 percent of citizens selected it. In Pennsylvania the default option was the cheaper one, and only 20 percent opted for the more expensive plan.[21]

Why does the status quo bias exist? Because the status quo represents what we already have (and have to give up in order to change), versus what we *might* have once we choose, which is far riskier. Why should this be? Because of the *endowment effect*.

Endowment Effect

The psychology underlying the status quo bias is what economist Richard Thaler calls the *endowment effect*, or *the tendency to value what we own more than what we do not own*. In his research on the endowment effect, Thaler has found that owners of an item value it roughly twice as much as potential buyers of the same item. In one experiment, subjects were given a coffee mug valued at $6.00 and they were asked what they would take for it. The average price below which they would not sell was $5.25. Another group of subjects were asked how much they would be willing to pay for the same mug and gave an average price of $2.75.[22]

Ownership endows value by its own virtue, and nature has endowed us to hold dear what is ours. Why? Evolution. The endowment effect begins with the natural propensity for animals to mark their territories and defend them through threat gestures and even physical aggression if necessary, thereby declaring the equivalent of private ownership to what was once a public good. The evolutionary logic runs like this: once a territory is declared taken by one animal, would-be trespassers have to invest considerable energy and risk grave bodily injury in attempts to acquire the property for themselves, so there is an endowment effect. We are more willing to invest in defending what is already ours than we are to take what is someone else's. Dogs, for example, will invest more energy in defending a bone from a challenger than they will in absconding with some other dog's bone. The endowment effect with property ownership has a direct and obvious connection to *loss aversion*, where we are twice as motivated to avoid the pain of loss as we are to seek the pleasure of gain. Evolution has wired us to care more about what we already have than what we might possess, and here we find the evolved moral emotion that undergirds the concept of private property.

Beliefs are a type of private property—in the form of our private thoughts with public expressions—and therefore the endowment effect applies to belief systems. The longer we hold a belief, the more we have invested in it; the more publicly committed we are to it, the more we endow it with value and the less likely we are to give it up.

Framing Effects

How beliefs are framed often determines how they are assessed, and this is called the *framing effect*, or *the tendency to draw different conclusions based on how data are presented*. Framing effects are especially noticeable in financial decisions and economic beliefs. Consider the following thought experiment presented in two different frames for the same financial problem:

1. Phones Galore offers the new Techno phone for $300; five blocks away FactoryPhones has the same model half off for $150. Do you make the short trip to save $150? Sure you would, right?

2. Laptops Galore offers the new SuperDuper computer for $1,500; five blocks away FactoryLaptops has the same model discounted to $1,350. Do you make the short trip to save $150? Nah, why bother?

In research where subjects are offered such choices, most people would take the trip in the first scenario but not the second, even though the amount saved is the same! Why? The framing changes the perceived value of the choice.

Framing effects can be found in both political and scientific beliefs. Here is a classic thought experiment with real-world implications: You are a contagious disease expert at the Centers for Disease Control and you have been told that the United States is preparing for the outbreak of an unusual Asian disease that is expected to kill six hundred people. Your team of experts has presented you with two programs to combat the disease:

Program A: Two hundred people will be saved.

Program B: There is a one-third probability that all six hundred people will be saved, and a two-thirds probability that nobody will be saved.

If you are like the 72 percent of the subjects in an experiment that presented this scenario, you chose Program A. Now consider another set of choices for the same scenario:

Program C: Four hundred people will die.

Program D: There is a one-third probability that nobody will die, and a two-thirds probability that all six hundred people will die.

Even though the net result of the second set of choices is precisely the same as the first, subjects switched preferences, from 72 percent for Program A to 78 percent for Program D. The framing of the question led to the shift in preference. We prefer to think in terms of how many people we may save instead of how many people will die—the "positive frame" is preferred over the "negative frame."[23]

Anchoring Bias

Lacking some objective standard to evaluate beliefs and decisions—which is usually not available—we grasp for any standard on hand, no matter how seemingly subjective. Such standards are called *anchors*, and this creates the *anchoring effect*, or *the tendency to rely too heavily on a past reference or on one piece of information when making decisions*. The comparison anchor can even be entirely arbitrary. In one study subjects were asked to give the last four digits of their Social Security numbers, and then asked to estimate the number of physicians in New York City. Bizarrely, people with higher Social Security numbers tended to give higher estimates for the number of docs in Manhattan. In a related study, subjects were shown an array of items to purchase—a bottle of wine, a cordless computer keyboard, a video game—and were then told that the value of the items was equal to the last two digits of their Social Security numbers. When subsequently asked the maximum price they would be willing to pay, subjects with high Social Security numbers consistently said that they would be willing to pay more than those with low numbers. With no objective anchor for comparison, this random anchor influenced them arbitrarily.

Our intuitive sense of the anchoring effect and its power leads negotiators in corporate mergers, representatives in business deals, and even disputants in divorces to begin from an extreme initial position in order to set the anchor high for their side.

Availability Heuristic

Have you ever noticed how many red lights you encounter while driving when you are late for an appointment? Me, too. How does the universe know that I left late? It doesn't, of course, but the fact that most of us notice more red lights when we are running late is an example of the *availability heuristic*, or the tendency to assign probabilities of potential outcomes based on examples that are immediately available to us, especially those that are vivid, unusual, or emotionally charged, which are then generalized into conclusions upon which choices are based.[24]

For example, your estimation of the probability of dying in a plane crash (or lightning strike, shark attack, terrorist attack, and so on) will be directly related to the availability of just such an event in your world, especially your exposure to it in mass media. If newspapers and especially television cover an event there is a good chance that people will overestimate the probability of that event happening.[25] An Emory University study, for example, revealed that the leading cause of death in men—heart disease—received the same amount of media coverage as the eleventh-ranked cause: homicide. In addition, drug use—the lowest-ranking risk factor associated with serious illness and death—received as much attention as the second-ranked risk factor of poor diet and lack of exercise. Other studies have found that women in their forties believe they have a 1 in 10 chance of dying from breast cancer, while their real lifetime odds are more like 1 in 250. This effect is directly related to the number of news stories about breast cancer.[26]

Representative Bias

Related to the availability bias is the *representative bias*, which, as described by its discoverers, psychologists Amos Tversky and Daniel Kahneman, means: "an event is judged probable to the extent that it represents the essential features of its parent population or generating process." And, more generally, "when faced with the difficult task of judging probability or frequency, people employ a limited number of heuristics which reduce these judgments to simpler ones."[27] The following thought experiment has become a classic in cognitive studies. Imagine that you are looking to hire someone for your company and you are considering the following candidate for employment:

> Linda is thirty-one years old, single, outspoken, and very bright. She majored in philosophy. As a student, she was deeply concerned with issues of discrimination and social justice and participated in antinuclear demonstrations.

Which is more likely? 1. Linda is a bank teller. 2. Linda is a bank teller and is active in the feminist movement.

When this scenario was presented to subjects, 85 percent chose the second option. Mathematically speaking, this is the wrong choice, because the probability of two events occurring together will always be less than the probability of one occurring by itself. And yet most people get this problem wrong because they fall victim to the representative fallacy, in which the descriptive terms presented in the second option seem more representative of the description of Linda.[28]

Hundreds of experiments reveal time and again that people make snap decisions under high levels of uncertainty, and they do so by employing these various rules of thumb to shortcut the computational process. For example, policy experts were asked to estimate the probability that the Soviet Union would invade Poland and that the United States would then break off diplomatic relations. Subjects gave this a probability of

4 percent. Meanwhile, another group of policy experts was asked to esti-mate the probability just that the United States would break off diplomatic relations with the Soviet Union. Although the latter was more likely, these experts gave it a smaller probability of happening. The experi-menters concluded that the more detailed two-part scenario seemed more representative of the actors involved.

Inattentional Blindness Bias

Arguably one of the most powerful of the cognitive biases that shape our beliefs is captured in the biblical proverb "There are none so blind as those who will not see." Psychologists call this *inattentional blindness,* or *the tendency to miss something obvious and general while attending to something special and specific.* The now-classic experiment in this bias has subjects watching a one-minute video of two teams of three players each, one team donning white shirts and the other black shirts, as they move about one another in a small room tossing two basketballs back and forth. The assigned task is to count the number of passes made by the white team. Unexpectedly, after thirty-five seconds a gorilla enters the room, walks directly through the farrago of bodies, thumps his chest, and exits nine seconds later.

How could anyone miss a guy in an ape suit? In fact, in this remarkable experiment by psychologists Daniel Simons and Christopher Chabris, 50 percent of subjects did not see the gorilla, even when asked if they noticed anything unusual.[29] For many years now I have incorporated the gorilla DVD into my public lectures, asking for a show of hands of those who did not see the gorilla. Out of the more than one hundred thousand people I have shown it to over the years, fewer than half saw the gorilla during the first viewing. (I show the clip a second time with no counting and every-one sees it.) I was able to decrease the figure even more by telling audiences that one gender is more accurate than the other at counting the passes, but I won't tell them which gender so as not to bias the test. This really makes people sit up and concentrate, causing even more to miss the gorilla.

Most recently, I filmed a special on gullibility for *Dateline NBC* with the host Chris Hansen, in which we reconstructed a number of classic psychological experiments that demonstrate many of these cognitive biases, one of which was inattentional blindness. Instead of a gorilla, how-ever, we had Chris Hansen himself walk right through the middle of a

Figure 12. Would You See the Gorilla?
Inattentional blindness is *the tendency to miss something obvious and general while attending to something special and specific.* The now-classic experiment in this bias has subjects watching a one-minute video of two teams of three players each, one team donning white shirts and the other black shirts, as they move about one another in a small room tossing two basketballs back and forth. The assigned task is to count the number of passes made by the white team. Unexpectedly, after thirty-five seconds a gorilla enters the room, walks directly through the farrago of bodies, thumps its chest, and exits nine seconds later. In this remarkable experiment by psychologists Daniel Simons and Christopher Chabris, 50 percent of the subjects did not see the gorilla, even when asked if they noticed anything unusual. PHOTO COURTESY OF DANIEL SIMONS AND CHRISTOPHER CHABRIS, "GORILLAS IN OUR MIDST: SUSTAINED INATTENTIONAL BLINDNESS FOR DYNAMIC EVENTS," *PERCEPTION* 28 (1999): 1059–74, AND THE LAB WEB PAGE OF DANIEL SIMONS: http://www.theinvisiblegorilla.com.

---

room in which there was a studio audience of people who thought that they were trying out for an NBC reality show. We arranged for a real New York basketball team to participate, but when I saw how small the room was and how close the audience members would be to the area where Chris would walk across the stage, I became concerned that the effect would not work. So I instructed our basketball players to really ham up their dribbling and passing and to emulate the Harlem Globetrotters with some very animated and vocal play. As well, I divided the studio audience into two groups, one of whom would count the number of passes by the white-shirted players and the other half the number of passes by the black-shirted players. Finally, I had them count the passes out loud. The effect was nearly complete. Only a couple of people noticed something unusual, and not one person in the audience saw that it was

Chris Hansen who walked across the stage, stopped, twirled around, and exited the stage. The audience was shocked when I explained what had just happened and brought Chris out to greet them.

Experiments such as these reveal a hubris in our powers of perception, as well as a fundamental misunderstanding of how the brain works. We think of our eyes as video cameras and our brains as blank tapes to be filled with percepts. Memory, in this flawed model, is simply rewinding the tape and playing it back in the theater of the mind. This is not at all what happens. The perceptual system, and the brain that analyzes its data, are deeply influenced by the beliefs it already holds. As a consequence, much of what passes before our eyes may be invisible to a brain focused on something else. In fact, eye trackers have been used to monitor subjects watching the film, and those who missed the gorilla were looking right at it.

Biases and Beliefs

Our beliefs are buffeted by a host of these and additional cognitive biases that I will briefly mention here (in alphabetical order):

Authority bias: the tendency to value the opinions of an authority, especially in the evaluation of something we know little about.

Bandwagon effect: the tendency to hold beliefs that other people in your social group hold because of the social reinforcement provided.

Barnum effect: the tendency to treat vague and general descriptions of personality as highly accurate and specific.

Believability bias: the tendency to evaluate the strength of an argument based on the believability of its conclusion.

Clustering illusion: the tendency to see clusters of patterns that, in fact, can be the result of randomness; a form of patternicity.

Confabulation bias: the tendency to conflate memories with imagination and other people's accounts as one's own.

Consistency bias: the tendency to recall one's past beliefs, attitudes, and behaviors as resembling present beliefs, attitudes, and behaviors more than they actually do.

Expectation bias / experimenter bias: the tendency for observers and especially for scientific experimenters to notice, select, and publish data that agree with their expectations for the outcome of an experiment, and to not notice, discard, or disbelieve data that appear to conflict with those experimental expectations.

False-consensus effect: the tendency for people to overestimate the degree to which others agree with their beliefs or that will go along with them in a behavior.

Halo effect: the tendency for people to generalize one positive trait of a person to all the other traits of that person.

Herd bias: the tendency to adopt the beliefs and follow the behaviors of the majority of members in a group in order to avoid conflict.

Illusion of control: the tendency for people to believe that they can control or at least influence outcomes that most people cannot control or influence.

Illusory correlation: the tendency to assume that a causal connection (correlation) exists between two variables; another form of patternicity.

In-group bias: the tendency for people to value the beliefs and attitudes of those whom they perceive to be fellow members of their group, and to discount the beliefs and attitudes of those whom they perceive to be members of a different group.

Just-world bias: the tendency for people to search for things that the victim of an unfortunate event might have done to deserve it.

Negativity bias: the tendency to pay closer attention and give more weight to negative events, beliefs, and information than to positive.

Normalcy bias: the tendency to discount the possibility of a disaster that has never happened before.

Not-invented-here bias: the tendency to discount the value of a belief or source of information that does not come from within.

Primacy effect: the tendency to notice, remember, and assess as more valuable initial events more than subsequent events.

Projection bias: the tendency to assume that others share the same or similar beliefs, attitudes, and values, and to overestimate the probability of others' behaviors based on our own behaviors.

Recency effect: the tendency to notice, remember, and assess as more valuable recent events more than earlier events.

Rosy retrospection bias: the tendency to remember past events as being more positive than they actually were.

Self-fulfilling prophecy: the tendency to believe in ideas and to behave in ways that conform to expectations for beliefs and actions.

Stereotyping or generalization bias: the tendency to assume that a member of a group will have certain characteristics believed to represent the group without having actual information about that particular member.

Trait-ascription bias: the tendency for people to assess their own personality, behavior, and beliefs as more variable and less dogmatic than those of others.

Bias Blind Spot

The *bias blind spot* is really a meta-bias in that it is grounded in all the other cognitive biases. It is *the tendency to recognize the power of cognitive biases in other people but to be blind to their influence upon our own beliefs*. In one study conducted by Princeton University psychologist Emily Pronin and her colleagues, subjects were randomly assigned high or low scores on a "social intelligence" test. Unsurprisingly, those given the high marks rated the test fairer and more useful than those receiving low marks. When asked if it was possible that they had been influenced by the score on the test, subjects responded that other participants had been far more biased than they were. Even when subjects admit to having a bias, such as being a member of a partisan group, this "is apt to be accompanied by the insistence that, in their own case, this status . . . has been uniquely *enlightening*—indeed, that it is the *lack* of such enlightenment that is making those on the other side of the issue take their misguided position," said Pronin. In a related study at Stanford University, students were asked to compare themselves to their peers on such personal qualities as friendliness and selfishness. Predictably, they rated themselves higher. Yet, even when the subjects were warned about the *better-than-average bias* and asked to reevaluate their original assessments, 63 percent claimed that their initial evaluations were objective, and 13 percent even claimed to be too modest![30]

The Middle Land of Belief

Now that we have drilled deep into the brain to examine the cognitive biases of belief, let us pull back for a broader view of what I call the *Middle Land* of belief.

Imagine these two series of twenty-five heads (H) and tails (T) coin flips and guess which series best represents randomness:

THTHTHTHTHTHTHTHTHTHTHTHT

HHHTTHTTTHTHHHHTTHHTTTTTH

Most people would say that the first series of alternative heads and tails looks the most random, whereas, in fact, both computer simulations and actual coin-flipping experiments generate something much more like the second series (try it yourself). When subjects are asked to imagine flipping a coin and are then instructed to write down the sequence of outcomes, their guesses are highly nonrandom. That is, their string of Ts and Hs more closely resembles the predictable first string above and not the less predictable and more (but not perfectly) random second string.

This fact goes a long way toward explaining the apparent nonrandom guessing in ESP experiments that paranormal researchers claim as evidence for psychic power. In fact, in their analysis of ESP research over the past century, Peter Brugger and Kirsten Taylor have redefined ESP as *effect of subjective probability*, noting that scientists have now conclusively demonstrated what typically happens in research in which one subject tries to determine or anticipate the thoughts or actions of a second subject using paranormal means. When the second subject is instructed to randomly perform some task (such as raising or lowering an arm), the sequence is not going to be random. Over time the second subject will develop a predictable pattern that the first subject will unconsciously learn.[31] This effect is called *implicit sequence learning*, and it has plagued paranormal research for over a century as researchers continue to fail to control for it. As the mathematician Robert Coveyou once quipped: "Random number generation is too important to be left to chance."[32]

The reason that our folk intuitions so often get it wrong is that we evolved in what the evolutionary biologist Richard Dawkins calls *Middle World*—a land midway between short and long, small and large, slow and fast, young and old. Out of alliterative preference, I call it *Middle Land*. In the Middle Land of space, our senses evolved for perceiving objects of middling size—between, say, grains of sand and mountain ranges. We are not equipped to perceive atoms and germs, on one end of the scale, or galaxies and expanding universes, on the other end. In the Middle Land of speed, we can detect objects moving at a walking or running pace, but the glacially slow movement of continents (and glaciers) and the bogglingly fast speed of light are literally imperceptible. Our Middle Land time scales range from the psychological "now" of three seconds in duration to the few decades of a human lifetime, far too short to witness evolution, continental drift, or long-term environmental changes. Our Middle Land folk numeracy leads us to pay attention to and remember short-term trends, meaningful coincidences, and personal anecdotes.

Additional random processes and our folk numeracy about them abound. Hollywood studio executives often fire successful producers after a short run of box-office bombs, only to watch the subsequent films under production during the producer's reign become blockbusters. Athletes who appear on *Sports Illustrated*'s cover typically experience career downturns, not because of a superstitious jinx but because of the "regression to the mean." The exemplary performance that landed them on the cover in the first place is a low-probability event that is difficult to repeat, and thus they "regress" back to their normal performance levels.

Extraordinary events do not always require extraordinary causes. Given enough time and opportunity, they can happen by chance. Understanding this can help us overcome our Middle Land propensity to find patterns and agents that are not actually there. Embrace the random. Find the pattern. Know the difference.

Science as the Ultimate Bias-Detection Machine

The study of cognitive biases has revealed that humans are anything but the Enlightenment ideal of rational calculators carefully weighing the evidence for and against beliefs. And these biases are far reaching in their effects. A judge or jury assessing evidence against a defendant, a CEO evaluating information about a company, or a scientist weighing data in favor of a theory will undergo the same cognitive temptations to confirm what is already believed.

What can we do about it? In science we have built-in self-correcting machinery. In experiments, strict double-blind controls are required, in which neither the subjects nor the experimenters know the experimental conditions during the data-collection phase. Results are vetted at professional conferences and in peer-reviewed journals. Research must be replicated in other labs unaffiliated with the original researcher. Disconfirming evidence, as well as contradictory interpretations of the data, must be included in the paper. Colleagues are rewarded for being skeptical. Nevertheless, scientists are no less vulnerable to these biases, so such precautions must be vigorously enforced, especially by the scientists themselves, because if you don't seek contradictory data against your theory or beliefs, someone else will, usually with great glee and in a public forum.

How this method of science developed historically and how it works today are the subject of the final chapters and epilogue of this book.

13

Geographies of Belief

THROUGHOUT THIS JOURNEY INTO THE BELIEVING BRAIN WE HAVE seen how we are not the rational calculators and logic machines that the Enlightenment philosophers who launched the Age of Reason envisioned. We are, in fact, subject to a host of factors that shape our beliefs. Patternicity ensures that we will seek and find patterns in both meaningful and meaningless noise. Agenticity drives us to infuse those patterns with meaning and intentional agents to explain why things happen as they do. These meaningful patterns form the core of our beliefs, for which our brains employ a host of cognitive biases that continually confirm our beliefs as true, and our understanding of reality is dependent upon those beliefs. To reiterate my thesis: *beliefs come first, the explanations for the beliefs follow.*

How, then, can we tell the difference between true and false patterns? How can we discern the difference between real and imaginary agents? How can we avoid the cognitive bias pitfalls that so burden our rationality? The answer is science. A brief tour through what I am calling the *geographies of belief* reveals that despite the subjectivity of our psychologies, relatively objective knowledge is available through the tools of science. The story of how those tools were created is a halting journey of exploration of the world and our place in it.

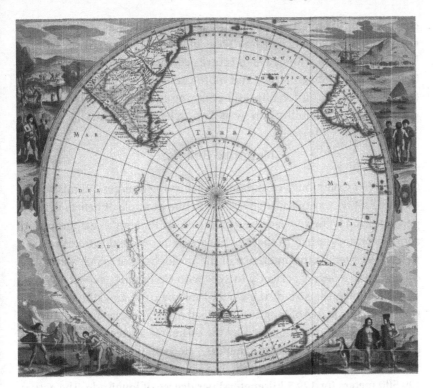

Figure 13. *Terra Australis Incognita*

Terra incognita are two of the most important words ever penned on the geography of belief, embodying the mental space of unlimited exploration—a story without end. They appear on this map, *Terra Australis Incognita*, by Hendrik Hondius, 1657. COURTESY OF DIXON LIBRARY, STATE LIBRARY OF NEW SOUTH WALES, AUSTRALIA.

Terra Incognita

The belief engine drives all forms of perception in all fields of knowledge, and there are few more dramatic examples than those from the history of exploration. Geographical maps shape cognitive maps, and vice versa. When Claudius Ptolemaeus of Alexandria—better known to history as Ptolemy—penned the words *Terra Australis Incognita* at the bottom of his second-century CE world map, he unwittingly also provided a cognitive map that shaped exploration for more than 1,500 years by freeing humanity from the constraints of a dogged and dogmatic commitment to certainty. The knowledge that there was still undiscov-ered land—codified in Latin as *terra incognita*—led explorers to new heights of adventure and gave to future generations an earth (and even-tually a cosmos) much larger and more variegated than ever imagined. (See figure 13.) An uncertain and doubting mind leads to fresh world visions and the possibility of new and ever-changing realities.[1]

Negative Beliefs

Christopher Columbus's confidence in achieving a successful mission to the Far East by way of sailing west is a prime example of beliefs driving perceptions. His first voyage was premised on Ptolemy's cartographical coordinates for the length that the Euro-Asian continent extends east, as well as the overall circumference of the world, both of which were mis-calculated to a degree perfectly in sync with Columbus's expectations.

To compute the size of the earth, Ptolemy used an estimate of 500 stadia per one degree of longitude, instead of the more accurate figure of 700 stadia per degree employed by the estimable ancient Greek geogra-pher and mathematician Eratosthenes. A stadium is about 185 meters, so 500 stadia equals 92,500 meters (or 92.5 kilometers) and 700 stadia equals 129,500 meters (or 129.5 kilometers) per degree of longitude. The actual circumference of the earth is 40,075 kilometers at the equator. Ptolemy's calculations estimated it to be about 33,300 kilometers, or 17 percent too small. Add to this Columbus's use of Marinus of Tyre's estimate on the high side of the length that the Euro-Asian land mass stretched eastward (thereby leaving less water to sail across), plus the fact that the land routes from Europe to China and India had become politically unstable after the fall of Constantinople in 1453, and Columbus's plan to sail west to get to the east was actually quite reasonable. (Sailing down the coast of Africa, around the Cape of Good Hope, and east to India and China had never been successfully completed and was considered potentially problematic at best and disastrous at worst.) Thus, in one of the most prescient coin-cidences in the history of serendipitous discovery, after sailing a little more than 5,000 kilometers westward across the "Ocean Sea" (the Atlan-tic) on his maiden voyage, Columbus encountered land in the exact place where he had calculated the Indies would be, and thus he dubbed the people he engaged there "Indians."[2]

Why did Columbus not immediately realize he was not in Asia? Surely the flora and fauna and people he discovered were nothing at all like what Marco Polo had reported from his land excursions eastward from Europe where he had met the Great Khan and absorbed Asian culture. The answer can be found in the dual problem of *perception* and *cognition*, or *data* and *theory*. What threw Columbus off was coarse-grained data coupled with incorrect theory. Marco Polo's reports of Asia were sketchy at best, allowing ample wiggle room for interpreting New World data as Old World facts. Plus, there was no theory of a New World, so in Columbus's mind when he made first contact with the New World on that fateful day in October 1492, where else could he be *but* Asia?

Because of the power of the paradigm to shape perceptions, Columbus's cognitive map told him what he was seeing. When his men dug up some common garden rhubarb, *Rheum rhaponticum* (used in pies), for example, the ship's surgeon determined that it was *Rheum officinale*, the medicinal Chinese rhubarb. The native American plant gumbo-limbo was mistaken for an Asiatic variety of the mastic evergreen tree that yields resin used to make lacquer, varnish, and adhesives. The South American *nogal del país* nut was classified as the Asian coconut, or at least what Marco Polo had described as such. Columbus deemed a plant with the aroma of cinnamon to be that valuable Asian spice. After first touching land in San Salvador, Columbus then sailed to Cuba, bringing with him some San Salvadorian captives to help with communications with the Cuban natives, who told him that there was gold to be found at "Cubanacan"—the middle of Cuba—which Columbus heard as "El Gran Can," or the Great Khan. When Columbus touched down again in Cuba during his second voyage, he recorded his navigation along what he thought were the shores of the Mangi kingdom in southern China, which had been described by Marco Polo. And so it went for all four voyages to "the Indies," with Columbus never once doubting where he was, despite never meeting the Great Khan. Such is the power of belief. New data pouring in through old paradigms only reinforced his confidence that he was where he believed he was—on the eastern boundary of the Old World, not the eastern edge of the New World.[3]

The power of the paradigm was witnessed again shortly after Columbus's epic voyages when Ferdinand Magellan set out to circumnavigate the globe in 1519. Once it was established that there was a continental land mass between Europe and Asia, explorers, cartographers, and scholars had two great unanswered geographical questions: (1) Is there a "northern passage" through or around the North American continent linking the Atlantic and Pacific oceans that ships sailing west from Europe could traverse and save months of travel time? (2) Is there really a great southern land mass, the *Terra Australis Incognita* of Ptolemy's imagination? This second question became the provocation for a slew of *negative discoveries*—looking for X but finding Y.

Naval surveyor James Cook secured the headship for these voyages on the premise that he would seek out this unknown territory until he would "discover it or fall in with the eastern side of the land discovered by Tasman and now called New Zealand." (Abel Janszoon Tasman also discovered the large island off the southeastern tip of Australia, which now bears his name—Tasmania.) There was putative evidence for the existence of this lost continent. The mysterious territory was reportedly first sighted by Marco Polo, later by Spanish and French voyagers, and most recently by the pirate Edward Davis. The continent was estimated to be as large as Asia and loaded with precious gems and minerals. Lush tropical surroundings were reportedly dotted with temples, and the people traveled about the land on the backs of elephants. It was an eighteenth-century El Dorado, the Shangri-la of the South Pacific.[4]

Prior to Cook, many adventurers crusaded for such voyages of negative discovery. Maupertuis cajoled Frederick the Great into financing a trip. In 1756, Charles de Brosses of Dijon published his *Histoire des Navigations aux Terres Australes*, in which he developed the theory that this continent must exist to counterbalance the weight of the landmasses of the Northern Hemisphere and prevent the earth from toppling over. To modern ears this sounds positively daffy because we know that the earth is not "floating" in any medium that would cause it to "right" itself, as an out-of-balance log might do in a pond of water. But, in fact, it was long believed—right up through the early part of the twentieth century—that the earth was, in fact, floating in an invisible substance called *the ether*.

A decade later, in 1766, a Scotsman named John Callander published a book ambitiously entitled *Terra Australis Cognita*. Callander proposed immediate colonization of this no longer incognito new continent. The following year the chief hydrographer to the British East India Company, Alexander Dalrymple, wrote his *Account of the Discoveries Made in the South Pacific Ocean*, reiterating the "global equilibrium theory" and offering precise latitude and longitude figures for the land that he estimated contained more than fifty million inhabitants. He insisted that its wealth would far exceed that of the American colonies, which would free England from the political and economic tribulations those troublemaker Americans were stirring up. Dalrymple believed that since he was so well informed about this southern land he should be given command of an expeditionary force. He would be the new (and last, he believed) Columbus. Since Dalrymple was not a naval officer, the command of Britain's voyage of discovery went to the virtually unknown forty-year-old Cook, who was savvy enough to include scientists among his crew members, thereby making his explorations among the greatest in the history of science. In the process of seeking out the unknown land of the south, Cook found, charted, and explored just about everything but the mythical land, including Tahiti, New Zealand, Tasmania, Australia, the Great Barrier Reef, Tonga, Easter Island, New Caledonia, New Guinea, the Sandwich Islands, and, finally, what *Terra Australis Incognita* would turn out to actually be—Antarctica.[5]

In the end, what was known on the map mattered less than what was unknown, for it is undiscovered country that drives exploration and innovation, placing *terra incognita* at the very heart of science.

Look Through the Tube

During this age of positive exploration and negative discovery, other geographies of belief with their own unknown territories were opening up to human exploration. In 1609, the Italian mathematician and astronomer Galileo Galilei turned toward the heavens a modified version of the telescope first invented by the Dutch spectacle maker Hans Lippershey, who originally created it for much more earthly matters, such as viewing the flags and contents of merchant vessels approaching port. At this time astronomy was at something of a standstill. With the exception of the sun and the moon, the unaided human eye was inadequate for observing astronomical bodies in any detail much beyond a point of light. Galileo improved the Lippershey "looker" with a larger lens and a greater magnifying eyepiece, pointed it upward, and made a number of startling observations.

Galileo noted, for example, that there were satellites orbiting Jupiter, that Venus had phases, and that there were mountains on the moon and spots on the sun. He even discerned that the Milky Way—the blurry belt of light cinched across the waist of the sky—actually comprised an uncountable number of individual stars. The discovery of Jupiter's moons was particularly significant in that it was evidence that the earth was not the center of *everything*, giving support to Copernicus's heliocentric theory, which Galileo had already committed himself to believe even before he could prove it. Moreover, Galileo's telescopic discoveries of mountains casting shadows on the moon, along with those pesky sunspots, posed a problem for Aristotelian cosmology, which held that all objects in space must be perfectly round and perfectly smooth.

The telescope provided an Archimedean point from which worldviews could be moved, but not everyone was eager to pick up the new fulcrum. Galileo's eminent senior colleague at the University of Padua, Cesare Cremonini, was so committed to Aristotelian cosmology that he refused to even look through the tube. In fact, Cremonini was skeptical that there were even any heavenly bodies to see through it, concluding that it was all a parlor trick: "I don't believe that anyone but he saw them, and besides, that looking through glasses would make me dizzy. Enough, I don't want to hear any more about it. But what a pity that Mr. Galileo has gotten involved in these entertainment tricks."[6] Cremonini's allegiance to Aristotle was due, in no small part, to the fact that the Catholic Church had wedded the uncontested authority of scripture (via the great thirteenth-century Augustinian scholar St. Thomas Aquinas) to the undeniable wisdom of Aristotle. Cremonini's fidelity was to "the philosopher," as he explained during the Inquisition: "I cannot and do not wish to retract my exposition of Aristotle because this is how I understand him, and I am paid to present him as I understand him, and, were I not to do so, I would be obliged to give back my pay."[7] Now *that* is loyalty to the company, and the Catholic Church was unquestionably the largest and most powerful corporate entity of its day.

Those who did look through Galileo's tube could not believe their eyes—literally. One of Galileo's colleagues reported that the instrument worked for terrestrial viewing but not celestial, because "I tested this instrument of Galileo's in a thousand ways, both on things here below and on those above. Below, it works wonderfully; in the sky it deceives one. I have as witnesses most excellent men and noble doctors . . . and all have admitted the instrument to deceive." A professor of mathematics at the Collegio Romano was convinced that Galileo had put the four moons of Jupiter inside the tube and that he, too, could show others such a marvel given the opportunity to "first build them into some glasses." Galileo was practically apoplectic in his frustration: "As I wished to show the satellites of Jupiter to the Professors in Florence, they would see neither them nor the telescope. These people believe there is no truth to seek in nature, but only in the comparison of texts."[8]

In Galileo's mind, the marring of the sun with spots and the moon with mountains sounded the death knell of Aristotelian cosmology. Aristotelian scholastics (also known as *Peripatetics*, or those who "think while pacing," an activity popular among Greek philosophers) tried desperately to "preserve the appearances" of the unblemished and incorruptible heavens, but Galileo was convinced it was only a matter of time, as he noted in sardonic anticipation in a 1612 letter: "I presume that these innovations will be the funeral and the finish of, or the last judgment on, pseudo-philosophy; signs of it have already appeared in the Moon and in the Sun. I am expecting to hear of great proclamations on this subject by the Peripatetics who will wish to preserve the immortality of the heavens. I do not know how it can be saved and preserved."[9] Partial preservation of the heavens came in 1616 when Galileo was granted permission to employ the Copernican system only for mathematical convenience to calculate planetary orbits. But he was warned both verbally and in writing that he was not to profess the sun-centered system as literally true.

Nevertheless, contrarian that he was, and operating under the assumption that his previous good standing with Cardinal Maffeo Barberini— now Pope Urban VIII—would grant him some leeway, in 1632 Galileo published his most famous work, *Dialogue Concerning the Two Chief World Systems, Ptolemaic and Copernican*, an unmistakable defense of the Copernican sun-centered system. Galileo's book was a masterpiece of literature, set down in the style of a dialogue between two proponents, one a supporter of the earth-centered geocentric theory and the other a champion of the sun-centered heliocentric system. The book's protagonist, a supporter of the geocentric model, was named "Simplicio" and bore a striking resemblance to the incumbent Pope Urban VIII, whom Galileo characterized as an irrational fool. *Dialogue* is a systematic attack on Aristotelian physics and cosmology, and on the Peripatetic reliance on authority over observation.

Unsurprisingly, Urban VIII was incensed, not only because Galileo had violated the restraint of 1616 on teaching the Copernican system as real, but also because the scientist had ridiculed the pope's own preferred position on the ongoing Ptolemaic-Copernican controversy. In

August 1632, the Holy Office prohibited further publication and sales of *Dialogue*. Shortly thereafter, the pope ordered Galileo to stand trial before the Inquisition in Rome in 1633, where he was found guilty of "vehement suspicion of heresy." In the penalty phase of the trial, the court decreed: "We condemn you to formal imprisonment in this Holy Office at our pleasure."[10] The now-aged astronomer formally renounced his sin:

> I have been pronounced by the Holy Office to be vehemently sus-
> pected of heresy—that is to say, of having held and believed that
> the sun is the center of the world and immoveable, and that the
> earth is not the center, and moves. Therefore, desiring to remove
> from the minds of your eminences, and of all faithful Christians,
> this strong suspicion reasonably conceived against me, with sin-
> cere heart and unfeigned faith I abjure, curse, and detest the afore-
> said errors and heresies.[11]

Given Galileo's commitment to observation over authority, what legend has him saying next (although apocryphal), fits his character so well that it should have been spoken: "*Eppur si muove*," "And yet it moves." When the legend becomes fact, print the legend.

This is, in fact, what happened to the legend that Galileo was tortured and jailed for his beliefs. Because the church did not release the documents detailing precisely what was done with Galileo, but did release statements that said Galileo would be subject to "rigorous examination" (which at the time everyone knew meant torture), people naturally assumed that Galileo was tortured and jailed for his beliefs.[12] In reality, because of Galileo's fame and the respect he held among so many prominent people in power, and especially because of his recantation, the court granted him a "salutary penance" performed "for the spiritual benefit of former heretics who had returned to the faith," and he was thereafter confined to what amounted to a very comfortable house arrest. He could leave the confines of the building and even go to visit his daughter in a nearby convent. Nevertheless, *Dialogue* was banned and Galileo was prohibited from ever again teaching the Copernican system.[13] Remarkably, Galileo's *Dialogue* remained on the Catholic Church's *Index of Prohibited Books* until 1835, and it was not until 1992 that Pope John Paul II exonerated Galileo with an official apologia that reveals how belief systems can and do change once they are decoupled from unchanging dogmas, even if it takes three and a half centuries to do so:

> Thanks to his intuition as a brilliant physicist and by relying on different arguments, Galileo, who practically invented the experimental method, understood why only the sun could function as the centre of the world, as it was then known, that is to say, as a planetary system. The error of the theologians of the time, when they maintained the centrality of the Earth, was to think that our understanding of the physical world's structure was, in some way, imposed by the literal sense of Sacred Scripture. Let us recall the celebrated saying attributed to Baronius, *"Spiritui Sancto mentem fuisse nos docere quomodo ad coelum eatur, non quomodo coelum gradiatur."* ["It was the Holy Spirit's intent to teach us how one goes to heaven, not how the heavens go."][14]

Why did redemption take so long? Galileo's own words in a 1615 letter to the grand duchess dowager Christina, with whom he had been corresponding about his heretical ideas in support of Copernicus, provide some insight: "Methinks that in the discussion of natural problems we ought not to begin at the authority of places of Scripture; but at sensible experiments and necessary demonstrations."[15]

Methinks Galileo knew perfectly well what he was doing—and what the consequences would be—by prodding these old Aristotelians into looking through his tube.

The Battle of the Books

The allegiance to the authority of both scripture and Aristotle made it very difficult for the scholars of Galileo's time to accept his observations— and especially the inductions he drew from them—as true. And he knew it. This is why Galileo commented in his book *Bodies in Water*, with epigrammatic poignancy, "The authority of Archimedes was of no more importance than that of Aristotle; Archimedes was right because his conclusions agreed with experiment."[16] Four centuries later, the physicist Richard Feynman echoed Galileo's principle in his observation about determining if your theory is right or wrong: "If it disagrees with experiment, it is wrong. In that simple statement is the key to science. It doesn't make any difference how beautiful your guess is, how smart you are, who made the guess, or what his name is. If it disagrees with experiment, it's wrong. That's all there is to it."[17]

What Galileo reflected in his observations was one end of a spectrum that grew out of the Scientific Revolution that had begun more than a century before and culminated in a battle of the books: *the book of authority* versus *the book of nature*. Andreas Vesalius's dissections of the human body in his 1543 *On the Fabric of the Human Body*, William Gilbert's geological observations on magnets and the earth in his 1600 *On the Magnet and Magnetic Bodies, and on the Great Magnet the Earth*, and William Harvey's tracking of the motion of the heart and blood in his 1628 *Anatomical Exercise on the Motion of the Heart and Blood in Animals* were all books of nature that challenged the ancient books of authority, in which scribes copied copies of copies originally set down centuries before, with little real-world fact checking.

The Scientific Revolution was revolting against the Catholic Church and its reliance on holy scripture (in Latin no less) as interpreted by authorities in a rigid ecclesiastical hierarchy. This is, in part, why the Catholic Church reacted so violently to the Protestant Reformation—Martin Luther said it was acceptable for everyone to read the Bible in the vernacular, that anyone can have a relationship with God directly without a priestly intermediary, and that such rigid hierarchies were unnecessary. This set the stage for later cultural and political battles between conservatives and liberals that have carried forward to this day.

How did the book of authority maintain its grip on the human imagination? An example can be found in the first-century CE Roman writer Dioscorides' work, *De Materia Medica*, the foremost classical source of botanical terminology and the leading pharmacological text for the next 1,600 years. *De Materia Medica* presented thorough descriptions of more than six hundred plants that the author collected while traveling with the armies of Emperor Nero and became the foundation of late medieval herbals when it was translated into seven languages and distributed throughout Europe. After Dioscorides' death, however, his disciples studied Dioscorides instead of nature. In time, copyists copying copies created a whole new nature that had little correspondence to reality. Leaves were drawn on branches for symmetry. Enlarged roots and stem systems were added to fill in oversized folio pages. Publishers used stock blocks of wood carved individually for roots, trunks, branches, and leaves, and combined them into composite illustrations of trees that existed nowhere in the world. Copyists' fancy and imagination became the norm. The "barnacle-tree," for example, was believed to actually grow barnacles; the "tree-of-life" was enveloped by a serpent with a woman's head; and the Narcissus plant grew tiny human figures. So powerful was Dioscorides' influence over the ages that late in the sixteenth century the chair of botany at the University of Bologna was conferred with the title "Reader of Dioscorides."[18]

The power of the book of authority is well exemplified in the illustrations in figure 14. The half-man / half-beast creature is "the true picture of the Lamia," from Edward Topsell's 1607 work *The Historie of Fourefooted Beastes*. The half-man / half-plant creature is the plant "Mandragora," more commonly known today as a mandrake (in the nightshades family), originally printed in a 1485 German book, *Herbarius*. Who ever saw such creatures? No one. But once they were printed in volumes that were copied endlessly from century to century without anyone checking the original sources—much less nature—they became reified as species in God's creation. Empirical observation and verification did not inhabit the cognitive space in the medieval mind. By contrast, the woodcut illustration of two artist-naturalists from Leonhart Fuchs's 1542 *De Historia Stirpium* (*The History of Plants*) reveals a phase transition from the book of authority to the book of nature. Instead of copiers copying copies made from previous copies, naturalists went outdoors to check with nature instead, and that meant the extinction of Lamia and Mandragora (although Bigfoot and the Loch Ness Monster live on in our imaginations).[19]

This battle of the books involves two different ways of thinking—two belief engines, as it were. The book of authority is grounded in *deduction*—the process of making specific statements from a generalized conclusion, or arguing from the general to the specific, *from theory to data*. The book of nature is grounded in *induction*—the process of drawing generalized conclusions from specific statements, or arguing from the specific to the general, *from data to theory*. It would be oversimplified and unrealistic to describe any one person or tradition as practicing pure induction or pure deduction, for none of us operates in a vacuum without inputs from many sources, and it is impossible to operate without both modes of thinking. Data and theory go hand in hand. Nevertheless, there are periods in the history of science when one has been emphasized more than

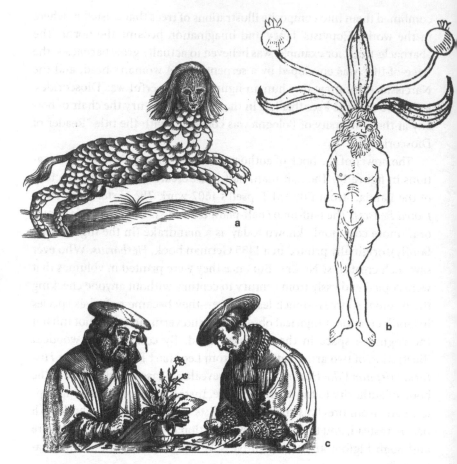

Figure 14. The Book of Authority Triumphs over the Book of Nature

So powerful was the tradition of respecting the authority of the ancients, that "naturalists" were little more than scribes who copied the copies of previous copiers from some long-ago original source. The half-man / half-beast creature called the "Lamia" (a) and the half-man / half-plant creature called the "Mandragora" (b) were both staples of sixteenth- and seventeenth-century works. The two artist-naturalists sketching an actual plant (c) mark a sea change in the shift from the book of authority to the book of nature. Lamia from Edward Topsell's 1607 *The Historie of Fourefooted Beastes*. Mandragora from the 1485 German *Herbarius*. The artist-naturalists from Fuchs's 1542 *De Historia Stirpium*. All are reprinted from ALAN DEBUS, *MAN AND NATURE IN THE RENAISSANCE* (CAMBRIDGE: CAMBRIDGE UNIVERSITY PRESS, 1978), PP. 36, 44, 45.

the other, and Galileo and his fellow revolutionaries were butting up against a deep tradition in deduction.

The pull of Aristotelian logic tied to deductive reasoning was compelling and hard to overcome. In the early 1600s, for example, while Galileo was making his first telescopic observations, it was suggested that space consisted literally of nothing—a vacuum. But how, then, would the planets move through it? According to Aristotle, an object moved through air or space by "impetus," in which air or "ether" passes and envelops the object, thereby pushing it from behind and giving it thrust. Just as an arrow moves through the atmosphere by the air enveloping it and pushing it from behind, so, too, do planets move through space with the ether surrounding them and pushing them from behind. Without the ether no thrust could exist to push a planet through space. The planets move, ergo no vacuum. Ether, thereafter, became the fifth element—along with earth, water, air, and fire—and belief in it persisted all the way into the twentieth century, until the experiments on the speed of light by physicists Albert Michelson and Edward Morley were fully accepted. Such is the endurance of belief, even in the sciences.

In 1620, a staunch challenge to Aristotle's deductive methodology was proffered by the English philosopher Francis Bacon in his book *Novum Organum*. This "new instrument" was the empirical or observational method. Rejecting both the unempirical tradition of scholasticism and the Renaissance quest to recover and preserve ancient wisdom, Bacon sought a blend of sensory data and reasoned theory, with emphasis on data and caution about theory. Ideally, he proposed, one should begin with observations then formulate a general theory from which logical predictions could be made. Bacon outlined how the mind works in this regard:

> There are and can be only two ways of searching into and discovering truth. The one flies from the senses and particulars to the most general axioms, and from these principles, the truth of which it takes for settled and immovable, proceeds to judgment and to the discovery of middle axioms. The other derives axioms from the senses and particulars, rising by a gradual and unbroken ascent, so that it arrives at the most general axioms last of all. This is the true way, but as yet untried.[20]

Impeding Bacon's goal, however, were psychological barriers that colored clear judgment of the facts, of which he identified four types: *idols of the cave* (individual peculiarities), *idols of the marketplace* (limits of language), *idols of the theater* (preexisting beliefs), and *idols of the tribe* (inherited foibles of human thought): "Idols are the profoundest fallacies of the mind of man. Nor do they deceive in particulars . . . but from a corrupt and crookedly-set predisposition of the mind; which doth, as it were, wrest and infect all the anticipations of the understanding." The power of beliefs to drive our observations and conclusions is profound: "The human understanding when it has once adopted an opinion . . . draws all things else to support and agree with it. And though there be a greater number and weight of instances to be found on the other side, yet these it either neglects and despises . . . in order that by this great and pernicious predetermination the authority of its former conclusions may remain inviolate." This is a superb example of the *confirmation bias*, which we saw in the previous chapter is where we look for and find confirmatory evidence for what we already believe and either ignore or rationalize disconfirming evidence. Everyone does it.

What is the answer to the problem of the idols? Science. Bacon's *Novum Organum* was part of a larger project he called *Instauratio Magna*, or the "Great Restoration." (See figure 15.) This was a plan to reorganize philosophy and the sciences, starting by challenging the authority of Aristotle with the new instrument of science. With the impudence only a man of Bacon's stature could muster, he boldly proposed that "there was but one course left . . . to try the whole thing anew upon a better plan and to commence a total reconstruction of sciences, arts, and all human knowledge, raised upon the proper foundations." Bacon suggested, "As water will not ascend higher than the level of the first springhead from whence it descendeth, so knowledge derived from Aristotle and exempted from liberty of examination will not rise again higher than the knowledge of Aristotle."[21]

The debate over the relative strengths and roles of induction and deduction in science continued for centuries and remains with us to this day. When Charles Darwin was coming of intellectual age and developing his theory of evolution, for example, the pendulum had swung over to the side of induction, and there was much handwringing among philosophers of science over what it was and how it was used in science. Although definitions varied, induction was roughly understood to mean

Figure 15. Francis Bacon's Great Restoration through the Exploration of Science

Frontispiece from Francis Bacon's 1620 *Instauratio Magna*, or "Great Restoration" through the *Novum Organum*, or new instrument of science. The ships represent the tools of scientific knowledge that carry the explorers (scientists) past the Pillars of Hercules (literally, the Strait of Gibraltar; figuratively, the gates of the great unknown). THE FRONTISPIECE FROM FRANCIS BACON, *INSTAURATIO MAGNA*, 1620, IS FROM E. L. EISENSTEIN, *THE PRINTING REVOLUTION IN EARLY MODERN EUROPE* (NEW YORK: CAMBRIDGE UNIVERSITY PRESS, 1983), P. 258.

arguing from the specific to the general, from data to theory. In 1830, however, astronomer John Herschel argued that induction was reasoning from the known to the unknown. In 1840, philosopher of science William Whewell insisted that induction was the superimposing of concepts on facts by the mind, even if they were not empirically verifiable. In 1843, philosopher John Stuart Mill claimed that induction was the discovery of general laws from specific facts, but that they had to be verified empirically. Johannes Kepler's discovery of the laws of planetary motion, for example, was considered to be a classic case study of induction. For Herschel and Mill, Kepler discovered these laws through careful observation and induction. For Whewell, the laws were self-evident truths that could have been known a priori and verified later by observation. By the 1860s, as the theory of evolution was gaining momentum and converts, Herschel and Mill carried the day on induction as observation, not so much because they were right and Whewell was wrong, but because empiricism was becoming integral to the understanding of how good science is done. This is, in part, what caused Darwin to delay publication of *On the Origin of Species*—he wanted to compile copious data for his theory before going public.[22]

The Power and Poverty of Pure Empiricism

All intellectual movements swing like pendulums through mental space, oscillating between extremes then settling into an ever more narrow groove of ideational range. So it was for the battle of the books as the extremes of fluctuation between authority and empiricism stabilized over time, and where today we (hopefully) recognize the importance of both data and theory. It was Galileo who first discovered the principle of the pendulum, so it is with some irony that I employ the metaphor here. As important as his empirical discoveries were to overthrowing the authoritative dogma of centuries past, when it came to his observations of the planet Saturn, Galileo succumbed to his own cognitive limitations and imagination.

After observing Saturn—the most distant planet of his day—through his tiny telescope, Galileo wrote to his astronomical colleague Johannes Kepler, "*Altissimum planetam tergeminum observavi*," "I have observed that the farthest planet is threefold." He then explained what he meant:

"This is to say that to my very great amazement Saturn was seen to me to be not a single star, but three together, which almost touch each other." He saw Saturn not as a planet with rings as we see it today in even the tiniest of home telescopes, but as one large sphere surrounded by two smaller spheres, thus accounting for its oblong shape.

Why did Galileo—champion of observation and induction—make this mistake? Having praised empiricism as the sine qua non of science, we must now admit its limitative effects. Galileo's error is instructive for an understanding of the interplay of data and theory, and when it came to Saturn, Galileo lacked them both. *Data*: Saturn is twice as far away as Jupiter, thus what few photons of light there were streaming through the cloudy glass in his little tube made resolution of the rings problematic at best. *Theory*: There was no theory of planetary rings. It is at this intersection of nonexistent theory and nebulous data that the power of belief is at its zenith and the mind fills in the blanks. Like Columbus before him, Galileo went to his grave believing not what his eyes actually saw but what his model of the world told him he was seeing. It was literally a case of *I wouldn't have seen it if I hadn't believed it.*

Galileo could not "see" the rings of Saturn, either directly or theoretically, but he certainly saw something, and herein lies the problem. *Altissimum planetam tergeminum observavi.* As the late Harvard evolutionary theorist and historian of science Stephen Jay Gould noted in his insightful commentary on the Galileo Saturn affair: "He does not advocate his solution by stating 'I conjecture,' 'I hypothesize,' 'I infer,' or 'It seems to me that the best interpretation. . . .' Instead, he boldly writes 'observavi'—I have *observed*. No other word could capture, with such terseness and accuracy, the major change in concept and procedure (not to mention ethical valuation) that marked the transition to what we call 'modern' science."[23]

Over time Galileo returned to Saturn often, and although he never saw the same thing twice, he stuck steadfastly with his original observation and conclusion. In his 1613 book on sunspots, he wrote: "I have resolved not to put anything around Saturn except what I have already observed and revealed—that is, two small stars which touch it, one to the east and one to the west." Challenged by a fellow astronomer who suggested that perhaps it was one oblong object rather than three spheres, Galileo boasted of his own superior observational skills of "the shape and distinction of the three stars imperfectly seen. I, who have observed it a thousand times at different periods with an excellent instrument, can assure you that no change whatever is to be seen in it."

The next time he pointed his tube to Saturn just before publication of his sunspot book, however, Galileo saw something rather different.

> But in the past few days I returned to it and found it to be solitary, without its customary supporting stars, and as perfectly round and sharply bounded as Jupiter. Now what can be said of this strange metamorphosis?. . . . Was it indeed an illusion and a fraud with which the lenses of my telescope deceive me for so long—and not only me, but many others who have observed it with me? . . . I need not say anything definite upon so strange and unexpected an event; it is too recent, too unparalleled, and I am restrained by my own inadequacy and the fear of error.[24]

Nevertheless, Galileo concluded in the book that despite this new data his original theory about what he saw was correct. Why? The answer may be found in the visual presentation of the data.

The great scholar of the visual display of quantitative information Edward Tufte notes in his 2006 book, *Beautiful Evidence*, with the accompanying page from Galileo's 1613 sunspot book (see figure 16), that "Galileo reported his discovery of Saturn's unusual shape as *2 visual nouns* that compare clear and murky telescopic views. In Galileo's work *Istoria e dimostrazioni intorno alle macchie solari* (1613), words and images combine to become simply evidence rather than different modes of evidence." The translation of the text in figure 16 accompanied by the two tiny drawings of Saturn reads: "The shape of Saturn is thus ∞ as shown by perfect vision and perfect instruments, but appears thus ⊂⊃ where perfection is lacking, the shape and distinction of the three stars being imperfectly seen." Tufte describes this sentence as "one of the best analytical designs ever" because it represented "Saturn as evidence, image, drawing, graphic, word, noun."[25] Despite his more recent observations that the "three stars" had become "solitary" and "as perfectly round and sharply bounded as Jupiter," Galileo's image, drawing, graphic, word, and noun were congealed into evidence that his original observations

ta imperfezzione dello ſtrumento, ò dell'occhio del riguardan-
te,perche ſendo la figura di Saturno così ◯◯◯ ,come moſtra-
no alle perfette viſte i perfetti ſtrumenti , doue manca tal
perfezzione appariſce così ◯◯ non ſi diſtinguendo perfetta-
mente la ſeparazione , e figura delle tre ſtelle ; ma io che mil-
le volte in diuerſi tempi con eccellente ſtrumento l'hò riguar-
dato, poſſo aſſicurarla , che in eſſo non ſi è ſcorta mutazione
alcuna, e la ragione ſteſſa fondata ſopra l'eſperienze,che hauia-

The shape of Saturn is thus ◯◯◯ as shown by perfect vision and perfect
instruments, but appears thus ◯◯ where perfection is lacking, the shape
and distinction of the three stars being imperfectly seen.

Figure 16. Galileo's Saturn as "Evidence, Image, Drawing, Graphic, Word, Noun"

The page from Galileo's 1613 book on sunspots, in which he returns to the
consideration of the Saturn enigma, concluding once again that he was
right in the first place that Saturn was a three-bodied object. Source: GALI-
LEO GALILEI, *ISTORIA E DIMOSTRAZIONI INTORNO ALLE MACCHIE SOLARI*
(ROME, 1613), P. 25. AS REPRODUCED IN EDWARD TUFTE, *BEAUTIFUL EVI-
DENCE* (CHESHIRE, CONN.: GRAPHICS PRESS, 2006), P. 49.

were correct. Galileo never fully retreated from his first definitive conclusion.

The solution to the Saturn problem is equally instructive of the *data-
theory* dialogue in the narrative of belief. It wasn't until 1659—half a
century after Galileo's observations—that Dutch astronomer Christiaan
Huygens published the solution in his great work *Systema Saturnium*,
one of the finest visual displays of both data and theory in the history of
science. In figure 17 we see on display thirteen interpretations of Saturn
produced by astronomers from 1610 (Galileo) to 1650 (Fontana and others), all wrong.

To our *data-theory* duo we should add *presentation* of the data and
theory. In many ways, presentation is everything in understanding how
beliefs are born, reinforced, and changed, because humans are so visually oriented as primates who once depended on three-dimensionality to
navigate through dense arboreal environs. The *data-theory-presentation*

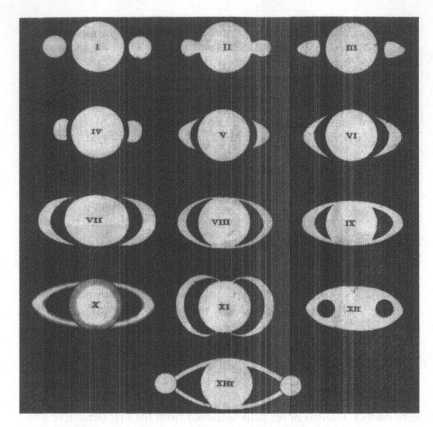

Figure 17. Christiaan Huygens's Catalogue of Errors

Dutch astronomer Christiaan Huygens solved the Saturn enigma in his 1659 work *Systema Saturnium*, in which he included this visual catalogue of the thirteen most prominent theories of Saturn, including those of I. Galileo, 1610; II. Scheiner, 1614; III. Riccioli, 1641 or 1643; IV–VII. Hevel, theoretical forms; VIII–IX. Riccioli, 1648–1650; X. Divini, 1646–1648; XI. Fontana, 1636; XII. Biancani, 1616; Gassendi, 1638, 1639; XIII. Fontana and others, 1644, 1645. Note the first image from Galileo's observation of Saturn from which he concluded: "I have observed that the farthest planet is threefold." Source: CHRISTIAAN HUYGENS, *SYSTEMA SATURNIUM* (THE HAGUE, 1659), FOLDOUT PLATE AT PP. 34–35. AS REPRODUCED IN EDWARD TUFTE, *VISUAL EXPLANATIONS* (CHESHIRE, CONN.: GRAPHICS PRESS, 1997), P. 107.

triad is on exquisite display in figure 18, in which Huygens takes those two-dimensional Saturns, blows them up into 3-D, and puts them in motion around the sun. It is a marvelous presentation of both data and theory, incorporating Copernicus's theory that the sun is at the center of the solar system instead of the earth (as in Ptolemaic cosmology), Kepler's first law that planetary orbits are elliptical instead of circular (as in Aristotelian cosmology), and Kepler's third law that the inner planets revolve around the sun faster than the outer planets.

Here we see the sun-Earth-Saturn system from above—an Archimedean point outside the solar system that grants a new perspective—with Saturn set in motion on its glacially slow 29.5-Earth-years-long orbit. About 1.8 Earth-years elapse between each of the 32 Saturns in the diagram. The effect is to show that Saturn will appear different to Earthbound observers at different times of the Earth year. This explains why in the course of half a century so many keen-eyed astronomers saw so many different Saturns, including a Saturn with no rings at all. Twice each

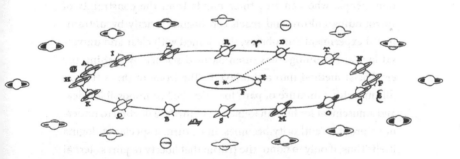

Figure 18. Saturn in 3-D and in Motion

The *data-theory-presentation* triad is on exquisite display here, in which Huygens takes those two-dimensional Saturns seen in Figure 17, blows them up into 3-D, and puts them in motion around the sun. It is a marvelous presentation of both data and theory, incorporating Copernicus's theory that the sun is at the center of the solar system instead of the earth (as in Ptolemaic cosmology), Kepler's first law that planetary orbits are elliptical instead of circular (as in Aristotelian cosmology), and Kepler's third law that the inner planets revolve around the sun faster than the outer planets. SOURCE: CHRISTIAAN HUYGENS, *SYSTEMA SATURNIUM* (THE HAGUE, 1659), P. 55. AS REPRODUCED IN EDWARD TUFTE, *VISUAL EXPLANATIONS* (CHESHIRE, CONN.: GRAPHICS PRESS, 1997), P. 108.

Saturn-year the rings appear edge on to Earth-bound observers. Edward Tufte eloquently describes the power of this visual explanation: "Huygens presents a series of still images in order to depict motion. To resolve such discontinuous spatial representations of continuous temporal activity, viewers must interpolate between images, closing up the gaps. Imaginative and original, this display is a classic, an exemplar of information design."[26]

The Saturn enigma and its ultimate solution reveals the interplay between *data, theory,* and *presentation,* between *induction, deduction,* and *communication,* between what we *see,* what we *think,* and what we *say.* We cannot untangle the three, for the mind engages them all to produce knowledge on which we act in the world. The Saturn affair demonstrates, in the master rhetorician Stephen Jay Gould's words, both "the power and poverty of pure empiricism." How? Gould's answer is one of the most eloquent ever penned on this contentious issue:

> The idea that observation can be pure and unsullied (and therefore beyond dispute)—and that great scientists are, by implication, people who can free their minds from the constraints of surrounding culture and reach conclusions strictly by untrammeled experiment and observation, joined with clear and universal logical reasoning—has often harmed science by turning the empiricist method into a shibboleth. The irony of this situation fills me with a mixture of pain for a derailed (if impossible) ideal and amusement for human foibles—as a method devised to undermine proof by authority becomes, in its turn, a species of dogma itself. Thus, if only to honor the truism that liberty requires eternal vigilance, we must also act as watchdogs to debunk the authoritarian form of the empiricist myth—and to reassert the quintessentially human theme that scientists can work only within their social and psychological contexts. Such an assertion does not debase the institution of science, but rather enriches our view of the greatest dialectic in human history: the transformation of society by scientific progress, which can only arise within a matrix set, constrained, and facilitated by society.[27]

In the 1920s, four centuries after Galileo changed the geography of knowledge of the world and its immediate environs in space, a cosmological matrix of data, theory, and presentation came together in a new pattern that completely changed the way we view the cosmos and our place in it. As bold a pattern shatterer as he was, Galileo could never have imagined just how inconceivably vast and vacuous the heavens would turn out to be. How that new pattern was discovered, delineated, doubted, debated, and ultimately determined to be correct provides us with a final example of how science works to adjudicate disputes over conflicting patterns, and how we can avoid the trap that belief-dependent realism holds for us if we do not employ the tools of science.

14

Cosmologies of Belief

ON A CLEAR NIGHT AWAY FROM CITY LIGHTS, IF YOU HAVE REASONably good eyesight you can just barely make out a fuzzy patch of light near the constellation Cassiopeia (the W-shaped pattern of stars), especially if you look a little to the side of it so that the photons that left the Andromeda galaxy 2.5 million years ago land on the periphery of your retina where the dim-light sensitive rods are located. On October 6, 1923, astronomer Edwin Hubble, wielding the one-hundred-inch Hooker telescope atop Mount Wilson in the San Gabriel Mountains above the Los Angeles basin—at the time the largest light-gathering instrument in the world—confirmed that this and many of the other cloudy images he had been focusing in his eyepiece were not nebulae *within* the Milky Way galaxy as many astronomers believed, but were, in fact, separate galaxies— "island universes" as they were romantically called—and that the universe is bigger than anyone imagined . . . a lot bigger.

What Hubble confirmed after centuries of debate is that our star is not merely a grain of sand among a hundred billion grains on a single beach; in fact, there are hundreds of billions of beaches, each one of which contains hundreds of billions of grains of sand. The story of how this remarkable discovery was made demonstrates how science works in practice: not only how it requires an elegant blend of data, theory, and presentation as seen in the Galileo story, but also how scientific disputes are resolved and what happens to previously accepted theories rendered obsolete by new observations. In the world of macroscience there are few targets of observation more nebulous than the cosmic nebulae that have perplexed observers for so long. The final resolution of their nature would result in a dramatic shift in our understanding of the large-scale structure of the universe . . . and beyond.

Lookback Time

When you look out into space the distances are so enormous that you are looking back into time; appropriately, astronomers call this *lookback time*. Light travels at a speed of about 186,000 miles per second, or about 671 million miles an hour. It takes light 1.3 seconds to travel from the moon to Earth, 8.3 minutes from the sun to Earth, and 4.4 years from our closest stellar neighbor Alpha Centauri to Earth. Thus, when I said that the light from the Andromeda galaxy left 2.5 million years ago, I was using a *lookback time* reference because it is 2.5 million light-years away. Geologists call such long time spans *deep time*. *Lookback time*, *deep time* . . . by any other name it dwarfs the imagination of creatures that live a scant four score years.

When it comes to such astronomically distant objects as galaxies, the naked eye could not help early astronomers grasp the nature of the nebula, and thus it is that humanity had to wait until modern optics could provide us with the observational tools needed to see such enormous distances. With one exception. On that clear night away from city lights, after you've found Andromeda, scan the rest of the celestial sphere and you will see a thick band of splotchy light that stretches across the entire sky. This is the Milky Way galaxy, and the problem of determining its nature is compounded by the fact that we're in the middle of it with no way to step off of our observer's platform for an Archimedean big-picture perspective. Ever since Galileo was able to discern individual stars in that band of light with his crude telescope, astronomers have debated its nature, where we live in relation to it, and if those other misty forms in the sky are similar to or different from the one in which we live.

Some astronomers speculated that a force made the stars orient themselves in a band across the sky, and that this structure rotated around the sun just like the planets. In 1750, an English watchmaker and teacher named Thomas Wright published his theory of the Milky Way in a book entitled *An Original Theory; or, New Hypothesis of the Universe*, in which he presciently conjectured that an observer's orientation in space determines the perception of what is observed. He concluded that the Milky

Way was a shell of stars on which our solar system resided, such that looking flat across the shell one sees lots of stars, but in looking up or down away from the shell one sees mostly empty space.[1] That's a close approximation to what we observe, only we now know that the Milky Way is a flat disc, like a Frisbee, and our solar system sits about three-quarters of the way out from its center. If you look "through" the disc—along the thick plane, that is—you see lots of stars, and these then appear as a band across the night sky. When you look away from the band you are looking either up or down from the disc.

Islands in the Sky

Such conjectures, however prescient in hindsight, gained little footing on the intellectual landscape until the great Prussian philosopher Immanuel Kant turned his perceptual powers skyward—if only in his mind's eye—when he suggested that the elliptical-shaped "nebulous stars" believed by many astronomers to be nearby were actually discs of countless stars very far away: "I easily persuaded myself that these stars can be nothing else than a mass of many fixed stars. On account of their feeble light, they are removed to an inconceivable distance from us." But why do some nebulae appear round, others elliptical shaped, and still others as a flat plane? Were these different objects entirely, or are they the same species of objects viewed at different angles? Kant reasoned his way to a nearly correct answer: "[I]f such a world of fixed stars is beheld at such an immense distance from the eye of the spectator situated outside of it, then this world will appear under a small angle as a patch of space whose figure will be circular if its plane is presented directly to the eye, and elliptical if it is seen from the side or obliquely."

These nebulae became known as Kant's "island universes," and he waxed poetic about them in his 1755 book *Universal Natural History and Theory of the Heavens*: "The infinitude of the creation is great enough to make a world, or a Milky Way of worlds, look in comparison with it what a flower or an insect does in comparison with the earth." As for the Milky Way itself, Kant outlined his theory in his usual insightful manner:

> Just as the planets in their system are found very nearly in a com-
> mon plane, the fixed stars are also related to their positions, as
> nearly as possible, to a certain plane which must be conceived as

drawn through the whole heavens, and by their being very closely massed in it they present that streak of light which is called the Milky Way. I have become persuaded that because this zone, illuminated by innumerable suns, has almost exactly the form of a great circle, our sun must be situated quite near this great plane. In exploring the causes of this arrangement, I have found the view to be very probable that the so-called fixed stars may really be slow moving, wandering stars of a higher order.[2]

The Great Debate

Kant's theory of the heavens set the stage for a multicentury debate between those who thought that the nebulae were stellar systems within our own galaxy of stars (the "nebular hypothesis"), and those who believed that they represented separate galaxies at great distances (the "island universe theory"). As retold by Timothy Ferris in his classic work *Coming of Age in the Milky Way*, as well as by Gale Christianson in his biography *Edwin Hubble: Mariner of the Nebulae*, and most recently by Marcia Bartusiak in her splendid history of *The Day We Found the Universe*, it was this debate that Edwin Hubble adjudicated at Mount Wilson on that fateful day in October 1923.[3]

In 1781, a comet chaser named Charles Messier published a catalog of the nebulae, primarily as a means of distinguishing these fixed blurry dots from the moving wispy comets he was searching for.[4] This became the definitive compendium of nebulae and is still in use today because historical nomenclature holds precedence in science (in the same way that we still use Carl von Linné's eighteenth-century pre-Darwinian binomial nomenclature for identifying organisms—for example, *Homo sapiens*). Messier's catalog gave grist to the telescopic mill. The great astronomer William Herschel, after his remarkable discovery of Uranus, ramped up the search by turning his twenty-foot tube with its twelve-inch mirror to the objects Messier said were not moving. "I have looked farther into space than ever a human being did before me," he boasted. He was able to resolve individual stars within the blotches, proving that there were island universes after all![5] Kant was right.

Not so fast. It turns out that Herschel was not imaging distant galaxies. He was looking at globular clusters—collections of stars in or near the Milky Way galaxy that astronomers differentiated from nebulae without discernible individual stars. Herschel correctly identified the Orion Nebula as an interstellar cloud of gas within our galaxy in the process of giving birth to new stars. As well, in 1790 Herschel imaged "a most singular Phaenomenon!": "a star of about the eighth magnitude, with a faint luminous atmosphere" in which "the star is perfectly in the center and the atmosphere is so diluted, faint and equal throughout, that there can be no surmise of its consisting of stars; nor can there be a doubt of the evident connection between the atmosphere and the star."[6] It was a planetary nebula—a star within our galaxy that is shedding its outer gaseous layer. This was evidence against Kant's island universe theory and in favor of the nebular hypothesis. By the 1790s Herschel had cataloged more than a thousand new nebulae and stellar clusters. Despite the wide variety of nebula types that he imaged, and over the voices of many skeptical colleagues, Herschel pronounced: "These curious objects, not only on account of their number, but also in consideration of their great consequence, are no less than whole sidereal systems" that "may well outvie our Milky-Way in grandeur."[7]

Conflicting Patterns of Data

With the hindsight bias, of course, we know how the story turns out. It is easy to rummage around in the dustbin of history and pull out those who were ahead of their time, which is what I've been doing thus far, but with two centuries left in the story astronomers had obviously not solved the riddle of the nebulae. An additional problem arises at this point: in a sense both theories were correct. On the one hand, there are lots of local phenomena within our galaxy that appear as fuzzy patches in the night sky: comets, gaseous clouds, globular clusters of stars, open clusters of stars, planetary nebulae, ancient nova and supernova stars that blew up and left only shells of gas, and so on. On the other hand, the vast majority of Messier's catalog objects labeled as nebulae are, in fact, island universes—galaxies of stars—enormous distances away from the Milky Way galaxy. The problem in distinguishing between the two categories of celestial objects comes down to better data and refined theory. The latter followed the former, and the former depended directly on improvements in telescope technology.

In the 1830s, an Irish nobleman named William Parsons, third earl of Rosse, constructed a thirty-six-inch telescope. Through his eyepiece he managed to barely discern spiral arms in M51—the fifty-first object in Messier's catalog—which took everyone by surprise because even those who believed in the island universe theory had no notion of what the structure of these other galaxies (let alone our own) might be. The Whirlpool galaxy, as it came to be known, seemed to indicate movement through arms coiled around a central axis that very much resembled a whirlpool, from whence its name.[8] In 1846, a supporter of the island universe theory named John Nichol suggested that some of the nebulae "are situated so deep in space that no ray from them could reach our Earth, until after traveling through the intervening abysses, during centuries whose number stuns the imagination."[9] In Nichol's imagination that number could be as high as thirty million years. This was a stunning figure to contemplate given that the prevailing worldview among the public at the time was a biblical age no older than ten thousand years. Privately, many scientists had their doubts, but none could have known how shy of the mark their educated guesses were—off, as it turns out, by orders of magnitude of very deep lookback time.

Once again we are getting ahead of ourselves in singling out our prognosticating champions of truth. There were other lines of evidence piling up against the island universe theory, and none more powerful than what was being imaged through a new device capable of discerning the elementary constituents of light. As Isaac Newton demonstrated back in the seventeenth century, if you pass white light through a glass prism it can be spread out into its component colors. Over the centuries scientists discovered that if you magnify a band of those colors you can see vertical lines that appear to represent the elements in the substance of the object that is generating the light. For example, if you heat up an element so hot that it burns bright enough to give off light, pass this light through a prism, and magnify it, you will find a characteristic set of lines that represent that element and no other—always and everywhere.

This device is called a *spectroscope*, and it was first employed by a German optics technician named Joseph von Fraunhofer, who attached a crude spectroscope to his telescope and noticed that similar patterns of lines appeared in the spectra of the sun, moon, and the other planets, which followed from the fact that the moon and planets are reflecting sunlight. But when Fraunhofer analyzed other stars he found different line patterns. Was the light from the stars coming from a different source? A few decades later a physicist named Robert Bunsen (of "Bunsen burner" fame) imaged a local fire through his spectroscope and found barium and strontium in the flames. Others followed, recording spectra of all manner of heated elements, and thus was born the technology of spectroscopy and the science of astrophysics. By cataloging the characteristic lines for elements on Earth, astronomers could then turn their spectroscopes (yoked to their telescopes) to the stars—and eventually the nebulae—in order to determine their composition.

In 1861, a physicist named Gustav Kirchhoff imaged the closest star to the earth—the sun—and found lines matching those of sodium, calcium, magnesium, iron, chromium, nickel, barium, copper, and zinc. On August 29, 1864, an English amateur astronomer named William Huggins turned a spectroscope to the light coming from the bright stars Betelgeuse and Aldebaran, where he identified iron, sodium, calcium, magnesium, and bismuth, confirming that the sun was just another star, and, alternatively, that the stars are the same species of celestial object as the sun. But then Huggins confused the debate when he did a spectroscopic analysis of one of Herschel's planetary nebulae and found only one distinct line.

> At first I suspected some displacement of the prism, and that I was looking at a reflection of the illuminated slit . . . then the true interpretation flashed upon me. The riddle of the nebulae was solved. The answer, which had come to us in the light itself, read: Not an aggregation of stars, but a luminous gas. Stars after the order of our own sun, and of the brighter stars, would give a different spectrum; the light of this nebula had clearly been emitted by a luminous gas.[10]

"The Nebular Hypothesis Made Visible"

With this new data the pendulum was swinging back in favor of the nebulae as internal galactic structures; perhaps, some speculated, they were stars and planetary systems under development. Demonstrating the power of this concept to drive percepts, in 1888 the relatively new technology of astrophotography was introduced at the Royal Astronomical Society's annual meeting with a dramatic photograph of Andromeda, which was declared by astronomers as "The nebular hypothesis made visible!" The mighty Andromeda was once again relegated to our galactic suburbs. Even the discovery of a nova in Andromeda, which later would be additional proof of its extragalactic origin, was reinterpreted through the lens of the nebular hypothesis as an anomaly—the very fact that it outshone the entire nebula "with the energy of some fifty million suns," wrote one astronomer, meant that it was simply impossible that this could be an exploding star in a distant galaxy. Instead, it was suggested that it could be "the sudden transformation of the nebula into a star," and thus the nebular hypothesis remained intact. "The question whether nebulae are external galaxies hardly any longer needs discussion," declared the astronomer Agnes Clerke in her definitive 1890 work, *The System of the Stars*. "It has been answered by the progress of discovery. No competent thinker, with the whole of the available evidence before him, can now, it is safe to say, maintain any single nebula to be a star system of coordinate rank with the Milky Way."[11]

At this point we would do well to remember Arthur C. Clarke's first law: "When a distinguished but elderly scientist states that something is possible, he is almost certainly right. When he states that something is impossible, he is very probably wrong."[12] As our account shifts into the twentieth century we will find that the progress of discovery supported Clarke over Clerke, beginning with an 1899 spectroscopic analysis of the Andromeda nebula by the German astronomer Julius Scheiner. Scheiner compared Andromeda to the spectra of the Orion nebula, which by then was determined to be a nearby cloud of interstellar gas. Andromeda's spectra more closely resembled that of an enormous cluster of stars and not just a cloud of gas. To test this hypothesis, in 1908 an astronomer at the Lick Observatory near San Jose, California, named Edward Fath measured the spectra of globular clusters and noted the similarity with the spectrum of Andromeda. Game, set, and match, as far as Fath was concerned: "The hypothesis that the central portion of a nebula like the famous one in Andromeda is a single star may be rejected at once, unless we wish to modify greatly the commonly accepted ideas as to what constitutes a star."[13] But since there was as yet no accurate and reliable means to measure the distance to such celestial objects, Fath could not discern whether Andromeda represented a nearby globular cluster or a distant island universe.

"Weighty evidence in favor of the well known 'island universe' theory"

The final pieces of the puzzle in this celestial mystery were put together in California, first at the Lick Observatory and finally at Mount Wilson, the first two mountaintop observatories in the world that were, in their day, on the cutting edge of peering into deep space and lookback time. In the late nineteenth century, a phenomenally wealthy industrialist named James Lick, in search of the biggest and boldest monuments to which he could attach his name, pledged $1 million to build an observatory on Mount Hamilton in the Diablo mountain range just inland from San Jose. There he erected the "Great Lick Refractor," a thirty-six-inch piece of glass mounted at the end of a jaw-droppingly long tube that remains to this day one of the most beautiful astronomical instruments ever constructed, a true concours d'elegance of science. But this telescope—one of the last of the great refractors ever built—was mainly employed in the study of planets and stars, which had come to consume astronomical careers. So when the observatory hired a young upstart astronomer who specialized in spectroscopy named James Keeler, he was sent across the valley to another peak where a secondary dome housed an inelegant workaday reflector telescope with a thirty-six-inch mirror and skeletal struts instead of a tube.

The transition between the old and the new—between the refracting lens and the reflecting mirror—was more than symbolic. (See figure 19.) The size of a lens is restricted by its weight because it can only be supported around the edge. Over time it may begin to sag and distort. A mirror, however, can be fully buttressed from beneath, and so a reflecting telescope can be made large enough to gather those precious few photons of light arriving from the far reaches of the universe. The Crossley, so named for the wealthy textile manufacturer who bought it in 1885 and then donated it to the Lick Observatory, had another advantage for the spectroscopist: glass lenses discriminately absorb some wavelengths more than others, limiting the scope and quality of spectroscopic analysis, whereas a mirror reflects all wavelengths equally, providing a truer portrait of the contents of the mysterious nebulae.[14]

One of the first long-term exposures Keeler made with the Crossley was of the controversial M51 Whirlpool galaxy, which stunned even the most conservative of astronomers with its obvious spiral shape implying

Figure 19. Lick Observatory's Telescope and the Mysterious Nebulae It Revealed

a

a. The Crossley telescope at the Lick Observatory contains a thirty-six-inch mirror at the bottom and a secondary mirror at the top of its tube, that together reflect the focused light into an eyepiece or spectroscope on the side of the tube. Through this instrument James Keeler was able to image thousands of nebulae. PHOTOGRAPH BY THE AUTHOR.

b. One such nebula was NGC 891 (the 891st object in the New General Catalog of deep space objects), which, when examined more closely, was discovered to include many other nebulae, from which Keeler concluded that they are separate "island universes" outside of the Milky Way galaxy. The close-up image with individual nebulae identified with arrows and the three bright stars corresponds to the upper right corner of the wide-angle photograph of galaxy NGC 891. COURTESY OF THE LICK OBSERVATORY.

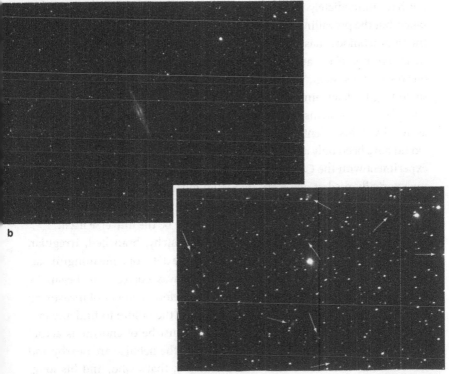

b

motion, along with internal structure in the form of distinct arms. As an added bonus, the four-hour exposure revealed seven other previously unknown nebulae, hinting that there were many more out there than anyone had previously imagined. Over time the Messier catalog had been vastly surpassed by the New General Catalog (NGC), which featured thousands of nebulae. As Keeler wheeled the Crossley around the sky snapping long exposures of this and that NGC object, he began to see a pattern of flattened discs of spiraling arms swirling around a bright center. In the background were countless more not-yet-cataloged tiny splotches of light. It was what we would today call a *fractal pattern*: with each increase of magnification for a particular patch of sky, a similar pattern emerged of scattered nebulae behind the primary target of the viewfinder. Extrapolating from his data set, an average of three nebulae per square degree in the sky, Keeler estimated that there were at least 120,000 of these celestial sphinxes, but he privately suspected that there were many more than this, perhaps an order of magnitude more.

Again, with hindsight we wonder how Keeler and his colleagues could not have immediately inferred spiral arms of countless stars at great distance, but the prevailing theory of star formation at that time was of a contracting nebulous mass that rotated as it contracted, thus giving planets their common plane and direction of revolution about a star, as we see in our own solar system. This is a problem in pattern detection and hypothesis testing to determine if the nebulae patterns represent developing star and planetary systems within our galaxy or island universe galaxies far away. Given his talents for both astrophotography and spectroscopy, it would have been only a matter of time before Keeler conducted a definitive experiment with the Crossley to determine which pattern was real, but he unexpectedly died at age forty-two in August 1900, so that task went to Heber Curtis throughout the 1910s, in the race against the Mount Wilson astronomers for the prize that would ultimately be the universe itself.

Curtis cataloged nebulae by adjective—patchy, branched, irregular, elongated oval, symmetrical—and searched the data for a meaningful pattern that would indicate which hypothesis was correct. He began by rephotographing spirals shot by Keeler years before in hopes of measuring rotation. When he found none he concluded, "the failure to find any evidence of rotation would indicate that they must be of enormous actual size, and at enormous distances from us." Or, the nebulae are nearby and not rotating. Who could tell? George Ritchey, that's who, and his long-time-exposure photograph of NGC 6946 in 1917 from the new Hale sixty-inch reflector telescope at Mount Wilson—named after the astronomer George Ellery Hale, who was in the habit of building the world's largest telescopes and had bagged another one here—revealed a nova that had flared up when compared to earlier photographs of that same object. Comparing this nova to the 1885 nova in Andromeda revealed that it was 1600 times dimmer, which Ritchey took to mean that it was 1600 times farther away. Unless, of course, there are different types of nova, some brighter and some dimmer—which there are—so more data and better theory were still needed. Curtis went to work, photographing nebulae previously imaged and comparing the plates in search of new dots of light. He found them, concluding that one in particular had to be at least twenty million light-years away, which led him to note, "The novae in spirals furnish weighty evidence in favor of the well known 'island universe' theory."[15]

This might have settled the issue, were it not for the fact that there was no reliable method for measuring distance out that far. As the British astronomer A. C. Crommelin noted in his 1918 comprehensive paper weighing the evidence for and against the island universe theory: "Whether true or false, the hypothesis of external galaxies is certainly a sublime and magnificent one. Instead of a single star-system it presents us with thousands of them, some large and conspicuous, others faint and small through their awful remoteness. Our conclusions in science must be based on evidence, and not on sentiment. But we may express the hope that this sublime conception may stand the test of further examination."[16]

Red Shifts and Variable Stars

The "sublime conception" of island universes, however, was not quite ready for prime time. The great British astrophysicist James Jeans developed a model for the evolution of solar systems that looked remarkably like what astronomers thought they were seeing in the nebulae. This model included stars that passed nearby a nebulous cloud, stirring up the particles into spiral shapes that would eventually coalesce into planets. At the Lowell Observatory in Arizona, the colorful and influential astronomer Percival Lowell threw his not inconsiderable weight behind the nebular hypothesis and was steadfastly confident that the fuzzy patches represented solar systems in formation. To bolster his belief he ordered his young charge Vesto Slipher to spectrographically analyze the nebulae to detect the characteristic lines of planets that he strongly suspected would be found within these faint structures, along with their radial velocity—how fast the nebulae were moving toward or away from us. This latter set of measurements would turn out to be the undoing of Lowell's theory.

In a marathon light-gathering night in September 1912, Slipher imaged Andromeda for 13.5 hours. The spectrographic plate revealed that there was a displacement of the spectral lines toward the blue end of the spectrum.[17] By now astronomers had determined that the shifting of spectral lines toward the blue meant that an object is moving toward us, and if shifted toward the red the object is receding away from us. This is the so-called Doppler effect, discovered by the Austrian physicist Christian Doppler, who noted that waves of light moving toward an observer will be squashed and therefore shifted toward the higher-frequency blue end of the spectrum, and if moving away the waves will be stretched and therefore shifted toward the lower-frequency red end of the spectrum. Andromeda was blue shifted. *Really* blue shifted—to the tune of three hundred kilometers per second by Slipher's calculations, which put Andromeda astronomically far beyond the range of motion ever measured of individual stars. How could an object moving this fast be located within the Milky Way?

Additional spectral shifts confirmed Slipher's initial finding. Nebula M81 was measured at one thousand kilometers per second—three times the speed of Andromeda—and it was moving away from us. By 1914, Slipher had more than a dozen nebular speeds, all within the range measured for Andromeda and M81—about twenty-five times faster than the average stellar velocity—and most receding from us. With these speeds, and the estimated size of the Milky Way, it seemed clear to many astronomers that these nebulae could not be within the Milky Way. The island universe theory was gaining momentum, and the seeds of the expanding universe theory were being sown.

What was needed to close out the debate was a reliable distance measurement, which was created in the early 1900s by Henrietta Swan Leavitt at Harvard, who began her career as a volunteer and worked her way up to being a "computer"—a woman who calculated figures for the all-male staff astronomers. She finally carved out a milestone career in astronomy for her work on Cepheid variable stars, which became the standard distance-measurement objects that Hubble noted on his photographic plate in 1923. Cepheid variables—named for the specimen discovered in the Cepheus the King constellation—vary in brightness over the course of days, weeks, or months, and they do so in a highly predictable manner: the brighter the variable the longer its period. Since Leavitt discovered these Cepheids in the Small Magellanic Cloud—those glowing patches in the southern sky first noted by Ferdinand Magellan during his circumnavigation of the globe—it meant that all the stars within that satellite galaxy were the same distance from us. Their periodicity was a direct measurement of their real luminosity and not an effect of varying distances.

Cepheid variables became the "standard candle" of light-distance measurement. If you have a particular type of candle for which all flames are the same size and brightness, and you discover some to be half as bright or a quarter as bright or an eighth as bright as the standard candle nearby, you can reasonably infer that they are four, eight, or sixteen times as far away. Once the distance to a Cepheid variable could be reliably established through such tried-and-true methods as parallax (how much the background stars shift behind the target stars when comparing images taken from one side of Earth's orbit to those taken from the other side six months later), then finding Cepheids in nebulae that are X times dimmer means that they are X times farther away. If Cepheid variables could be found inside nebulae at distances much greater than the size of the Milky Way, that would confirm that these stars are located in nebulae well outside of our galaxy and validate the island universe theory.

The "Big Galaxy" Hypothesis and the Mysterious Rotating Nebulae

There was one more line of evidence against the island universe hypothesis, and that was the work of the great cosmologist Harlow Shapley on the size of the Milky Way. Shapley began by gathering data on globular clusters from the one-hundred-inch Hooker telescope recently unveiled as the world's largest atop Mount Wilson. By 1920 Shapley concluded that these stellar globes circle about the center of the Milky Way like wasps swarming about a nest. Since it had by now been determined that the sun is nowhere near the center of the Milky Way, Shapley increased the estimated size of the Milky Way by an order of magnitude, from 30,000 light-years to 300,000 light-years across. He called it his "big galaxy" hypothesis, and it was a galaxy easily large enough to accommodate all celestial objects—including those pesky nebulae—into the known universe. If Shapley was right, there is only one island universe and we're in it, along with the nebulae. To test his hypothesis, Shapley returned to the data on whether or not nebulae rotate. If they do, then they cannot be that far away, because an object whose rotational motion was detectable over a course of only a few years at that distance would mean that it was rotating faster than the speed of light, which is not possible. Because some astronomers thought that they had detected just such motion in Andromeda, Shapley concluded that it could be no farther away than about 20,000 light-years.

Measurement of the rotational speeds of the nebulae began in earnest by the Dutch astronomer Adriaan van Maanen on the sixty-inch Hale telescope at Mount Wilson in 1915. Using a stereoscopic viewfinder that alternated two identical photographic plates shot at different times, Van Maanen compared photographs of spiral nebulae taken in 1899, 1908, and 1914 to his most recent photographs. Scanning the images for anything that moved or any rotational change from one year to the next, Van Maanen thought he saw motion in M101—the Pinwheel nebula—which he estimated was completing one full revolution every 85,000 years. If M101 was an island universe at a vast distance, this would mean that the stars on the nebula's edge would be rotating faster than the speed of light, which Einstein had recently proven was impossible. Ergo, M101—and by extension the other spiral nebulae—were nearby and well within Shapley's newly reconstituted 300,000-light-years-across Milky Way. Shapley wrote Van Maanen: "Congratulations on the nebulous results! Between us we have put a crimp in the island universes, it seems,—you by bringing the spirals in and I by pushing the Galaxy out."[18]

Since the theories were in conflict, the rub was in the data, which Heber Curtis at the Lick challenged. He attempted to measure nebular rotational motion himself but couldn't. Where Van Maanen thought he saw rotational periods of 160,000 years for M33, 45,000 years for M51, and 58,000 years for M81, Curtis saw no motion at all. How can this be? Either nebula are rotating or they are not, right? Herein lies a problem in patternicity and how the mind fills in the details when the data do not speak for themselves, which they rarely do. Measuring nebular rotation was incredibly tedious work in which error measurement could easily exceed the measurement of motion itself, leading to a completely erroneous conclusion. It would be like estimating the speed of a car at 30 mph, ± 30 mph. This, it would seem, was what happened. As improvements in measuring quality increased, motion of the nebula decreased . . . until it disappeared entirely.

"VAR!"

Enter Edwin Hubble, one of the grandest characters in the long and colorful history of astronomy, who cultivated a British air of aristocracy even though he was from Missouri. Hubble arrived at Mount Wilson shortly after the magnificent new one-hundred-inch Hooker telescope (see figure 20) came online, with a capacity to discern a candle at a distance of five thousand miles. Hubble's considerable intellect and ambition

Figure 20. The Mount Wilson 100-inch Telescope That Solved the Riddle of the Nebulae

The one-hundred-inch Hooker telescope atop Mount Wilson in the San Gabriel Mountains in Southern California, where Edwin Hubble demonstrated once and for all that the mysterious nebulae were not small nearby gaseous objects within the Milky Way galaxy, but were instead "island universes"—galaxies—similar in structure to our own but very far away.
PHOTOGRAPH BY AUTHOR.

were afforded the technology to adjudicate once and for all the great debate between the nebular hypothesis and the island universe theory.

The year 1923 was Hubble's annus mirabilis, starting with several months of classifying and cataloging familiar nebulae, followed by the discovery of fifteen variable stars in NGC 6822, eleven of which were Cepheid variables. Hubble employed the new standard candles to compute the nebula's distance at 700,000 light-years, well beyond even Shapley's 300,000-light-years-across "big galaxy." On October 4, Hubble photographed a number of nebulae, including Andromeda. During the next day's detailed laboratory analysis of the plates he thought he spotted a nova, maybe three. His attention heightened, he rephotographed Andromeda the next night and confirmed: "nova suspected." Hubble then went to the archives to compare the plate with those shot previously, and there on the new plate he scratched in "N" for nova—new star—for three specs of light. Triple-checking his plate, Hubble realized that one of the dots was not new; it was, in fact, a variable star—a Cepheid variable, no less! Hubble wrote in the logbook for the one-hundred-inch telescope, "On this plate (H335H), three stars were found, 2 of which were novae, and 1 proved to be a variable, later identified as a Cepheid—the first to be recognized in M31."[19] On the plate Hubble crossed out the "N" and scratched in "VAR!" The date on the plate reads "6-Oct 1923." (See figure 21.) This is the day the universe changed.

Over the next several months Hubble returned to Andromeda and tracked the light curve for his Cepheid, which varied over 31.415 days, from which he computed that the star was seven thousand times brighter than our sun. Yet it was barely noticeable on a photographic plate after hours of light-gathering exposure, which could mean only one thing: Andromeda was very, *very* far away. Hubble wrote to Shapley (who was now at Harvard): "You will be interested to hear that I have found a Cepheid variable in the Andromeda Nebula (M31). I have followed the nebula this season as closely as the weather permitted and in the last five months have netted nine novae and two variables."[20] Using the same technique that Shapley had used for measuring globular clusters and the size of the Milky Way, Hubble calculated that Andromeda was at least a million light-years away. If true, this would mean that Andromeda was an island universe.

Shapley was slow to see the new data in the same way Hubble did, telling him that he found Hubble's letter to be "the most entertaining

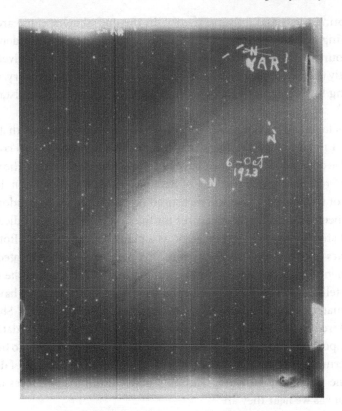

Figure 21. The Photograph That Changed the Universe
Edwin Hubble's photograph of Andromeda in which he identified Cepheid variable stars—used for measuring distance—that allowed him to compute that this nebula was too far away to be located within the Milky Way, and therefore must itself be an "island universe." COURTESY OF MOUNT WILSON OBSERVATORY.

piece of literature I have seen for a long time," and cautioned him that Cepheids with variable periods longer than twenty days may not be reliable indicators of distance. Hubble responded with more data, imaging nine variable stars in NGC 6822, then another dozen variables in Andromeda, three of which were the coveted Cepheids, plus another fifteen variables in M33, M81, and M101. In another letter to Shapley, Hubble opted for diplomacy in gently nudging his colleague and erstwhile competitor to shift paradigms: "the straws are all pointing in one direction and it will do no harm to begin considering the various possibilities involved" in accepting the island universe theory. In the end Shapley came around, showing Hubble's letter to a Harvard astronomy graduate student and pronouncing: "Here is the letter that has destroyed my universe."[21] Shortly thereafter, Shapley championed the island universe theory, abandoning his previously held belief in light of this new and unmistakable data.

As for Adriaan van Maanen's data of the nebular rotations that convinced not a few astronomers that the nebular hypothesis was correct, Hubble concluded that it must have been measurement error: "The problem of reconciling the two sets of data has a certain fascination, but in spite of this I believe that the measured rotations must be abandoned. I have been examining the measures for the first time and the indications point steadily to a magnitude error as a plausible explanation. Rotation appears to be a forced interpretation."[22] A perplexed and frustrated Van Maanen went back to his astronomical plates and recrunched the numbers, telling Shapley, "I cannot find a flaw in M33, for which I have the best material. They seem to be as consistent as possibly can be." Shapley countered with a diplomatic comparison between two sets of data and corresponding theories: "I am completely at a loss to know what to believe concerning those angular motions; but there seems to be no way of doubting the Cepheids, providing Hubble's period-luminosity curves are as definite as we hear they are."

They were, and years later when Shapley was asked in an interview why he defended Van Maanen's rotational data for so long, he responded in the third person: "They wonder why Shapley made this blunder. The point . . . is that Van Maanen was his friend and he believed in friends." An admirable trait to be sure, which can even cloud the judgment of data-hardened scientists, but in the end data and theory must trump belief and friendship.

The great debate over the celestial nebulae serves as a classic study in the history of science demonstrating that, in time, disputes are settled and debates are resolved through higher-quality data and more comprehensive theory. Perhaps science does not progress as quickly as we might like, and scientists often cling to cherished theories long after the data indicates that they should (especially when yoked to friendship), but eventually change does come, paradigm shifts are made, revolutions are

undertaken, and cumulative progress is made toward a greater understanding of the true nature of nature.

Where do we go from the island universe theory? What can there be
beyond island galaxies populating an expanding universe?

Science and the Greatest Unsolved Mystery

There is one mystery I will concede has proven to be a knotty one for science, and that is the matter of how our universe came to be. The mystery
is presented in two general ways, one impossible to answer and the other
potentially (but not yet) answerable. In the first configuration, the question is asked, *What existed before our universe began?* Or *Why is there
something rather than nothing?*

Phrasing the questions in this way is not only unscientific, it is non-
sensical, along the lines of asking *What time was it before time began?* Or
What is north of the North Pole? Asking why there is something rather
than nothing presumes "nothing" is the natural state of things out of
which "something" needs an explanation. Maybe "something" is the natural state of things and "nothing" would be the mystery to be solved. As
the physicist Victor Stenger noted: "Current cosmology suggests that no
laws of physics were violated in bringing the universe into existence. The
laws of physics themselves are shown to correspond to what one would
expect if the universe appeared from nothing. There is something rather
than nothing because something is more stable."[23]

The theist's answer to the problem of existence is that God existed
before the universe and subsequently brought it into existence out of
nothing (ex nihilo) in a single creation moment as described in Genesis.
But the very conception of God existing *before* the universe and *then*
creating it implies a time sequence. In both the religious and scientific
worldviews, time began with the big bang creation of the universe, so
God would have to exist outside of space and time, which means that as
finite beings delimited by living in a finite universe we cannot possibly
know anything about such a supernatural entity, unless he became a
natural being and entered our world to perform miracles.

In any case, in this conception of the mystery we are limited by language and cognition: because our brains are finite and limited we cannot
really grasp what "infinity" or "nothing" or "eternity" really mean, and
such thought experiments result in paradoxes that dissolve into tautologies,

along the lines of defining *gravity* as the tendency of objects to attract one another, and then explaining that objects attract one another because of *gravity*.[24] It is paradoxical to think of the universe as giving birth to time and space and then asking what there was before the universe. It is tautological to define God as the creator of the universe, and then explain the universe as a creation of God. These language and cognition conundrums cannot lead us to a satisfactory answer to the question. This limerick by the physicist George Gamow well captures the paradox:

> *There was a young fellow from Trinity*
> *Who took [the square root of infinity]*
> > *But the number of digits*
> > *Gave him the fidgets;*
> *he dropped Math and took up Divinity.*

The second configuration of the mystery gives scientists something to work with: *Why is our universe so finely tuned to enable stars, planets, life, and intelligence to arise?* This is known as the *fine-tuning problem*, and in my opinion it is the best argument that theists have for the existence of God. Even nonreligious scientists are stunned by the odd configuration of numbers that had to be just so or else life could not exist. Sir Martin Rees, Britain's astronomer royal, in his book *Just Six Numbers* outlined the problem, noting that, "our emergence from a simple Big Bang was sensitive to six 'cosmic numbers'" that are "well tuned" for the emergence of matter and life.[25] Here are the six numbers:

1. Ω (omega) = 1, the amount of matter in the universe: if Ω was greater than 1 it would have collapsed long ago and if Ω was less than 1 no galaxies would have formed.
2. ε (epsilon) = .007, how firmly atomic nuclei bind together: if epsilon were .006 or .008, matter as we know it could not exist as it does.
3. D = 3, the number of dimensions in which we live: if D were 2 or 4, life could not exist.
4. N = 10^{39}, the ratio of the strength of electromagnetism to that of gravity: if it had just a few less zeros the universe would be too young and too small for life to evolve.

5. Q = 1/100,000, the fabric of the universe: if Q were smaller the universe would be featureless and if Q were larger the universe would be dominated by giant black holes.

6. λ (lambda) = 0.7, the cosmological constant, or "antigravity" force that is causing the universe to expand at an accelerating rate: if λ were larger it would have prevented stars and galaxies from forming.

The fine-tuning of these six numbers (there are more, but these are the big ones) that make life possible is sometimes explained by the "anthropic principle," most prominently stated by physicists John Barrow and Frank Tipler in their 1986 book *The Anthropic Cosmological Principle*: "It is not only man that is adapted to the universe. The universe is adapted to man. Imagine a universe in which one or another of the fundamental dimensionless constants of physics is altered by a few percent one way or the other? Man could never come into being in such a universe. That is the central point of the anthropic principle. According to the principle, a life-giving factor lies at the center of the whole machinery and design of the world."[26] The anthropic principle troubles scientists because of its antithesis, known as the "Copernican principle," which states that we are not special. Intelligent design theorists, creationists, and theologians hold that this fine-tuning is evidence for intelligent design by a deity, and the anthropic principle is their hypothesis. I suggest that there are at least six alternatives to this hypothesis that better support the Copernican principle hypothesis.[27]

1. The universe is not so finely tuned for life since the vast majority of the universe is empty space, and what little matter there is—in the form of stars and planets—is mostly inhospitable to life.

2. The idea that the universe is finely tuned for us is a problem in *cosmic chauvinism*, a grander variant of what Carl Sagan called "carbon chauvinism," or the belief that life cannot be based on anything other than carbon. By rejecting cosmic chauvinism we see that the universe is not finely tuned for us, we are finely tuned for it. It is difficult for us to conceive of how a different physics could produce different forms of life, but it could. Science has had only four centuries to study the nature of life; evolution has had

four billion years to create life. Evolution is smarter than science. It is too provincial of us to say that we know for sure that life could not evolve under a different set of laws.

3. Such numbers as the speed of light and Planck's constant are, on one level, arbitrary numbers that can be configured in different ways so that their relationship to the other constants are not so coincidental or mysterious. As well, such constants may be *inconstant* over vast spans of time, varying from the big bang to the present, making the universe finely tuned only now but not earlier or later in its history. Physicists John Barrow and John Webb call these numbers the "inconstant constants," and have demonstrated how in particular the speed of light, gravitation, and the mass of the electron have, in fact, been inconstant over time.[28]

4. There may be an underlying principle behind the six magic numbers that will be found when the grand unified theory of physics is discovered and constructed. Instead of six mysterious numbers, there will just be one. Until we have a comprehensive theory of physics that connects the quantum world of subatomic particles to the cosmic world of general relativity, we do not yet know enough about the nature of our universe to make the leap to something beyond nature. Caltech cosmologist Sean Carroll notes:

> Possibly general relativity is not the correct theory of gravity, at least in the context of the extremely early universe. Most physicists suspect that a quantum theory of gravity, reconciling the framework of quantum mechanics with Einstein's ideas about curved spacetime, will ultimately be required to make sense of what happens at the very earliest times. So if someone asks you what really happened at the moment of the purported Big Bang, the only honest answer would be: "I don't know."[29]

That grand unified theory of everything will itself need an explanation, but it may be explicable by some other theory we have yet to comprehend out of our sheer ignorance at this moment in the history of science.

5. As a historian of science I strongly suspect that there are grander vistas still to be discovered by astronomers and cosmologists that will change the nature of the problem altogether, from explaining the nature and origin of the universe to explaining something else entirely. Consider the sequence of our visage of the cosmos over the past millennia: from the ancient Babylonians' Earth-centered cosmology with a canopy of stars rotating around it that was picked up by the Hebrews and solidified by Aristotle's model of a motionless Earth, to the medieval worldview of Earth at the center and the stars and planets rotating close by on their crystal spheres, to the sixteenth-century Copernican revolution that put Earth in motion and the stars far away, to William Herschel's eighteenth-century conjecture that the fuzzy patches in the sky were "island universes," to Edwin Hubble's twentieth-century discovery that those nebulae were not in the Milky Way galaxy but were actually galaxies of immense size and distance expanding away from a big bang beginning, to the twenty-first-century finding that the universe is expanding at an accelerating rate, to . . . what?

6. Based on the history of astronomy, and other converging lines of evidence and logic, I would like to make the case for a *multiverse*, in which our universe—which was born in a big bang and will most likely expand forever and die with a whimper—is just one of many bubble universes all with different laws of nature.[30] Those universes with the six magic numbers will generate matter, which coalesces into stars, some of which collapse into black holes and a singularity, the same entity out of which our universe may have sprung. Thus, universes like ours give birth to baby universes with those same six numbers, some of which develop intelligent life smart enough to discover this Darwinian process of cosmic evolution. A *multiverse* containing a multitude of universes fits this historical trajectory of expanding cosmic horizons and reinforces the long-standing Copernican principle that we are but fleeting actors on this planetary proscenium.

Of course, we should apply the rules of science and skepticism to the multiverse hypothesis as vigorously as we would any other. Are there any good reasons to believe in a multiverse? There are, and the models come in a variety of flavors that, in keeping with the pattern of numeration above, I'll classify into six types.

1. *The eternal-return multiverse.* This form of multiverse arises out of an eternal boom-and-bust cycle of expansion and contractions of the universe, with our universe just one "episode" of the bubble's eventual collapse and re-expansion in an eternal cycle. Cosmologist Sean Carroll argues "that space and time did exist before the Big Bang; what we call the Bang is a kind of transition from one phase to another." As such, he says, "there is no such thing as an initial state, because time is eternal. In this case, we are imagining that the Big Bang isn't the beginning of the entire universe, although it's obviously an important event in the history of our local region."[31] This multiverse seems unlikely because all the evidence to date shows that our universe is not only still expanding, but its expansion is accelerating. There does not appear to be enough matter in our universe to halt the expansion and bring it back into a big crunch that could launch it back into a new bubble out of another big bang.[32]

2. *Multiple-creations multiverse.* In the theory of inflationary cosmology, the universe sprang into existence from a bubble nucleation of space-time, and if this process of universe creation is natural then there may be multiple bubble nucleations that give rise to many universes that expand but remain separate from one another without any causal contact between them. If such causally disconnected universes existed, however, there is no way to get information from them, and so this is an inherently untestable hypothesis and thus is no better than the anthropic principle hypothesis.[33]

3. *The many-worlds multiverse.* This type of multiverse is derived out of the "many worlds" interpretation of quantum mechanics, in which there are an infinite number of universes in which every possible outcome of every possible choice that has ever been available, or will be available, has happened in one of them. This multiverse is grounded in the bizarre findings of the famous "double-slit" experiment in which light is passed through two slits and forms an interference pattern of waves on a back surface (like throwing two stones in a pond and watching the concentric wave patterns

interact, with crests and troughs adding and subtracting from one another). The spooky part comes when you send single photons of light one at a time through the two slits—they still form an interference wave pattern even though they are not interacting with other photons. How can this be? One answer is that the photons are interacting with photons in other universes! In this type of multiverse—sometimes configured as "parallel universes"—you could meet your doppelgänger, and depending on which universe you entered, your parallel self would be fairly similar or dissimilar to you, a theme that has become a staple of science fiction. This version of the multiverse, in my opinion, doesn't pass the smell test. The idea of there being multiple versions of me and you out there—and in an infinite multiverse model there would be an infinite number of us—just seems prima facie absurd and even less likely than the theistic alternative.

4. *The multidimensional string theory multiverse.* A multidimensional multiverse may come about when a three-dimensional "brane" (a membranelike structure on which our universe exists) moves through higher-dimensional space and collides with another brane, the result of which is the energized creation of another universe.[34] A related multiverse is derived through string theory, which by at least one calculation allows for 10^{500} possible worlds, all with different self-consistent laws and constants.[35] That's a 1 followed by 500 zeros possible universes (recall that 1 followed by 12 zeros is a trillion!). If true, it would be miraculous if there were not intelligent life in a number of them. Victor Stenger created a computer model that analyzes what just 100 different universes would be like under constants different from our own, ranging from five orders of magnitude above to five orders of magnitude below their values in our universe. Stenger found that long-lived stars of at least one billion years—necessary for the production of life-giving heavy elements—would emerge within a wide range of parameters in at least half of the universes in his model.[36]

5. *Quantum foam multiverse.* In this model, universes are created out of nothing, but in the scientific version of ex nihilo the nothing of the vacuum of space actually contains quantum foam, which may fluctuate to create baby universes. In this configuration, any quantum object in any quantum state may generate a new universe,

each one of which represents every possible state of every possible object.[37] This is Stephen Hawking's explanation for the fine-tuning problem that he himself famously presented in the 1990s:

> Why is the universe so close to the dividing line between collapsing again and expanding indefinitely? In order to be as close as we are now, the rate of expansion early on had to be chosen fantastically accurately. If the rate of expansion one second after the big bang had been less by one part in 10^{10}, the universe would have collapsed after a few million years. If it had been greater by one part in 10^{10}, the universe would have been essentially empty after a few million years. In neither case would it have lasted long enough for life to develop. Thus one either has to appeal to the anthropic principle or find some physical explanation of why the universe is the way it is.[38]

Hawking's collaborator Roger Penrose layered on even more mystery when he noted that the "extraordinary degree of precision (or 'fine tuning') that seems to be required for the Big Bang of the nature that we appear to observe . . . is one part in $10^{10^{123}}$ at least." Penrose suggested two pathways to an answer: either it was an act of God, "or we might seek some scientific/mathematical theory."[39] Hawking opted for the second with this explanation: "Quantum fluctuations lead to the spontaneous creation of tiny universes, out of nothing. Most of the universes collapse to nothing, but a few that reach a critical size, will expand in an inflationary manner, and will form galaxies and stars, and maybe beings like us."[40]

6. *The natural selection multiverse.* For my money the best multiverse model is that proffered by American cosmologist Lee Smolin, who adds a Darwinian component to an evolving cosmos in which there is a "natural selection" of differentially reproducing bubble universes. Smolin thinks that, like its biological counterpart, there might be a selection from different "species" of universes, each containing different laws of nature. Universes like ours will have lots of stars, which means they will have lots of black holes that collapse into singularities, a point at which infinitely strong gravity causes matter to have infinite density and zero volume. Many

cosmologists today believe that our universe began with a big bang out of a singularity, so it is reasonable to conjecture that collapsing black holes create new baby universes out of these singularities. Baby universes with laws of nature similar to ours will be bio-philic, whereas universes with radically different laws of nature that disallow stars, cannot have black holes, and thus will not hatch any baby universes, will go extinct. The long-term result of this cosmic evolutionary process would be a preponderance of universes like ours, so we should not be surprised to find ourselves in a universe suitable for life.[41]

How can we test the multiverse hypothesis? The theory that new universes can emerge from collapsing black holes may be illuminated through additional knowledge about the properties of black holes. Other bubble universes might be detected in the subtle temperature variations of the cosmic microwave background radiation left over from the big bang of our own universe, and NASA recently launched a spacecraft constructed to study this radiation. Another way to test these theories might be through the Laser Interferometer Gravitational Wave Observatory (LIGO) that is designed to detect exceptionally faint gravitational waves. If there are other universes, perhaps ripples in gravitational waves will signal their presence. Maybe gravity is such a relatively weak force (compared to electromagnetism and nuclear forces) because some of it "leaks" out to other universes. Maybe.

⌖

In late 2010, Stephen Hawking and Leonard Mlodinow presented their answer to the biggest of Big Questions ("Why is there something rather than nothing?," "Why do we exist?," and "Why this particular set of laws and not some other?") in their book, *The Grand Design*. They approach the problem from what they call "model-dependent realism," based on the assumption that our brains form models of the world from sensory input, that we use the model most successful at explaining events and assume that the models match reality (even if they do not), and that when more than one model makes accurate predictions, "we are free to use whichever model is most convenient." Employing this method, the authors explain, "it is pointless to ask whether a model is real, only whether it agrees with observation." The two models that describe light discussed above—the wave/particle models—serve as an example of model-dependent realism, where each model agrees with certain observations but neither one is sufficient to explain all observations. Hawking and Mlodinow explain the results of the double-slit experiment through the model developed by Richard Feynman called "sum over histories," in which every particle in the double-slit experiment takes every possible path that it can, and thus it interacts with itself in its different histories (instead of interacting with particles in other universes in the alternate model presented above).

To model the entire universe, Hawking and Mlodinow employ "M-theory," an extension of string theory that includes eleven dimensions (ten of space and one of time) and incorporates all five current string theory models. As in Feynman's "sum-over-histories" model of light, Hawking and Mlodinow propose that the universe itself takes every possible path—experiences all possible histories—and this results in the most multiple multiverse imaginable. "In this view, the universe appeared spontaneously, starting off in every possible way," Hawking and Mlodinow explain. "Most of these correspond to other universes. While some of those universes are similar to ours, most are very different. In fact, many universes exist with many different sets of physical laws." Although, as we saw, some people call these different universes the multiverse, Hawking and Mlodinow claim that "these are just different expressions of the Feynman sum over histories." Employing multiple models to explain multiple universes as nothing more than one system with multiple histories, Hawking and Mlodinow conclude, "For these reasons M-theory is the only candidate for a complete theory of the universe. If it is finite—and this has yet to be proved—it will be a model of a universe that creates itself."[42]

How can a universe create itself? The answer has to do with the total energy of the universe, which Hawking and Mlodinow state must be constant and always remain zero. Since it costs energy to create a body such as a star or planet, locally there are non-zero energy imbalances. "Because gravity is attractive, gravitational energy is negative: One has to do work to separate a gravitationally bound system, such as the earth and the moon," the authors explain. "This negative energy can balance the positive energy needed to create matter." But how do entire universes arise? "On the scale of the entire universe, the positive energy of the matter can be balanced by the negative gravitational energy, and so there is no restriction on the creation of whole universes. Because there is a law like gravity, the universe can and will create itself from nothing. . . . Spontaneous creation is the reason there is something rather than nothing, why the universe exists, why we exist." Although the authors admit that the theory has yet to be confirmed by observation, if it is, then no creator explanation is necessary because the universe creates itself. I call this *auto ex nihilo.*

At present there is no positive evidence for the multiverse hypothesis, but neither is there positive evidence for the traditional answer to the question: God. For both hypotheses we are left with the reductio ad absurdum question of *What came before the multiverse or God?* If God is defined as that which does not need to be created, then why can't the multiverse be defined as that which does not need to be created? Perhaps both are eternal and need no creation explanation. In any case, we have only negative evidence along the lines of "I can't think of any other explanation," which is no evidence at all.

If there is one lesson that the history of science has taught us, it is that it is arrogant to think that we now know enough to know that we cannot know. So for the time being it comes down to cognitive and emotional preference: an answer with only negative evidence or no answer at all. *God, multiverse,* or *unknown.* Which one you choose depends on your own belief journey and how much you want to believe.

The Truth Is Out There

WHEN I CALL MYSELF A *SKEPTIC* I SIMPLY MEAN THAT I TAKE A SCIEN-tific approach to the evaluation of claims. Science is skepticism and sci-entists are naturally skeptical. Scientists have to be skeptical because most claims turn out to be false. Weeding out the few kernels of wheat from the substantial pile of chaff requires extensive observation, careful experimentation, and cautious inference to the best conclusion.

What makes science so potent is that there is a well-defined method for getting at answers to questions about the world—a world that is real and knowable. Where philosophy and theology depend on logic and rea-son and thought experiments, science employs empiricism, evidence, and observational experiments. It is the only hope we have of avoiding the trap of belief-dependent realism.

Science and the Null Hypothesis

Science begins with something called a *null hypothesis*. Although statis-ticians mean something very specific about this (having to do with com-paring different sets of data), I am using the term *null hypothesis* in its more general sense: the hypothesis under investigation is not true, or null, until proven otherwise. A null hypothesis states that X does not cause Y. If you think X does cause Y then the burden of proof is on you to provide convincing experimental data to reject the null hypothesis.

The statistical standards of proof needed to reject the null hypothesis are substantial. Ideally, in a controlled experiment, we would like to be at least 95 to 99 percent confident that the results were not due to chance before we offer our provisional assent that the effect may be real. Everyone is familiar with the process already through news stories about the FDA approving a new drug after extensive clinical trials. The trials to which they refer involve sophisticated methods to test the claim that Drug X (say, a statin drug) causes Disease Y (say, heart disease linked to cholesterol) to decrease. The null hypothesis states that statins do not reduce heart disease by lowering cholesterol. Rejecting the null hypothesis means that there was a statistically significant difference in rates of heart disease between the experimental group receiving the statins and the control group that did not.

Here is a relatively simple example of how this method of statistical significance works in relation to the null hypothesis to answer this question: can a psychic using ESP alone determine whether a playing card from a deck is red or black? Psychics typically claim that they can do this, but in my experience what people *say* they can do and what they can *actually* do are not always the same. How can we test this claim? If we place the cards down onto a table one by one with the psychic stating either red or black for each card, how many correct hits would the psychic need in order for us to conclude that the card color determinations were not due to chance? In this scenario, the null hypothesis is that the psychic will do no better than chance, and thus to reject the null hypothesis we will need to establish a figure for the number of correct hits needed in each round. By chance, we would expect the psychic to get about half correct. In a deck of 52 cards, half of which are red and half of which are black, random guessing or flipping a coin will produce, on average, 26 correct hits.

Of course, as anyone who has flipped coins for fun knows, 10 flips do not necessarily always result in 5 heads and 5 tails. There are streaks and deviations from symmetry—6 heads and 4 tails, or 3 heads and 7 tails— all within the realm of chance. Or as anyone who has gambled at a roulette wheel knows, sometimes red comes up more than black, or vice versa, without any violations of chance and randomness. In fact, we count on such asymmetrical streaks in our betting schemes and hope that we're disciplined enough to walk away from the table during a temporary deviation from chance in our favor before the odds swing the other way.

So we can't just test our psychic on one short series of card guesses, because by chance the psychic may be expected to get a series of hits. We need to run multiple trials, in which some rounds may result in slightly below chance (say, 22, 23, 24, or 25 hits) and other rounds may result in slightly above chance (say, 27, 28, 29, or 30 hits). The variation may be even greater and still be due to nothing but chance. What we need to determine is the number by which we can confidently reject the null hypothesis. In this example, that number is 35. The psychic would need to get 35 correct hits out of a 52-card deck in order for us to reject the null hypothesis at the 99 percent confidence level. The statistical method by which this figure is derived need not concern us here.[1] The point is that even though 35 out of 52 doesn't sound like it would be that hard to obtain, in fact by chance alone it would be so unusual that we could confidently state ("at the 99 percent confidence level") that something else besides chance was going on here.

What might that be? It could be ESP. But it could be something else as well. Perhaps our controls were not tight enough. Maybe the psychic was getting the red/black information by some other normal (as opposed to paranormal) means of which we were not aware (such as the reflection of the card face in the table surface). Possibly the psychic was cheating, and we don't know how. I've seen James Randi do this very experiment with an entire deck of cards, resulting in two perfect piles of all red and all black cards. The magician Lennart Green shuffles and scrambles a deck of cards, fumbles with them for a while as if he's all thumbs, clumsily pushes them back together, then proceeds to deal out four winning poker hands or an entire sequence of a suit in order, all while blindfolded.[2] But Randi and Green are magicians and these are magic tricks. That I do not know how they are done does not make them real (paranormal) magic, and the fact that most scientists do not know how such magic tricks are done means we need to be even more vigilant in our controls when testing psychics, perhaps even attaching a magician to our research team. The argument from personal incredulity—if I can't explain it, then it must be true—does not hold water in science.

Even with all these controls in place, certainty still eludes science. The scientific method is the best tool ever devised to discriminate between true and false patterns, to distinguish between reality and fantasy, and to detect baloney, but we must always remember that we could be wrong. Rejecting the null hypothesis is not a warranty on truth, yet failure to reject the null hypothesis does not make the claim false. We must keep an open mind, but not so open that our brains fall out. Provisional truths are the best we can do.

Science and the Burden of Proof

The null hypothesis also means that the burden of proof is on the person asserting a positive claim, not on the skeptics to disprove it. I once appeared on *Larry King Live* to discuss UFOs (a perennial favorite of his), accompanied by a table full of UFOlogists (a five-to-one ratio of believers to skeptics seems to be the norm in television shows that cover such topics). Larry's questions for us skeptics typically miss this central tenet of science. ("Dr. Shermer, do you have an explanation for Mr. X's UFO sighting at three in the morning in Nowhere, Arizona?" If I don't, the assumption is that it must be extraterrestrial.) The burden of proof is not on the skeptics to disprove UFOs; it is on the UFO claimant to prove that it is extraterrestrial.

Although we cannot run a controlled experiment that would yield a statistical probability of rejecting the null hypothesis that aliens are not visiting Earth, proving that they are would be simple: show us an alien spacecraft or an extraterrestrial body. Until then, keep searching and get back to us when you have something. Unfortunately for UFOlogists, scientists cannot accept as definitive proof of alien visitation such evidence as blurry photographs, grainy videos, and anecdotes about spooky lights in the sky. Photographs and videos are often misperceived and can be easily doctored, and lights in the sky have many prosaic explanations: aerial flares, lighted balloons, experimental aircraft, helicopters, clouds, swamp gas, or even the planet Venus, which, if you are driving on an undulating highway away from city lights, really does appear to be a bright light following your car. Nor will government documents with redacted (blacked-out) paragraphs count as evidence for ET contact, because we know that governments keep secrets for a host of reasons having to do with military defense and national security. Yes, governments lie to their citizens, but lying about X does not make Y true. Terrestrial secrets do not equate to extraterrestrial cover-ups.

So many claims of this nature are based on *negative evidence*. That is, if science cannot explain X, then your explanation for X is necessarily true. Not so. In science lots of mysteries remain unexplained until further evidence arises, and problems are often left unsolved until another day. I recall a mystery in cosmology in the early 1990s whereby it appeared that there were stars older than the universe itself—the daughter was older than the mother! Thinking that I might have a hot story to write about that would reveal something deeply wrong with current cosmological models that I could publish in the inchoate days of *Skeptic* magazine, I first queried Caltech cosmologist Kip Thorne, who assured me that the discrepancy was merely a problem in the current estimates of the age of the universe and that it would resolve itself in time with more data and better dating techniques. It did, as so many problems in science eventually do. In the meantime, it's okay to say: "I don't know," "I'm not sure," and "Let's wait and see."

Science and the Convergence Method

To be sure, not all claims are subject to laboratory experiments and statistical tests. There are many historical and inferential sciences that require nuanced analyses of data and a *convergence of evidence* from multiple lines of inquiry that point to an unmistakable conclusion. Just as detectives employ the convergence of evidence technique to deduce who most likely committed a crime, scientists employ the method to deduce the likeliest explanation for a particular phenomenon. Cosmologists reconstruct the history of the universe through a convergence of evidence from cosmology, astronomy, astrophysics, spectroscopy, general relativity, and quantum mechanics. Geologists reconstruct the history of the earth through a convergence of evidence from geology, geophysics, and geochemistry. Archaeologists piece together the history of a civilization from pollen grains, kitchen middens, potsherds, tools, works of art, written sources, and other site-specific artifacts. Environmental scientists reconstruct climate history through the environmental sciences, meteorology, glaciology, planetary geology, geophysics, chemistry, biology, ecology, and others. Evolutionary biologists uncover and explain the history of life through geology, paleontology, botany, zoology, biogeography, comparative anatomy and physiology, genetics, etc.

Even though these inferential sciences do not fit the model of experimental laboratory sciences, hypothesis testing can be employed nevertheless. Indeed, scientists working within such historical sciences must test hypotheses in order to avoid the confirmation bias, the hindsight bias, and many other cognitive biases that will surely color their interpretations of the data. As Frank Sulloway noted at the end of his scientific treatise on the psychology of history: "When the mind is confronted with more information than it can absorb, it looks for meaningful (and usually confirmatory) patterns. As a consequence, we tend to minimize evidence that is incongruous with our expectations, causing the dominant worldview to bring about its own reaffirmation." In fact, Sulloway suggests that Charles Darwin may well be the greatest historian who ever lived because he went out of his way to test his hypotheses about the history of life, and this became the foundation of his work culminating in *On the Origin of Species*, which revolutionized his field, changing it from genteel speculations by amateur naturalists to the rigorous science it is today. Darwin even employed his new science to the history of his own life, as Sulloway explains: "Charles Darwin understood this human predilection for reaffirming the status quo. In his *Autobiography* he noted how quickly he tended to forget any fact that seemed to contradict his theories. He therefore made it a 'golden rule' to write this information down so that he would not overlook it. Like Darwin's golden rule, hypothesis testing overcomes certain limitations in how the human mind processes information."[3]

Science and the Comparative Method

How do you test a historical hypothesis? One way is through the *comparative method*, which was brilliantly employed by UCLA geographer Jared Diamond in his book *Guns, Germs, and Steel*, in which he explained the differential rates of development between civilizations around the globe over the past thirteen thousand years.[4] Why, Diamond asked, did Europeans colonize the Americas and Australia, rather than Native Americans and Australian Aborigines colonize Europe? Diamond rejected the hypothesis that inherited differences in abilities between the races precluded some groups from developing as fast as others. Instead Diamond proposed a biogeographical theory having to do with the availability of domesticated grains and animals to trigger the development of farming, metallurgy, writing, non-food-producing specialists, large populations, military and government bureaucracies, and other components that gave rise to Western cultures. Without these plants and animals, and a concatenation of other factors, none of these characteristics of our civilization could exist.

Employing the comparative method, Diamond compared Australia and Europe and noted that Australian Aborigines could not strap a plow to or mount the back of a kangaroo, as Europeans did the ox and horse. As well, indigenous wild grains that could be domesticated were few in number and located only in certain regions of the globe—those regions that saw the rise of the first civilizations. The east-west-oriented axis of the Euro-Asian continent lent itself to diffusion of domesticated grains and animals as well as knowledge and ideas, so Europe was able to benefit much earlier from the domestication process. By comparison, the north-south-oriented axis of the Americas, Africa, and the Asia-Malaysia-Australia corridor did not lend itself to such fluid transportation, and thus those areas already not well suited biogeographically for farming could not even benefit from diffusion. In addition, through constant interactions with domesticated animals and other peoples, Euro-Asians developed immunities to numerous diseases that, when brought by them in the form of germs to Australia and the Americas, along with their guns and steel, produced a genocide on a hitherto unseen scale. Furthermore, in less than one generation modern Australian Aborigines learned to fly planes, operate computers, and do anything that any European inhabitant of Australia can do. Comparatively, when European farmers were transplanted to Greenland they went extinct when their environment changed, not because their genes devolved.

Such comparative methods are the result of *natural experiments of history*, numerous examples of which Diamond presented in his 2010 book of that title, including a timely study comparing Haiti to the Dominican Republic. Both countries inhabit the same island but because of geopolitical differences one ended up dirt poor while the other flourishes.[5] What happened? This is a *natural experiment of borders*, similar to the one on the Korean peninsula. Adding a border between North and South Korea in 1945 resulted in dictatorship and poverty for North Korea, which in 2008 had an annual GDP of $13.34 billion and $555 per capita income, compared to South Korea's annual GDP of $929.1 billion and $19,295 per capita income. Think about how different your life would be if you made $555 a year compared to $19,295 a year, and you can feel the power of the comparative method. The border that divides the island of Hispaniola is striking: on one side the land is green and forested while on the other side the land is brown and treeless. Rain-laden weather fronts come from the east and dump their watery load on the eastern Dominican Republic side of the island, leaving the western side drier and with less fertile soils for agricultural productivity. Deforestation of the fewer trees on the Haitian side led to soil erosion, decreased soil fertility, loss of timber for the building industry and wood for charcoal fuel, heavier sediment loads in rivers, and decreased watershed protection, leading to lower hydroelectric power. This set up a negative feedback cycle of environmental degradation for Haiti.

Comparing the political history of the two sides of the island reveals a second set of factors at work. Christopher Columbus's brother Bartholomeo colonized Hispaniola in 1496 for Spain, establishing the capital at Santo Domingo on the egress of the Ozama River on the eastern side of the island. Two centuries later, during tensions between France and Spain, the Treaty of Ryswick in 1697 granted France dominion over the western half of the island, and the border was permanently established by the Treaty of Aranjuez in 1777. Because France was richer than Spain and slavery was an integral part of its economy, it turned western Hispaniola into a center of the slave trade with a population consisting of 85 percent slaves, compared to the eastern half under Spain with only 10–15 percent slaves. The raw numbers are staggering: about 500,000 slaves in the western side of the island compared to only 15,000 to 30,000 slaves in the eastern side. For a time, Haiti was richer than the Dominican Republic. For a time. But the slave economy led to a significantly greater population density that, coupled with France's hunger for timber from the island, led to rapid deforestation and subsequent environmental squalor. The fact that the Haitian slaves developed their own Creole language spoken by no one else in the world further isolated Haiti from the type of economic and cultural exchange that leads to prosperity.

When both the Haitians and Dominicans gained their independence in the nineteenth century, another comparative difference unfolded. Haitian slave revolts were violent and Napoléon's intervention to try to restore order resulted in the Haitians' deep distrust of Europeans. They wanted nothing to do with future trade and investments, imports and exports, or immigration and emigration, and so they did not benefit economically from these and other factors. By contrast, Dominican independence was relatively nonviolent, and it shuttled back and forth for decades between independence and control by Spain, which in 1865 decided that it did not want the territory. Throughout this period the

Dominicans spoke Spanish, developed exports, traded with European countries, and attracted European investors and a diverse immigrant population of Germans, Italians, Lebanese, and Austrians, who helped build a vibrant economy. Both countries succumbed to the power of evil dictators in the mid-twentieth century. Rafael Trujillo's control of the Dominican Republic involved considerable economic growth because of his desire to personally enrich himself, and this led to a vibrant export industry (most of which he owned); scientists and foresters were imported to help preserve the forests for Trujillo's personal use and profiteering through his logging companies. Haiti's dictator François "Papa Doc" Duvalier did none of this and instead further isolated the Haitians from the rest of the world.

Employing the comparison method with such natural experiments of history is no different from what sociologists and economists do in comparing natural experiments of society today. We cannot intentionally impoverish one group of people and then observe if their health, education, and crime rates change. But we can look around and find pockets of impoverished people in inner cities and then measure various factors and compare those to other socioeconomic classes. The process is as rigorous a scientific methodology as any to be found in the experimental sciences. Once an inferential or historical science is well established through the accumulation of positive evidence, it becomes a testable science.

Science and the Principle of Positive Evidence

The convergence of evidence method and the comparison method are routinely used by paleontologists and evolutionary biologists to test hypotheses about evolution, and the results accumulate in the form of positive evidence in support of the theory of evolution. For creationists to disprove evolution, they would need to unravel all these independent lines of evidence as well as construct a rival theory that can explain them better than the theory of evolution. They haven't, and instead employ only *negative evidence* in the form of "if evolutionary biologists cannot present a natural explanation of X, then a supernatural explanation of X must be true." Not so. The *principle of positive evidence* states that you must have positive evidence in favor of your theory and not just negative evidence against rival theories.

The principle of positive evidence applies to all claims. Skeptics are like people from Missouri, the Show-Me state. Show me positive evidence for your claim. Show me a Sasquatch body. Show me the archaeological artifacts from Atlantis. Show me a Ouija board that spells words with securely blindfolded participants. Show me a Nostradamus quatrain that predicted World War II or 9/11 *before* (not after) the fact (postdictions don't count in science because of the hindsight bias). Show me the evidence that alternative medicines work better than placebos. Show me an ET or take me to the mother ship. Show me the Intelligent Designer. Show me God. Show me and I will believe.

Most people (scientists included) treat the God question separate from all these other claims. They are right to do so as long as the particular claim in question cannot—even in principle—be examined by science. But what might that include? Most religious claims are testable, such as prayer positively influencing healing. In this case, controlled experiments to date show no difference between prayed-for and not-prayed-for patients. What would compel me to believe would be something unequivocal, such as a new limb growing on an amputee. Amphibians can do it. The new science of regenerative medicine appears on the verge of being able to do it. Surely an omnipotent deity could do it.

Science and Belief

We now come to the end of this narrative journey of belief, but it is really just the beginning of a new understanding of how the brain generates beliefs and reinforces them as truths. Of the many mysteries we have uncovered and questions we have tried to answer, one in particular stands out. *Homo rationalis*—that species of human who carefully weighs all decisions through cold, hard logic and rational analysis of the data—is not only extinct but probably never existed. Mr. Spock is science fiction. And it's a good thing, because people who have suffered brain damage to the emotional networks in their brains—particularly their limbic systems—find it nearly impossible to make even the simplest of decisions about the most mundane choices in life—which toothpaste to buy, for example: with so many brands and sizes and qualities and prices to consider, reason alone will leave you standing there in the store aisle, frozen in indecision. Analysis paralysis. An emotional leap of faith beyond reason is often required just to get through the day, let alone make the big decisions in life.

In the end, all of us are trying to make sense of the world, and nature has gifted us with a double-edged sword that cuts for and against. On one edge, our brains are the most complex and sophisticated information-processing machines in the universe, capable of understanding not only the universe itself but also the process of understanding. On the other edge, by the very same process of forming beliefs about the universe and ourselves, we are also more capable than any other species of self-deception and illusion, of fooling ourselves even while we are trying to avoid being fooled by nature.

In the end I want to believe. I also want to know. The truth is out there, and although it may be difficult to find, science is the best tool we have for uncovering it.

Ad astra per aspera![6]

Notes

Prologue: I Want to Believe

1. "Harris Poll Reveals What People Do and Do Not Believe," Harris, 2009, http://www.harrisinteractive.com/.
2. "Three in Four Americans Believe in Paranormal," Gallup, June 16, 2005, http://www.gallup.com/poll/16915/Three-Four-Americans-Believe-Paranormal.aspx. Similar percentages of belief were found in this 2005 Gallup Poll:

Psychic or spiritual healing	55 percent
Demon possession	42 percent
ESP	41 percent
Haunted houses	37 percent
Telepathy	31 percent
Clairvoyance (know past / predict future)	26 percent
Astrology	25 percent
Psychics are able to talk to the dead	21 percent
Reincarnation	20 percent
Channeling spirits from the other side	9 percent

3. "Paranormal Beliefs Come (Super)Naturally to Some," Gallup, November 1, 2005, http://www.gallup.com/poll/19558/Paranormal-Beliefs-Come-SuperNaturally-Some.aspx.
4. "Britons Report 'Psychic Powers,'" BBC News, May 26, 2006, http://news.bbc.co.uk/2/hi/uk_news/5017910.stm.
5. "Americans' Belief in Psychic Paranormal Phenomena Is Up Over Last Decade," Gallup News Service, June 8, 2001.
6. National Science Foundation, *Science Indicators Biennial Report*, 2002. The section on pseudoscience, "Science Fiction and Pseudoscience," is in chap. 7, "Science and Technology: Public Understanding and Public Attitudes," http://www.nsf.gov/statistics/seind02/c7/c7h.htm.
7. W. Richard Walker, Steven J. Hoekstra, and Rodney J. Vogl, "Science Education Is No Guarantee of Skepticism," *Skeptic* 9, no. 3 (2002): 24–25.
8. Stephen Hawking and Leonard Mlodinow, *The Grand Design* (New York: Bantam Books, 2010), 7.

Chapter 1: Mr. D'Arpino's Dilemma

1. The dialogues in this chapter are from an interview I recorded with Chick on Saturday, October 17, 2009, in person at my home in Altadena, California.
2. David L. Rosenhan, "On Being Sane in Insane Places," *Science* 179 (January 1973): 250–58.
3. The radio interview is on a cassette tape I've had for thirty-five years. Contrary to the expectations of the time that magnetic tape would not last more than two decades, it still sounds crystal clear.

Chapter 2: Dr. Collins's Conversion

1. Francis Collins, *The Language of God: A Scientist Presents Evidence for Belief* (New York: Free Press, 2007).
2. Interview conducted by phone on Friday, November 6, 2009.
3. The quote is inscribed on Kant's tomb and comes from his section on the moral law in his 1788 book *Critique of Practical Reason*: "Two things fill the mind with ever new and increasing admiration and awe, the oftener and more steadily we reflect on them: the starry heavens above me and the moral law within me. I do not merely conjecture them and seek them as though obscured in darkness or in the transcendent region beyond my horizon: I see them before me, and I associate them directly with the consciousness of my own existence." Accessible here: http://www.utsc.utoronto.ca/~sobel/Mystery_Glory/m_gStarry.pdf.
4. Quotes from Collins in this section are from *The Language of God*. Italicized quotes by Collins in the previous and subsequent sections are from my interview with him.

Chapter 3: A Skeptic's Journey

1. See Michael Shermer, *Why Darwin Matters: The Case Against Intelligent Design* (New York: Times Books, 2006). The central tenet of the book, most notably in the chapter on why conservatives and Christians should accept the theory of evolution, is that scientific theories describe the world as it really is, whereas religion describes the world as we would like to make it in terms of improving the human condition.
2. E-mail correspondence, November 22–23, 2009. That last qualifier is vintage Navarick humor. Interestingly, on the matter of internal states and mind, Navarick added: "However, like Skinner, I fully acknowledge the reality of private events ('conscious' experiences) that are directly sensed, like a toothache or internal speech. But I do not see these private events as adequate explanations of behavior."
3. See P. Edwards, "Socrates," in *Encyclopedia of Philosophy* (New York: Macmillan, 1967), 7:482.
4. "Books That Made a Difference in Readers' Lives," http://www.noblesoul.com/ore/books/rand/atlas/faq.html#Q6.4.
5. Brian Doherty, "She's Back," *Reason*, December 2009, http://reason.com/archives/2009/11/09/ayn-rand-is-back.
6. Jennifer Burns, *Goddess of the Market: Ayn Rand and the American Right* (New York: Oxford University Press, 2009), 286.
7. Nathaniel Branden, *Judgment Day: My Years with Ayn Rand* (Boston: Houghton Mifflin, 1989), 255–56.
8. Galambos never published his long-promised book in his lifetime, so my summary of his theory comes from my own extensive notes from the V-50 class and a series of three-by-five leaflets he printed called "Thrust for Freedom," numbered sequentially and presenting the definitions quoted here. In 1999, Galambos's estate issued

volume one of *Sic Itur Ad Astra (The Way to the Stars)*, a 942-page tome published by the Universal Scientific Publications Company Inc. Galambos's dream was to be a space entrepreneur and fly customers to the moon. In order to realize this dream he believed that space exploration had to be privatized, which meant that society itself, in its entirety, would have to be privatized.

9. As emblazoned in Latin on a plaque posted at the Panama Canal that also served as the institute's motto: *Aperire Terram Gentibus*.

10. Ludwig von Mises, *Human Action*, 3rd ed. (Chicago: Contemporary Books, 1966), 2.

11. These have never been published, and I have no intention of ever publishing them.

12. Friedrich A. von Hayek, *The Road to Serfdom* (Chicago: University of Chicago Press, 1944); Hayek, *The Constitution of Liberty* (Chicago: University of Chicago Press, 1960); Henry Hazlitt, *Economics in One Lesson* (New York: Harper and Brothers, 1946); Milton Friedman, *Free to Choose: A Personal Statement* (New York: Harcourt Brace Jovanovich, 1980).

13. Mises, *Human Action*, 860.

14. Freeman Dyson, "One in a Million," a review of *Debunked! ESP, Telekinesis, and Other Pseudoscience*, by Georges Charpak and Henri Broch, trans. Bart K. Holland, *New York Review of Books* 51, no. 5 (March 25, 2004).

15. I am here paraphrasing a line used by the comedian Bill Maher in his film *Religulous*, who is much funnier than I am when he makes that argument.

Chapter 4: Patternicity

1. Kevin R. Foster and Hanna Kokko, "The Evolution of Superstitious and Superstition-Like Behaviour," *Proceedings of the Royal Society B* 276, no. 1654 (2009): 31–37.

2. William D. Hamilton, "The Evolution of Altruistic Behavior," *American Naturalist* 97 (1963): 354–56; Hamilton, "The Genetical Evolution of Social Behavior," *Journal of Theoretical Biology* 7, no. 1 (1964): 1–52.

3. Michael Shermer, *The Science of Good and Evil* (New York: Times Books, 2003); Shermer, *The Mind of the Market* (New York: Times Books, 2008).

4. Foster and Kokko begin with a slightly different formula than mine—pb > c—where a belief may be held when the probability (p) of the benefit (b) is greater than the cost (c). For example, believing that the rustle in the grass is a dangerous predator when it is only the wind doesn't cost much, but believing that a dangerous predator is the wind may cost an animal its life. As Foster and Kokko note, we are very poor at estimating such probabilities (p). Since the cost (c) of believing that the rustle in the grass is a dangerous predator when it is just the wind is relatively low compared to the opposite, there would have been a beneficial (b) selection for believing that most patterns are real.

5. B. F. Skinner, "Superstition in the Pigeon," *Journal of Experimental Psychology* 38 (1948): 168–72.

6. Koichi Ono, "Superstitious Behavior in Humans," *Journal of the Experimental Analysis of Behavior* 47 (1987): 261–71.

7. Charles Catania and David Cutts, "Experimental Control of Superstitious Responding in Humans," *Journal of the Experimental Analysis of Behavior* 6, no. 2 (1963): 203–8.

8. Konrad Lorenz, *On Aggression*, trans. Marjorie Kerr Wilson (New York: Harcourt, Brace and World, 1966).

9. Edvard A. Westermarck, *The History of Human Marriage*, 5th ed. (London: Macmillan, 1921); Steven Pinker, *How the Mind Works* (New York: W. W. Norton, 1997).

10. Niko Tinbergen, *The Study of Instinct* (New York: Oxford University Press, 1951).

11. Vincent de Gardelle and Sid Kouider, "How Spatial Frequencies and Visual Awareness Interact During Face Processing," *Psychological Science*, November 2009, 1–9, http://pss.sagepub.com/content/early/2009/11/11/0956797609354064.full.pdf+html.
    For a slightly dissenting view in which facial recognition did not appear to be processed holistically, see this recent study: Yaroslav Konar, Patrick J. Bennett, and Allison B. Sekuler, "Holistic Processing Is Not Correlated with Face-Identification Accuracy," *Psychological Science*, December 2009, http://pss.sagepub.com/content/early/2009/12/16/0956797609356508.full.
    An article published just before this book went to press argues that the bizarreness of the reversed features is due to differential lighting, whether from the top down or bottom up, which would result in the reversed features showing a different shading than the rest of the face. And yet the effect is still evident in the Obama example I present here. See Zenobia Talati, Gillian Rhodes, and Linda Jeffrey, "Now You See It, Now You Don't: Shedding Light on the Thatcher Illusion," *Psychological Science*, January 2010, http://pss.sagepub.com/content/early/2010/01/08/0956797609357854.full.
12. Benjamin Libet, "Unconscious Cerebral Initiative and the Role of Conscious Will in Voluntary Action," *Behavior and Brain Sciences* 8 (1985): 529–66.
13. Irenäus Eibl-Eibesfeldt, *Ethology: The Biology of Behavior*, trans. Erich Kinghammer (New York: Holt, Rinehart and Winston, 1970).
14. Paul Ekman, *Emotions Revealed: Recognizing Faces and Feelings to Improve Communication and Emotional Life* (New York: Times Books, 2003).
15. S. Werner and H. Elke, "On the Function of Warning Coloration: A Black and Yellow Pattern Inhibits Prey-Attack by Naive Domestic Chicks," *Behavior Ecology and Sociobiology* 16 (1985): 249.
16. D. W. Pfennig, W. R. Harcombe, and K. S. Pfennig, "Frequency-Dependent Batesian Mimicry," *Nature* 410, no. 323 (March 15, 2001).
17. V. Sourjik and H. C. Berg, "Receptor Sensitivity in Bacterial Chemotaxis," *Proceedings of the National Academy of Science* 99, no. 1 (January 8, 2002): 123–27.
18. Niko Tinbergen, *Animal Behavior* (New York: Time Inc., 1965).
19. Deirdre Barrett, *Supernormal Stimuli: How Primal Urges Overran Their Evolutionary Purpose* (New York: W. W. Norton, 2010).
20. Ibid., 41.
21. Ibid., 122.
22. R. V. Exline and L. C. Winter, "Affection Relations and Mutual Gaze in Dyads," in *Affect, Cognition, and Personality: Empirical Studies*, ed. Silvan S. Tonkin and Carroll E. Inyard (New York: Springer, 1965).
23. J. B. Rotter, "Generalized Expectancies for Internal Versus External Control of Reinforcement," *Psychological Monographs* 80, no. 1 (1966): 1–28.
24. G. N. Marshall et al., "The Five-Factor Model of Personality as a Framework for Personality-Health Research," *Journal of Personality and Social Psychology* 67, no. 2 (August 1994): 278–86; J. Tobacyk and G. Milford, "Belief in Paranormal Phenomena: Assessment Instrument Development and Implications for Personality Functioning," *Journal of Personality and Social Psychology* 44, no. 5 (May 1983): 1029–37.
25. Bronislaw Malinowski, *Magic, Science, and Religion* (New York: Doubleday, 1954), 139–40.
26. Michael Shermer, *Why People Believe Weird Things: Pseudoscience, Superstition, and Other Confusions of Our Times* (New York: W. H. Freeman, 1997), 295–96.
27. These studies are cited in Jennifer A. Whitson and Adam D. Galinsky, "Lacking Control Increases Illusory Pattern Perception," *Science* 322 (October 3, 2008): 115–17.

28. Susan Blackmore and Rachel Moore, "Seeing Things: Visual Recognition and Belief in the Paranormal," *European Journal of Parapsychology* 10 (1994): 91–103.
29. J. Musch and K. Ehrenberg, "Probability Misjudgment, Cognitive Ability, and Belief in the Paranormal," *British Journal of Psychology* 93, no. 2 (May 2002): 169–77; Peter Brugger, Theodor Landis, and Marianne Regard, "A 'Sheep-Goat Effect' in Repetition Avoidance: Extra-Sensory Perception as an Effect of Subjective Probability?" *British Journal of Psychology* 81 (1990): 455–68.
30. Whitson and Galinsky, "Lacking Control Increases Illusory Pattern Perception."
31. Satoshi Kanazawa, "Outcome or Expectancy? Antecedent of Spontaneous Causal Attribution," *Personality and Social Psychology Bulletin* 18, no. 6 (1992): 659–68; B. Weiner, "'Spontaneous' Causal Thinking," *Psychological Bulletin* 97, no. 1 (1985): 74–84; H. H. Kelley, *Attribution in Social Interaction* (Morristown, N.J.: General Learning Press, 1971).
32. D. L. Hamilton and S. J. Sherman, "Perceiving Persons and Groups," *Psychological Review* 103, no. 2 (1996): 336–55.
33. This research, and many others like it, are nicely summarized in Ellen Langer's latest book, *Counterclockwise: Mindful Health and the Power of Possibility* (New York: Ballantine Books, 2009).
34. Association for the Treatment and Training in the Attachment of Children, http://www.ATTACh.org/.
35. Jean Mercer, Larry Sarner, and Linda Rosa, *Attachment Therapy on Trial: The Torture and Death of Candace Newmaker* (New York: Praeger, 2003). See also the Web site for Advocates for Children in Therapy, http://www.ChildrenInTherapy.org/.

Chapter 5: Agenticity

1. The concept of *agenticity* is derived, in part, from what the philosopher Daniel Dennett calls the *intentional stance*, whereby we predict the actions of others based on what we believe about their intention, although I take it much further. Dennett explains the concept this way: "First you decide to treat the object whose behavior is to be predicted as a rational agent; then you figure out what beliefs that agent ought to have, given its place in the world and its purpose. Then you figure out what desires it ought to have, on the same considerations, and finally you predict that this rational agent will act to further its goals in the light of its beliefs. A little practical reasoning from the chosen set of beliefs and desires will in most instances yield a decision about what the agent ought to do; that is what you predict the agent will do." Daniel Dennett, *The Intentional Stance* (Cambridge, Mass.: MIT Press, 1987).
2. I first introduced the concept of agenticity in my June 2009 column in *Scientific American*.
3. Bruce M. Hood, *Supersense: Why We Believe in the Unbelievable* (New York: Harper Collins, 2009), x.
4. Ibid., 183.
5. Ibid., 213.
6. Ibid., 214.
7. Ibid., 247–48.
8. Michael A. Persinger, *Neuropsychological Bases of God Beliefs* (New York: Praeger, 1987).
9. The show originally aired in 2000–2001. Clips from the series can be accessed on YouTube, keyword *Michael Shermer*.

10. The television segment on Michael Persinger and my participation in his experiment can be viewed at http://www.youtube.com/watch?v=nCVzz96zKA0.

11. Jon Ronson, *The Men Who Stare at Goats* (London: Picador, 2004).

12. You can see this and other striking visual and auditory illusions in my TED talk posted at http://www.skeptic.com/ under "Skepticism 101." There are entire Web pages dedicated to finding reverse lyrics and words in songs and speeches, for example, http://www.reversespeech.com/.

13. Such auditory priming and illusions have been studied scientifically by University of California–San Diego psychologist Diana Deutsch. For example, a repetitive tape loop of a two-syllable word educes different words and phrases in different people's minds, often depending on what they are thinking about at the moment they hear the repeated syllables. Diana Deutsch, "Musical Illusions," in *Encyclopedia of Neuroscience*, ed. Larry R. Squire (Boston: Elsevier, 2009), 5:1159–67.

14. Peter Suedfeld and Jane S. P. Mocellin, "The Sensed Presence in Unusual Environments," *Environment and Behavior* 19, no. 1 (January 1987): 33–52.

15. The complete poem and explanatory notes are available at http://www.bartleby.com/201/1.html.

16. John Geiger, *The Third Man Factor: The Secret of Survival in Extreme Environments* (New York: Weinstein Books, 2009).

17. Quoted in ibid., 84–85. Originally recounted in Charles A. Lindbergh, "33 Hours to Paris," *Saturday Evening Post*, June 6, 1953; and Lindbergh, *The Spirit of St. Louis* (New York: Charles Scribner's Sons, 1953).

18. Reinhold Messner and Horst Höfler, *Hermann Buhl: Climbing Without Compromise* (Seattle: The Mountaineers, 2000), 150.

19. Quoted in Geiger, *Third Man Factor*, 175–76.

20. William Laird McKinlay, *The Last Voyage of the Karluk: A Survivor's Memoir of Arctic Disaster* (New York: St. Martin's Press, 1976), 57.

21. James Allan Cheyne, "Sensed Presences in Extreme Contexts: A Review of *The Third Man Factor*," *Skeptic* 15, no. 2 (2009): 68–71.

22. The final ranking was as follows: (8) Hawaii Ironman Triathlon, (7) Badwater Ultramarathon 146-Mile Cross-Country Run, (6) La Traversée Internationale (25-mile swim), (5) Raid Gauloises Wilderness Competition, (4) U.S. Army's Best Ranger Competition, (3) Iditarod sled dog race, (2) Vendée Globe around-the-world sailing race, and (1) Race Across America.

23. I document these experiences, and many others, in Michael Shermer, *Race Across America: The Agonies and Glories of the World's Longest and Cruelest Bicycle Race* (Waco, Tex.: WRS Publishing, 1993).

24. Quoted in Daniel Coyle, "That Which Does Not Kill Me Makes Me Stranger," *New York Times*, February 5, 2006, http://www.nytimes.com/2006/02/05/sports/playmagazine/05robicpm.html.

25. Ryan Hudson, "The Iditarod, More Hallucinations Than Burning Man," SB Nation, March 16, 2010, http://www.sbnation.com/2010/3/16/1376103/iditarod-hallucination-2010-lance-mackey-newton-marshall.

26. Lew Freedman, *Anchorage Daily News*, March 19, 1993, quoted at http://www.helpsleddogs.org/remarks-mushersmistreatingdogs.htm#hallucinate.

27. Samuel M. McClure, David I. Laibson, George Loewenstein, and Jonathan D. Cohen, "Separate Neural Systems Value Immediate and Delayed Monetary Rewards," *Science* 306, no. 5695 (October 15, 2004): 503–7.

28. Antonio R. Damasio, *Descartes' Error: Emotion, Reason, and the Human Brain* (New York: Putnam, 1994); Ellen Peters and Paul Slovic, "The Springs of Action: Affective

and Analytical Information Processing in Choice," *Personality and Social Psychological Bulletin* 26, no. 12 (December 2000): 1465–75; Jon Elster, *Ulysses and the Sirens: Studies in Rationality and Irrationality* (New York: Cambridge University Press, 1979); Roy F. Baumeister, Todd F. Heatherton, and Dianne M. Tice, *Losing Control: How and Why People Fail at Self-Regulation* (San Diego: Academic Press, 1994); George Loewenstein, "Out of Control: Visceral Influences on Behavior," *Organizational Behavior and Human Decision Processes* 65, no. 3 (March 1996): 272–92; George F. Loewenstein and Jennifer Lerner, "The Role of Affect in Decision Making," in *Handbook of Affective Sciences*, ed. R. J. Davidson, K. R. Scherer, and H. H. Goldsmith (New York: Oxford University Press, 2003), 619–42.

29. Andy Clark, *Supersizing the Mind: Embodiment, Action, and Cognitive Extension* (New York: Oxford University Press, 2008).

30. Peter Brugger and Christine Mohr, "Out of the Body, but Not Out of Mind," *Cortex* 45 (2009): 137–40.

31. A. Newberg, E. D'Aquili, and V. Rause, *Why God Won't Go Away* (New York: Ballantine Books, 2001).

32. V. S. Ramachandran and Eric L. Altschuler, "The Use of Visual Feedback, in Particular Mirror Visual Feedback, in Restoring Brain Function," *Brain* 132, no. 7 (2009): 1693–1710.

33. Rama's TED talk about this research can be viewed here: http://www.ted.com/talks/vilayanur_ramachandran_on_your_mind.html.

34. Michael Gazzaniga, *The Ethical Brain* (New York: Dana Press, 2005), 150.

Chapter 6: The Believing Neuron

1. Richard Dawkins, *The Ancestor's Tale: A Pilgrimage to the Dawn of Evolution* (New York: Houghton Mifflin, 2004), 551–52.

2. There are many excellent books on neuroscience. Two recent ones I consult often are Joseph LeDoux, *Synaptic Self: How Our Brains Become Who We Are* (New York: Viking, 2002); and Christof Koch, *The Quest for Consciousness: A Neurobiological Approach* (Denver: Roberts and Company, 2004).

3. Gabriel Kreiman, Itzhak Fried, and Christof Koch, "Single Neuron Correlates of Subjective Vision in the Human Medial Temporal Lobe," *Proceedings of the National Academy of Sciences USA* 99, no. 12 (June 11, 2002): 8378–83.

4. James Olds and Peter Milner, "Positive Reinforcement Produced by Electrical Stimulation of Septal Area and Other Regions of Rat Brain," *Journal of Comparative and Physiological Psychology* 47 (1954): 419–27.

5. M. E. Olds and J. L. Fobes, "The Central Basis of Motivation: Intracranial Self-Stimulation Studies," *Annual Review of Psychology* 32 (January 1981): 523–74; M. P. Bishop, S. T. Elder, and R. G. Heath, "Intracranial Self-Stimulation in Man," *Science* 140, no. 3565 (April 26, 1963): 394–96.

6. Morten Kringelbach and Kent C. Berridge, eds., *Pleasures of the Brain* (New York: Oxford University Press, 2010).

7. Personal correspondence, January 10, 2010.

8. Peter Brugger and Christine Mohr, "The Paranormal Mind: How the Study of Anomalous Experiences and Beliefs May Inform Cognitive Neuroscience," *Cortex* 44, no. 10 (November/December 2008): 1291–98.

9. P. Reed, D. Wakefield, J. Harris, J. Parry, M. Cella, and E. Tsakanikos, "Seeing Non-Existent Events: Effects of Environmental Conditions, Schizotypal Symptoms, and Sub-Clinical Characteristics," *Journal of Behavior Therapy and Experimental Psychiatry* 39, no. 3 (September 2008): 276–91.

10. Christine Mohr, Theodor Landis, and Peter Brugger, "Lateralized Semantic Priming: Modulation by Levodopa, Semantic Distance, and Participants' Magical Beliefs," *Neuropsychiatric Disease and Treatment* 2, no. 1 (March 2006): 71–84.
11. Peter Krummenacher, Christine Mohr, Helene Haker, and Peter Brugger, "Dopamine, Paranormal Belief, and the Detection of Meaningful Stimuli," *Journal of Cognitive Neuroscience* 22, no. 8 (August 2010): 1–12.
12. J. K. Seamans and C. R. Yang, "The Principal Features and Mechanisms of Dopamine Modulation in the Prefrontal Cortex," *Progress in Neurobiology* 74, no. 1 (September 2004): 1–58.
13. Carl Sagan, *The Dragons of Eden: Speculations on the Evolution of Human Intelligence* (New York: Ballantine Books, 1977).
14. P. Brugger, A. Gamma, R. Muri, M. Schäfer, and K. I. Taylor, "Functional Hemispheric Asymmetry and Belief in ESP: Towards a 'Neuropsychology of Belief,'" *Perceptual and Motor Skills* 77, no. 3 (December 1993): 1299–308.
15. Ibid., 1299.
16. Personal correspondence, January 13, 2010. See also Andrea Marie Kuszewski, "The Genetics of Creativity: A Serendipitous Assemblage of Madness" (METODO Working Papers, no. 58, 2009), http://ssrn.com/abstract=1393603.
17. Anna Abraham, Sabine Windmann, Irene Daum, and Onur Güntürkün, "Conceptual Expansion and Creative Imagery as a Function of Psychoticism," *Consciousness and Cognition* 14, no. 3 (September 2005): 520–34.
18. Personal correspondence, January 13, 2010.
19. Ibid.
20. Kary Mullis, *Dancing Naked in the Mind Field* (New York: Random House, 1998), 5.
21. As I was finishing up this chapter I saw Kary at the TED 2010 conference and asked his permission to include our exchange, which he kindly granted, adding that my skepticism hadn't fazed his confidence in his beliefs in the least!
22. Michael Shermer, *In Darwin's Shadow: The Life and Science of Alfred Russel Wallace* (New York: Oxford University Press, 2002).
23. The historian of science Richard Milner offers this insight from Wallace that he applied to Mullis: "Alfred Russel Wallace, the great Victorian naturalist and evolutionist, wrote in his 1874 *Defense of Spiritualism* that indeed 'the pure dry air of California' was known to produce 'powerful and . . . startling manifestations.'" See Richard Milner, *Darwin's Universe: Evolution from A to Z* (Berkeley: University of California Press, 2009), 309–10. Of course, brain pattern filters have to operate in an environment, and as a native Californian firmly ensconced in L.A., I can attest that this is, indeed, La La Land.
24. M. I. Posner and G. J. DiGirolamo, "Executive Attention: Conflict, Target Detection, and Cognitive Control," in *The Attentive Brain*, ed. Raja Parasuraman (Cambridge, Mass.: MIT Press, 1998).
25. C. S. Carter, T. S. Braver, D. M. Barch, M. M. Botvinick, D. Noll, and J. D. Cohen, "Anterior Cingulate Cortex, Error Detection, and the Online Monitoring of Performance," *Science* 280, no. 5364 (1998): 747–49.
26. Daniel H. Mathalon, Kasper W. Jorgensen, Brian J. Roacha, and Judith M. Forda, "Error Detection Failures in Schizophrenia," *International Journal of Psychophysiology* 73, no. 2 (August 2009): 109–17. Although their data showed decreased error detection in schizophrenics compared to healthy subjects, they did not find decreased activity in the ACC of the schizophrenic patients. Some neuroscientists believe that the ACC is involved in a lot of cognition, not just error detection. See M. F. Rushworth, M. E. Walton, S. W. Kennerley, and D. M. Bannerman, "Action Sets

and Decisions in the Medial Frontal Cortex," *Trends in Cognitive Science* 8, no. 9 (September 2004): 410–17; M. F. Rushworth, T. E. Behrens, P. H. Rudebeck, and M. E. Walton, "Contrasting Roles for Cingulate and Orbitofrontal Cortex in Decisions and Social Behaviour," *Trends in Cognitive Sciences* 11, no. 4 (April 2007): 168–76.

27. "Natural-Born Dualists: A Talk with Paul Bloom," Edge Foundation Inc., May 13, 2004, http://www.edge.org/3rd_culture/bloom04/bloom04_index.html. See also Paul Bloom, *How Pleasure Works: The New Science of Why We Like What We Like* (New York: W. W. Norton, 2010).

28. Ibid. "Natural-Born Dualists."

29. Oliver Sacks, *The Man Who Mistook His Wife for a Hat and Other Clinical Tales* (New York: Summit Books, 1985).

30. Sacks describes this and other hallucinations and their causal explanations in his TED talk available here: http://www.ted.com/talks/oliver_sacks_what_hallucina tion_reveals_about_our_minds.html.

31. Ibid.

32. Helen L. Gallagher and Christopher D. Frith, "Functional Imaging of 'Theory of Mind,'" *Trends in Cognitive Sciences* 7, no. 2 (February 2003): 77–83.

33. Giacomo Rizzolatti, Luciano Fadiga, Vittorio Gallese, and Leonardo Fogassi, "Premotor Cortex and the Recognition of Motor Actions," *Cognitive Brain Research* 3, no. 2 (March 1996): 131–41.

34. L. Fogassi, P. F. Ferrari, B. Gesierich, S. Rozzi, F. Chersi, and G. Rizzolatti, "Parietal Lobe: From Action Organization to Intention Understanding," *Science* 308, no. 5722 (April 29, 2005): 662–67; V. Gallese, L. Fadiga, L. Fogassi, and G. Rizzolatti, "Action Recognition in the Premotor Cortex," *Brain* 119, no. 2 (1996): 593–609.

35. M. Iacoboni, R. P. Woods, M. Brass, H. Bekkering, J. C. Mazziotta, and G. Rizzolatti, "Cortical Mechanisms of Human Imitation," *Science* 286, no. 5449 (December 24, 1999): 2526–28; G. Rizzolatti and L. Craighero, "The Mirror-Neuron System," *Annual Review of Neuroscience* 27 (July 2004): 169–92.

It should be noted that the activity imaged in such fMRI studies is not the same as the recording of individual neurons in monkeys' brains. As University of Groningen psychologist Christian Keysers explained, "When we record signals from neurons in monkeys, we can really know that a single neuron is involved in both doing the task and seeing someone else do the task. With imaging, you know that within a little box about three millimeters by three millimeters by three millimeters, you have activation from both doing and seeing. But this little box contains millions of neurons, so you cannot know for sure that they are the same neurons—perhaps they're just neighbors." See Lea Winerman, "The Mind's Mirror," *Monitor on Psychology* 36, no. 9 (October 2005): 48, http://www.apa.org/monitor/oct05/mirror.html.

36. Vittorio Gallese and Alvin Goldman, "Mirror Neurons and the Simulation Theory of Mind-Reading," *Trends in Cognitive Sciences* 2, no. 12 (December 1998): 493–501.

37. L. Fogassi, P. F. Ferrari, B. Gesierich, S. Rozzi, F. Chersi, and G. Rizzolatti, "Parietal Lobe: From Action Organization to Intention Understanding," *Science* 308 (2005): 662–67.

38. Sam Harris, Sameer A. Sheth, and Marks S. Cohen, "Functional Neuroimaging of Belief, Disbelief, and Uncertainty," *Annals of Neurology* 63 (2007): 141–47.

39. Sam Harris, Jonas Kaplan, Ashley Curiel, Susan Bookheimer, Marco Iacoboni, and Mark Cohen, "The Neural Correlates of Religious and Nonreligious Belief," *PloS One* 4, no. 10 (2009): e0007272.

40. Personal correspondence, December 23, 2009.

41. I should note that there are good reasons to be cautious when extrapolating from brain scanning research studies such as those discussed in this book, most notably those employing fMRI scans, for a number of reasons, including five that I outlined in an article for *Scientific American Mind* (Michael Shermer, "Why You Should Be Skeptical of Brain Scans," *Scientific American Mind*, October/November 2008, 67–71): (1) The MRI tube is a very unnatural environment; it is claustrophobic, with the subjects' heads locked into place to prevent movement; (2) scans measure blood flow change, not neural activity, and there is a delayed reaction from neuronal firing to blood rushing to the area; (3) the brain scan colors are artificial and exaggerate the differences between activity in one area and surrounding areas, which are often very subtle; (4) brain scan findings are statistical compilations across many subjects, not any one person's brain; and (5) brain areas activate for various reasons. The neuroscientist Russell Poldrack told me: "It is tempting to look at one of those spots and say 'This is where X happens in your brain,' when in fact that area could be lighting up when involved in all sorts of tasks. Take the right prefrontal cortex that lights up when you do almost any difficult task. One way to think about it is in terms of networks, not modules. When you are engaged in thinking about money, there's a network of several different areas involved in communicating with each other in a particular way. So, the prefrontal cortex may be involved in a lot of different tasks. But in communication with specific other brain networks, it becomes active when engaged in one particular task, such as thinking about money."

Chapter 7: Belief in the Afterlife

1. Eric Lax, *On Being Funny: Woody Allen and Comedy* (New York: Charterhouse, 1975), 208.
2. Quoted in Garrison Keillor, *A Prairie Home Companion Pretty Good Joke Book* (New York: Highbridge Co., 2001), 13.
3. "Harris Poll Reveals What People Do and Do Not Believe," Harris 2009, http://www .harrisinteractive.com. These results confirm those of a 2007 Pew Forum survey showing that 74 percent of Americans believe in heaven with Mormons the largest cohort at 95 percent, black Protestant churchgoers at 91 percent, white Evangelicals at 86 percent, and Muslims (with or without seventy-two virgins) at 85 percent. On the other end of the belief spectrum, not counting atheists, agnostics, and secularists, only 51 percent of Hindus, 46 percent of Jehovah's Witnesses, 38 percent of Jews, and 36 percent of Buddhists believe that they will live on, not just in their apartments (pace Woody Allen), but in some ethereal place beyond their bodies. Tellingly, across the board only 59 percent believe in hell, demonstrating once again the power of wishful thinking. U.S. Religions Landscape Survey, "Summary of Key Findings," Pew Forum on Religion & Public Life, http://religions.pewforum.org/ pdf/report2religious-landscape-study-key-findings.pdf (N=35,000). The oddest finding in the Pew survey was that 12 percent of atheists and 18 percent of agnostics said that they believe in heaven and—consistent with the wishful thinking self-serving bias—there were lower percentages for belief in hell (10 percent for atheists, 12 percent for agnostics)! Hope springs eternal.
4. Helen L. Gallagher and Christopher D. Frith, "Functional Imaging of 'Theory of Mind,'" *Trends in Cognitive Sciences* 7, no. 2 (February 2003): 77.
5. Two recent books that use these lines of evidence are Deepak Chopra, *Life After Death: The Burden of Proof* (New York: Harmony Books, 2006); and Dinesh D'Souza, *Life After Death: The Evidence* (Washington, D.C.: Regnery Press, 2009).
6. Rupert Sheldrake, *A New Science of Life: The Hypothesis of Formative Causation*

(Los Angeles: J. P. Tarcher, 1981); Sheldrake, *The Presence of the Past: Morphic Resonance and the Habits of Nature* (New York: Harper Collins, 1988).

7. Rupert Sheldrake, *Seven Experiments that Could Change the World: A Do-It-Yourself Guide to Revolutionary Science* (New York: Riverhead Books, 1995).

8. Rupert Sheldrake, *The Sense of Being Stared At: And Other Aspects of the Extended Mind* (New York: Crown, 2003). See also Sheldrake's Web page experimental protocol, http://www.sheldrake.org/experiments/olt/start.html, and http://www.sheldrake .org/experiments/staring/staring_experiment.html.

   Sheldrake's papers on this research, giving the results of thousands of trials published in several journals, are also available in full text versions at http://www .sheldrake.org.

9. http://www.csicop.org/si/show/psychic_staring_effect_an_artifact_of_pseudo_ randomization/.

10. Richard Wiseman and Marilyn Schlitz, "Experimenter Effects and the Remote Detection of Staring," *Journal of Parapsychology* 61 (1997): 197–207.

11. The following ratings were made by me from the *Journal of Consciousness Studies* 12, no. 6 (2005), an open peer commentary on "Sheldrake and His Critics: The Sense of Being Glared At." Sheldrake provided two target articles on which fourteen peers commented. Sheldrake was then given the last word with a concluding article. Commentators, affiliations, and my ratings of their response to Sheldrake's target articles follow. Scale: 1 to 5 (1, critical; 2, mildly critical; 3, neutral; 4, mildly supportive; 5, supportive).

   Anthony Atkinson, lecturer in psychology, Durham University: 1
   Ian Baker, postgraduate researcher, Koestler Parapsychology Unit, Edinburgh: 4
   Susan Blackmore, visiting lecturer, psychology, University of West England: 1
   William Braud, professor, Global Programs, Institute of Transpersonal
      Psychology: 5
   Jean Burns, physicist, founding editor of the *Journal of Consciousness Studies*: 2
   Roger Carpenter, reader in oculomotor physiology, University of Cambridge: 1
   Chris Clarke, visiting professor of applied mathematics, University of
      Southampton: 3
   Ralph Ellis, professor of philosophy, Clark Atlanta University: 1
   David Fontana, visiting professor of transpersonal psychology, John Moores
      University: 5
   Christopher French, professor of psychology, University of London: 2
   Dean Radin, Institute of Noetic Sciences, president, Parapsychological
      Association: 5
   Marilyn Schlitz, director of research, Institute of Noetic Sciences: 4
   Stefan Schmidt, Institute of Environmental Medicine, University
      Hospital Freiburg: 2
   Max Velmans, professor of psychology, University of London: 3

12. Rupert Sheldrake, "Research on the Feeling of Being Stared At," *Skeptical Inquirer*, March/April 2000, 58–61.

13. Daryl J. Bem and Charles Honorton, "Does Psi Exist? Replicable Evidence for an Anomalous Process of Information Transfer," *Psychological Bulletin* 115 (1994): 4–18.

14. Ray Hyman, "Anomaly or Artifact? Comments on Bem and Honorton," *Psychological Bulletin* 115 (1994): 19–24.

15. Julie Milton and Richard Wiseman, "Does Psi Exist? Lack of Replication of an Anomalous Process of Information Transfer," *Psychological Bulletin* 125, no. 4 (July 1999): 387–91.

16. Daryl J. Bem, "Response to Hyman," *Psychological Bulletin* 15, no. 1 (1994): 25–27.

17. Provided at Hameroff's Web page: http://www.quantumconsciousness.org/.

18. Information about the film may be accessed at http://www.whatthebleep.com/.

19. I heard Gell-Mann use this term in the 1980s after he delivered a lecture by this title at Caltech, and it has caught on ever since. Because he earned his Nobel Prize in quantum physics he is eminently qualified to so judge claims made on its behalf.

20. Stuart Hameroff and Roger Penrose, "Orchestrated Reduction of Quantum Coherence in Brain Microtubules: A Model for Consciousness," in *Toward a Science of Consciousness—The First Tucson Discussions and Debates*, ed. S. R. Hameroff, A. W. Kaszniak, and A. C. Scott (Cambridge, Mass.: MIT Press, 1996), 507–40.

21. Victor Stenger, *The Unconscious Quantum: Metaphysics in Modern Physics and Cosmology* (Buffalo, N.Y.: Prometheus Books, 1995).

22. J. E. Whinnery and A. M. Whinnery, "Acceleration-Induced Loss of Consciousness: A Review of 500 Episodes," *Archives of Neurology* 47 (1990): 764–76.

23. K. Augustine, "Near-Death Experiences with Hallucinatory Features," *Journal of Near-Death Studies* 26, no. 1 (2007): 3–31.

24. James E. Whinnery, "Psychophysiologic Correlates of Unconsciousness and Near-Death Experiences," *Journal of Near-Death Studies* 15, no. 4 (1997): 231–58.

25. J. E. Whinnery, "Technique for Simulating G-Induced Tunnel Vision," *Aviation and Space Environmental Medicine* 50 (1979): 1076.

26. David E. Comings, *Did Man Create God? Is Your Spiritual Brain at Peace with Your Thinking Brain?* (Duarte, Calif.: Hope Press, 2008).

27. O. Blanke, S. Ortigue, T. Landis, and M. Seeck, "Neuropsychology: Stimulating Illusory Own-body Perceptions," *Nature* 419 (September 19, 2002): 269–70.

28. Newberg, Aquili, and Rause, *Why God Won't Go Away*.

29. Cosimo Urgesi, Salvatore M. Aglioti, Miran Skrap, and Franco Fabbro, "The Spiritual Brain: Selective Cortical Lesions Modulate Human Self-Transcendence," *Neuron* 65, no. 3 (2010): 309–19.

30. P. V. Lommel, R. V. Wees, V. Meyers, and I. Elfferich, "Near-Death Experience in Survivors of Cardiac Arrest: A Prospective Study in the Netherlands," *Lancet* 358, no. 9298 (2001): 2039.

31. Mark Crisplin, "Near-Death Experiences and the Medical Literature," *Skeptic* 14, no. 2 (2008): 14–15.

32. Marlene Dobkin de Rios, *Hallucinogens: Cross-cultural Perspective* (Albuquerque: University of New Mexico Press, 1984).

33. Richard Strassman, *DMT: The Spirit Molecule* (Rochester, Vt.: Park Street Press, 2001).

34. Comings, *Did Man Create God?*, 384–85.

35. For a general discussion of brain-generated psychological states and experiences, see Antonio Damasio, *The Feeling of What Happens: Body, Emotions, and the Making of Consciousness* (London: Vintage, 2000).

36. If PBS's Charlie Rose and his hour-long one-on-one interview style on a minimalist set is on one end of the interview spectrum, and Jerry Springer's circus sideshow is on the other end, Larry King hovers somewhere in the middle ground between salacity and solemnity.

37. All quotes in this section are from the complete transcript of the show available at http://transcripts.cnn.com/TRANSCRIPTS/0912/22/lkl.01.html.

38. Chopra, *Life After Life*, 222–23.
39. Ibid., 223.

Chapter 8: Belief in God

1. D. B. Barrett, G. T. Kurian, and T. M. Johnson, eds., *World Christian Encyclopedia: A Comparative Survey of Churches and Religions in the Modern World*, 2 vols. (New York: Oxford University Press, 2001).
2. U.S. Religions Landscape Survey, "Summary of Key Findings."
3. Charles Darwin, *The Descent of Man* (London: John Murray, 1871), 2:395.
4. Ibid., 1:163.
5. Ibid., 1:166.
6. Michael Shermer, *How We Believe* (New York: Times Books, 1999).
7. Donald E. Brown, *Human Universals* (New York: McGraw-Hill, 1991).
8. Chris Boehm, "Egalitarian Society and Reverse Dominance Hierarchy," *Current Anthropology* 34 (1993): 227–54; Boehm, *Hierarchy in the Forest: Egalitarianism and the Evolution of Human Altruism* (Cambridge, Mass.: Harvard University Press, 1999).
9. N. G. Waller, B. Kojetin, T. Bouchard, D. Lykken, and A. Tellegen, "Genetic and Environmental Influences on Religious Attitudes and Values: A Study of Twins Reared Apart and Together," *Psychological Science* 1, no. 2 (1990): 138–42.
10. N. G. Martin, L. J. Eaves, A. C. Heath, R. Jardine, L. M. Feingold, and H. J. Eysenck, "Transmission of Social Attitudes," *Proceedings of the National Academy of Sciences USA* 83 (1986): 4364–68.
11. L. J. Eaves, H. J. Eysenck, and N. G. Martin, *Genes, Culture and Personality: An Empirical Approach* (London: Academic Press, 1989), 385.
12. David E. Comings et al., "The DRD4 Gene and Spiritual Transcendence Scale of the Character Temperament Index," *Psychiatric Genetics* 10 (2001): 185–89.
13. Dean Hamer, *Living with Our Genes: Why They Matter More Than You Think* (New York: Anchor, 1999).
14. Dean Hamer, *The God Gene: How Faith Is Hardwired into Our Genes* (New York: Anchor, 2005).
15. Scholarly research on religion began in earnest in the late nineteenth century when anthropologists such as Edward Tylor and James Frazer argued that religious belief is an extension of primitive animism and superstitious magic. The psychologist Sigmund Freud viewed it as an obsessional neurosis, or an illusion of the mind. Sociologist Émile Durkheim claimed that religion represents the sacred part of the social structure, in contrast with Karl Marx's theory that it is largely a tool of alienation and an opiate of the masses. The historian of religion Mircea Eliade thought religion to be the most sacred part of the human psyche, while anthropologist E. E. Evans-Pritchard saw religion as society's "construct of the heart," which it needs as much as science's "construct of the mind." Anthropologist Clifford Geertz believed that religion is a cultural system of symbols that act to empower, give meaning, and provide motivation, whereas the renowned sociologists of religion Rodney Stark and William Bainbridge have suggested that religion is a form of economic exchange for goods and services unavailable through secular sources. See Edward B. Tylor, *Primitive Culture: Researches into the Development of Mythology, Philosophy, Religion, Language, Art, and Custom* (London: John Murray, 1871); James G. Frazer, *The Golden Bough: A Study in Magic and Religion* (New York: Macmillan, 1924); Sigmund Freud, *The Future of an Illusion*, trans. J. Strachey (New York: Norton, 1927, 1961); Émile Durkheim, *Elementary Forms of the Religious Life*, trans. J. W. Swain

(New York: Collier Books, 1912, 1961); Karl Marx, *The Marx-Engels Reader*, ed. R. C. Tucker (New York: W. W. Norton, 1869, 1978); Mircea Eliade, *The Sacred and the Profane: The Nature of Religion*, trans. W. R. Trask (New York: Harcourt Brace, 1957); E. E. Evans-Pritchard, *Theories of Primitive Religion* (Oxford, U.K.: Clarendon Press, 1965); Clifford Geertz, "Religion as a Cultural System," in *Anthropological Approaches to the Study of Religion*, ed. M. Banton (London: Tavistock Press, 1966); Rodney Stark and W. S. Bainbridge, *A Theory of Religion* (New Brunswick, N.J.: Rutgers University Press, 1987).

16. Thomas H. Huxley, *Collected Essays* (New York: D. Appleton and Co., 1894), 5: 237–38.

17. Arthur C. Clarke's third law states "Any sufficiently advanced technology is indistinguishable from magic." Clarke's first law: "When a distinguished but elderly scientist states that something is possible he is almost certainly right. When he states that something is impossible, he is very probably wrong." Clarke's second law: "The only way of discovering the limits of the possible is to venture a little way past them into the impossible." Clarke's first law was first published in "Hazards of Prophecy: The Failure of Imagination," an essay in his 1962 book *Profiles of the Future*. The second law was originally a derivative of the first and it became "Clarke's second law" later, after Clarke proposed the third law in a revised 1973 edition of *Profiles of the Future*. He said, "As three laws were good enough for Newton, I have modestly decided to stop there."

18. I first proposed Shermer's last law in Michael Shermer, "Shermer's Last Law," *Scientific American*, January 2002, 33. Since I do not believe in naming laws after oneself, as the good book warns: the last shall be first and the first shall be last.

19. Ray Kurzweil, *The Singularity Is Near* (New York: Penguin, 2006). See also http://singularity.com/.

20. Daniel G. Gibson et al., "Creation of a Bacterial Cell Controlled by a Chemically Synthesized Genome," *Science* 329, no. 5987 (July 2, 2010): 52–56.

21. Michio Kaku, *The Physics of the Impossible: A Scientific Exploration in the World of Phasers, Force Fields, Teleportation, and Time Travel* (New York: Anchor Books, 2009).

22. Michio Kaku, *Parallel Worlds: A Journey Through Creation, Higher Dimensions, and the Future of the Cosmos* (New York: Anchor Books, 2007).

23. Walter Isaacson, *Einstein: His Life and Universe* (New York: Simon and Schuster, 2007).

24. Ibid., 291.

25. For an excellent summary of Einstein's religious attitudes and belief in God, see Isaacson, *Einstein*, chap. 17, "Einstein's God."

26. Isaacson, *Einstein*, 386.

27. Ibid., 388.

28. Ibid., 335.

29. Michael Gilmore, "Einstein's God: Just What Did Einstein Believe About God?" *Skeptic* 5, no. 2 (1997): 62–64, http://www.theeway.com/skepticc/archives50.html.

30. Read the entire debate here: http://www.templeton.org/belief/debates.html#groopman.

Chapter 9: Belief in Aliens

1. You can read the book in its entirety at Joseph P. Firmage, *The Truth* (International Space Sciences Organization, 1999), http://www.bibliotecapleyades.net/ciencia/ciencia_thetruth.htm.

2. Jon Swartz, "CEO Quits Job Over UFO Views," *San Francisco Chronicle*, January 9, 1999, http://www.sfgate.com/cgi-bin/article.cgi?file=/chronicle/archive/1999/01/09/MN19158.DTL.

3. International Space Sciences Organization, http://orgs.tigweb.org/103.

4. Firmage, *Truth*, 237.

5. *The Truth*, condensed ed., pt. 4, UFOseek, http://www.ufoseek.org/part4.htm.

6. Ibid.

7. Swartz, "CEO Quits Job."

8. Firmage, *Truth*, pt. 2, "Teachers Have Taught Us."

9. Ibid., 229.

10. Carl Sagan, *The Demon-Haunted World: Science as a Candle in the Dark* (New York: Ballantine Books, 1996).

11. J. A. Cheyne, S. D. Rueffer, and I. R. Newby-Clark, "Hypnagogic and Hypnopompic Hallucinations During Sleep Paralysis: Neurological and Cultural Construction of the Nightmare," *Consciousness and Cognition* 8, no. 3 (1999): 319–37.

12. Richard J. McNally, Natasha B. Lasko, Susan A. Clancy, Michael L. Macklin, Roger K. Pitman, and Scott P. Orr, "Psychophysiological Responding During Script-Driven Imagery in People Reporting Abduction by Space Aliens," *Psychological Science* 15, no. 7 (2004): 493–97.

13. Richard McNally, *Remembering Trauma* (Cambridge, Mass.: Harvard University Press, 2003).

14. Susan A. Clancy, *Abducted: How People Come to Believe They Were Kidnapped by Aliens* (Cambridge, Mass.: Harvard University Press, 2005), 154.

15. Ibid., 150. See also Gregory L. Reece, *UFO Religion: Inside Flying Saucer Cults and Culture* (New York: Palgrave, 2007).

16. http://www.youtube.com/watch?v=X2_1DofIVqg.

17. For a highly readable and very entertaining account of people who search for aliens, from goofball fringers to hardcore scientists, see Joel Achenbach, *Captured by Aliens: The Search for Life and Truth in a Very Large Universe* (New York: Simon and Schuster, 1999).

18. The best single-volume summary covering all aspects of the question in a highly readable and yet scholarly treatment is Michael A. G. Michaud, *Contact with Alien Civilizations: Our Hopes and Fears About Encountering Extraterrestrials* (New York: Copernicus Books, 2007).

19. Stephen Webb, *If the Universe Is Teeming with Aliens . . . Where Is Everybody? Fifty Solutions to the Fermi Paradox and the Problem of Extraterrestrial Life* (New York: Copernicus Books, 2002).

20. You can watch the video here: http://www.youtube.com/watch?v=JKAXrmkx12g.

21. Personal correspondence, August 19, 2009.

22. This *progressivist bias*, in fact, is pervasive in nearly all evolutionary accounts and is directly challenged by counterfactual thinking. I once explained to my young daughter that polar bears are a good example of a transitional species between land and marine mammals, since they are well adapted for both land and marine environments. But this is not correct. Polar bears are not "becoming" marine mammals. They are not "transitioning" to anything. They are perfectly well adapted for doing just what they do. They may become marine mammals should, say, global warming melt the polar ice caps. Then again, they may just go extinct. In either case, there is no long-term drive for polar bears to progress to anything, since evolution creates immediate adaptations only for local environments. The same applies to our hominid ancestors.

23. Richard G. Klein, *The Human Career: Human Biological and Cultural Origins* (Chicago: University of Chicago Press), 367–493.
24. Richard Leakey, *The Origin of Humankind* (New York: BasicBooks, 1994), 134.
25. Klein, *Human Career*, 441–42.
26. Christopher Wills, *Children of Prometheus* (Reading, Mass.: Perseus Books, 1998), 143–45.
27. Shermer, *How We Believe.*
28. Klein, *Human Career*, 469.
29. Ian Tattersall, "Once We Were Not Alone," *Scientific American*, January 2000, 56–62.
30. Ian Tattersall, *The Fossil Trail: How We Know What We Think About Human Evolution* (New York: Oxford University Press, 1995), 212.
31. Leakey, *Origin of Humankind*, 132.
32. Ibid., 138.
33. Ibid., 20.
34. Tattersall, *Fossil Trail*, 246.
35. George Basalla, *Civilized Life in the Universe: Scientists on Intelligent Extraterrestrials* (New York: Cambridge University Press, 2006), 10–12.
36. Michael Shermer, *The Borderlands of Science: Where Sense Meets Nonsense* (New York: Oxford University Press, 2001).
37. David Swift, *SETI Pioneers: Scientists Talk About Their Search for Extraterrestrial Intelligence* (Tucson: University of Arizona Press, 1990), 57.
38. Frank Drake and Dava Sobel, *Is Anyone Out There? The Scientific Search for Extraterrestrial Intelligence* (New York: Delacorte, 1992), 160.
39. David Brin, "Shouting at the Cosmos ... Or How SETI Has Taken a Worrisome Turn into Dangerous Territory," 2006, http://www.davidbrin.com/.
40. Michael Crichton, "Aliens Cause Global Warming" (speech at the California Institute of Technology, January 17, 2003), http://www.crichton-official.com/.
41. Paul Davies, *Are We Alone? Philosophical Implications for the Discovery of Extraterrestrial Life* (New York: BasicBooks, 1995), 135.
42. Paul Davies, *The Eerie Silence: Renewing Our Search for Alien Intelligence* (New York: Houghton Mifflin, 2010), 192–93.
43. Swift, *SETI Pioneers*, 219.
44. Carl Sagan, *Contact* (New York: Pocket Books, 1986), 431.
45. Robert Plank, *The Emotional Significance of Imaginary Beings: A Study of the Interaction Between Psychopathology, Literature, and Reality in the Modern World* (Springfield, Ill.: Thomas, 1968).
46. Basalla, *Civilized Life*, 14.
47. Steven J. Dick, *Plurality of Worlds: The Origins of the Extraterrestrial Debate from Democritus to Kant* (New York: Cambridge University Press, 1982); Dick, *The Biological Universe: The Twentieth-Century Extraterrestrial Life Debate and the Limits of Science* (New York: Cambridge University Press, 1996).
48. Clancy, *Abducted*, 154.
49. Michael Shermer, "Deities for Atheists," *Science* 311 (March 3, 2006): 1244.
50. Personal correspondence, March 10, 2006.

Chapter 10: Belief in Conspiracies

1. Arthur Goldwag, *Cults, Conspiracies, and Secret Societies: The Straight Scoop on Freemasons, the Illuminati, Skull and Bones, Black Helicopters, the New World Order, and Many, Many More* (New York: Vintage Books, 2009).

2. Michael Shermer, *Denying History: Who Says the Holocaust Never Happened and Why Do They Say It?* (Berkeley: University of California Press, 2000).

3. Phil Molé, "9/11 Conspiracy Theories: The 9/11 Truth Movement in Perspective," eSkeptic, September 11, 2006, http://www.skeptic.com/eskeptic/06-09-11.

4. This claim is made by Jim Hoffman in his book *Waking Up from Our Nightmare: The 9/11/01 Crimes in New York City* (San Francisco: Irresistible/Revolutionary, 2004), and on his Web page, http://911research.wtc7.net/talks/towers/text/index.html.

5. Blanchard's entire analysis may be found on the Web site he edits: http://www.implosionworld.com.

6. The Web page World for 911 Truth, for example, dedicated an entire section to refuting me: http://world911truth.org/response-to-michael-shermer/.

7. My challenge to the 9/11 "truthers" and their response can be found here: http://trueslant.com/michaelshermer/2009/12/28/911-truthers-foiled-by-1225-attack/#comments.

Chapter 11: Politics of Belief

1. John T. Jost, Jack Glaser, Arie W. Kruglanski, and Frank J. Sulloway, "Political Conservatism as Motivated Social Cognition," *Psychological Bulletin* 129, no. 3 (2003): 339–75.

2. "Is Conservatism a Mild Form of Insanity?" *Psychology Today*, September 6, 2008, http://www.psychologytoday.com/blog/genius-and-madness/200809/is-political-conservatism-mild-form-insanity.

3. Julian Borger, "Study of Bush's Psyche Touches a Nerve," *Guardian*, August 13, 2003, http://www.guardian.co.uk/world/2003/aug/13/usa.redbox.

4. Jonathan Haidt, "What Makes People Vote Republican?" Edge Foundation Inc., September 9, 2008, http://www.edge.org/3rd_culture/haidt08/haidt08_index.html.

5. Arthur C. Brooks, *Who Really Cares? The Surprising Truth About Compassionate Conservatism* (New York: BasicBooks, 2007).

6. Daniel B. Klein and Charlotta Stern, "Professors and Their Politics: The Policy Views of Social Scientists," *Critical Review* 17, no. 3–4, 257–304.

7. Stanley Rothman, S. Robert Lichter, and Neil Nevitte, "Politics and Professional Advancement Among College Faculty," *Forum*, 2005, http://www.cmpa.com/documents/05.03.29.Forum.Survey.pdf.

8. John McGinnis, Matthew A. Schwartz, and Benjamin Tisdell, "The Patterns and Implications of Political Contributions by Elite Law School Faculty," *Georgetown Law Journal* 93 (2005): 1167–1212.

9. Tim Groseclose and Jeffrey Milyo, "A Measure of Media Bias," *Quarterly Journal of Economics*, November 2005, 1191–1237.

10. Donald Green, Bradley Palmquist, and Eric Schickler, *Partisan Hearts and Minds: Political Parties and the Social Identities of Voters* (New Haven: Yale University Press, 2002).

11. Jonathan Haidt, "The Moral Emotions," in *Handbook of Affective Sciences*, ed. R. J. Davidson, K. Scherer, and H. H. Goldschmidt (New York: Oxford University Press, 2003).

12. Jonathan Haidt, "The Emotional Dog and Its Rational Tail: A Social Intuitionist Approach to Moral Judgment," *Psychological Review* 108 (2001): 814–34.

13. Ibid. See also F. Cushman, L. Young, and M. Hauser, "The Role of Conscious Reasoning and Intuition in Moral Judgment: Testing Three Principles of Harm," *Psychological Science* 17, no. 12 (2006): 1082–89; and Moral Foundations Theory, http://www.moralfoundations.org/.

14. Ernst Fehr and Simon Gachter, "Altruistic Punishment in Humans," *Nature* 415 (2002): 137–40. See also R. Boyd and P. J. Richerson, "Punishment Allows the Evolution of Cooperation (Or Anything Else) in Sizable Groups," *Ethology and Sociobiology* 13 (1992): 171–95.

15. I outline this history and develop a theory around it in Shermer, *Science of Good and Evil*.

16. For an excellent summary of the evidence for our tribal nature and what we can do about it, see David Bereby, *Us and Them: The Science of Identity* (Chicago: University of Chicago Press, 2005).

17. L. J. Eaves, H. J. Eysenck, and N. G. Martin, *Genes, Culture and Personality: An Empirical Approach* (London: Academic Press, 1989). The correlation coefficient was .62. Squaring this number gives us an estimate of the percentage of variance accounted for by genetics, which is .384, or roughly 40 percent with error variance.

18. Thomas Sowell, *A Conflict of Visions: Ideological Origins of Political Struggles* (New York: BasicBooks, 1987), 24–25.

19. Steven Pinker, *The Blank Slate: The Modern Denial of Human Nature* (New York: Viking, 2002), 290–91.

20. I present this data in much greater detail in two of my books: *The Science of Good and Evil* and *The Mind of the Market*.

21. James Madison, "The Federalist No. 51: The Structure of the Government Must Furnish the Proper Checks and Balances Between the Different Departments," *Independent Journal*, February 6, 1788.

22. Abraham Lincoln, "First Inaugural Address," March 4, 1861, Bartleby.com, http://www.bartleby.com/124/pres31.html.

23. I just made up the word *idealpolitik*, but a quick Google search shows me that it is not original. Alas.

24. John Stuart Mill, *On Liberty* (New York: Penguin Books, 2006), 13.

25. Ibid., 7.

26. Timothy Ferris, *The Science of Liberty: Democracy, Reason, and the Laws of Nature* (New York: Harper, 2010), 262. This is an excellent treatise on the relationship of science and society.

27. Personal correspondence, March 18, 2010.

28. Ed Husain, *The Islamist: Why I Joined Radical Islam in Britain, What I Saw Inside, and Why I Left* (New York: Penguin, 2008).

29. Quoted in Marc Erikson, "Islamism, Fascism, and Terrorism," *Asia Times*, November 5, 2002, http://www.atimes.com/atimes/Middle_East/DK05Ak01.html.

30. Personal correspondence, March 18, 2010.

31. David Frum and Richard Perle, *An End to Evil: How to Win the War on Terror* (New York: Random House, 2004).

Chapter 12: Confirmations of Belief

1. Leonard Mlodinow, *The Drunkard's Walk: How Randomness Rules Our Lives* (New York: Vintage, 2009), 176–79.

2. Raymond Nickerson, "Confirmation Bias: A Ubiquitous Phenomenon in Many Guises," *Review of General Psychology* 2, no. 2 (1998): 175–220.

3. Mark Snyder, "Seek and Ye Shall Find: Testing Hypotheses About Other People," in *Social Cognition: The Ontario Symposium on Personality and Social Psychology*, ed. E. T. Higgins, C. P. Heiman, and M. P. Zanna (Hillsdale, N.J.: Erlbaum, 1981), 277–303.

4. John M. Darley and Paget H. Gross, "A Hypothesis-Confirming Bias in Labeling Effects," *Journal of Personality and Social Psychology* 44 (1983): 20–33.

5. Bonnie Sherman and Ziva Kunda, "Motivated Evaluation of Scientific Evidence" (paper presented at the annual meeting of the American Psychological Society, Arlington, Va., 1989).

6. Deanna Kuhn, "Children and Adults as Intuitive Scientists," *Psychological Review* 96 (1989): 674–89.

7. Deanna Kuhn, Michael Weinstock, and Robin Flaton, "How Well Do Jurors Reason? Competence Dimensions of Individual Variation in a Juror Reasoning Task," *Psychological Science* 5 (1994): 289–96.

8. D. Westen, C. Kilts, P. Blagov, K. Harenski, and S. Hamann, "The Neural Basis of Motivated Reasoning: An fMRI Study of Emotional Constraints on Political Judgment During the U.S. Presidential Election of 2004," *Journal of Cognitive Neuroscience* 18 (2006): 1947–58.

9. Baruch Fischhoff, "For Those Condemned to Study the Past: Heuristics and Biases in Hindsight," in Daniel Kahneman, Paul Slovic, and Amos Tversky, *Judgment Under Uncertainty: Heuristics and Biases* (New York: Cambridge University Press, 1982), 335–51.

10. John C. Zimmerman, "Pearl Harbor Revisionism," *Intelligence and National Security* 17, no. 2 (2002): 127–46.

11. Philip Tetlock, *Expert Political Judgment: How Good Is It? How Can We Know?* (Princeton, N.J.: Princeton University Press, 2005).

12. Geoffrey Cohen, "Party Over Policy: The Dominating Impact of Group Influence on Political Beliefs," *Journal of Personality and Social Psychology* 85 (2003): 808–82.

13. In Carol Tavris and Elliot Aronson, *Mistakes Were Made (But Not by Me): Why We Justify Foolish Beliefs, Bad Decisions, and Hurtful Acts* (New York: Mariner Books, 2008), 130–32. See also the Innocence Project, http://www.innocenceproject.org/.

14. M. Ross and F. Sicoly, "Egocentric Biases in Availability and Attribution," *Journal of Personality and Social Psychology* 37 (1979): 322–36; R. M. Arkin, H. Cooper, and T. Kolditz, "A Statistical Review of the Literature Concerning the Self-serving Bias in Interpersonal Influence Situations," *Journal of Personality* 48 (1980): 435–48; M. H. Davis and W. G. Stephan, "Attributions for Exam Performance," *Journal of Applied Social Psychology* 10 (1980): 235–48. For a general summary of the attribution bias see Carol Tavris and Carole Wade, *Psychology in Perspective*, 2nd ed. (New York: Longman/Addison Wesley, 1997).

15. R. E. Nisbett and L. Ross, *Human Inference: Strategies and Shortcomings of Social Judgment* (Englewood Cliffs, N.J.: Prentice-Hall, 1980).

16. Preliminary results of our study were originally published in Shermer, *How We Believe*.

17. The full data set and analysis will be published in Michael Shermer and Frank J. Sulloway, "Religion and Belief in God: An Empirical Study." In preparation.

18. Lisa Farwell and Bernard Weiner, "Bleeding Hearts and the Heartless: Popular Perceptions of Liberal and Conservative Ideologies," *Personality and Social Psychology Bulletin* 26, no. 7 (2000): 845–52.

19. Costs, deaths, casualties of Iraq war: "Home and Away: Iraq and Afghanistan War Casualties," CNN, http://www.cnn.com/SPECIALS/2003/iraq/forces/casualties/; Bush quote: http://mediamatters.org/research/200612220015.

20. William Samuelson and Richard Zeckhauser, "Status Quo Bias in Decision Making," *Journal of Risk and Uncertainty* 1 (1988): 7–59.

21. Samuelson and Zeckhauser, "Status Quo Bias in Decision Making"; Daniel Kahneman, J. L. Knetsch, and Richard H. Thaler, "Anomalies: The Endowment Effect, Loss Aversion, and Status Quo Bias," *Journal of Economic Perspectives* 5, no. 1 (1991): 193–206; E. J. Johnson, J. Hershey, J. Meszaros, and H. Kunreuther, "Framing, Probability Distortions, and Insurance Decisions," *Journal of Risk and Uncertainty* 7 (1993): 35–51.

22. Richard Thaler, Daniel Kahneman, and Jack Knetsch, "Experimental Tests of the Endowment Effect and the Coarse Theorem," *Journal of Political Economy*, December 1990, 1325–48.

23. Amos Tversky and Daniel Kahneman, "The Framing of Decisions and the Psychology of Choice," *Science* 211 (1981): 453–58; Tversky and Kahneman, "Rational Choice and the Framing of Decisions," *Journal of Business* 59, no. 4 (1986): 2; B. De Martino, D. Kumaran, B. Seymour, and R. S. Dolan, "Frames, Biases, and Rational Decision-Making in the Human Brain," *Science* 313 (2006): 684–87.

24. Amos Tversky and Daniel Kahneman, "Availability: A Heuristic for Judging Frequency and Probability," *Cognitive Psychology* 5 (1973): 207–32.

25. B. Combs and P. Slovic, "Newspaper Coverage of Causes of Death," *Journalism Quarterly* 56 (1979): 837–43.

26. Barry Glassner, *The Culture of Fear: Why Americans Are Afraid of the Wrong Things* (New York: BasicBooks, 1999).

27. Amos Tversky and Daniel Kahneman, "Availability: A Heuristic for Judging Frequency and Probability," in Kahneman, Slovic, and Tversky, *Judgment Under Uncertainty*, 163.

28. Amos Tversky and Daniel Kahneman, "Extension Versus Intuitive Reasoning: The Conjunction Fallacy in Probability Judgment," *Psychological Review* 90 (1983): 293–315.

29. Daniel J. Simons and Christopher Chabris, "Gorillas in Our Midst: Sustained Inattentional Blindness for Dynamic Events," *Perception* 28 (1999): 1059–74. You can watch the video clip at http://viscog.beckman.uiuc.edu/djs_lab/.

30. Emily Pronin, D. Y. Lin, and L. Ross, "The Bias Blind Spot: Perceptions of Bias in Self Versus Others," *Personality and Social Psychology Bulletin* 28 (2002): 369–81.

31. Peter Brugger and Kirsten I. Taylor, "ESP: Extrasensory Perception or Effect of Subjective Probability?" *Journal of Consciousness Studies* 10 (2003): 221–46. Brugger and Taylor demonstrate, in their words: "(1) as human subjects' guesses are highly nonrandom and (2) as no finite sequence of target alternatives is free of bias, above-chance matching of guesses to targets simply reflects the amount of sequential information common to both target and guess sequences."

32. Robert R. Coveyou, "Random Generation Is Too Important to Be Left to Chance," *Applied Mathematics* 3 (1969): 70–111.

Chapter 13: Geographies of Belief

1. John K. Wright, "Terrae Incognitae: The Place of the Imagination in Geography," *Annals of the Association of American Geographers* 37, no. 1 (1947): 1–15.

2. William D. Phillips and Carla Rahn Phillips, *The Worlds of Christopher Columbus* (New York: Cambridge University Press, 1992).

3. Christopher Columbus, *The Four Voyages: Being His Own Log-Book, Letters and Dispatches with Connecting Narratives*, ed. and trans. J. M. Cohen (New York: Penguin Classics, 1992).

4. Peter C. Mancall, *Travel Narratives from the Age of Discovery: An Anthology* (New

York: Oxford University Press, 2006); Ronald S. Love, *Maritime Exploration in the Age of Discovery, 1415–1800* (New York: Greenwood Press, 2006).

5. Nicholas Thomas, *Cook: The Extraordinary Voyages of Captain James Cook* (New York: Walker and Company, 2004).

6. Quoted in Giorgio de Santillana, *The Crime of Galileo* (New York: Time Inc., 1962), 28.

7. Quoted in Mario Biagioli, *Galileo Courtier: The Practice of Science in the Culture of Absolutism* (Chicago: University of Chicago Press, 1993), 236.

8. Quoted in De Santillana, *Crime of Galileo.*

9. For a recounting of Galileo's trials and tribulations with the church, see Richard Olson, *Science Deified and Science Defied* (Berkeley: University of California Press, 1982); and A. C. Crombie, *Augustine to Galileo* (Cambridge, Mass.: Harvard University Press, 1979).

10. Quoted in Maurice Finocchiaro, ed. and trans., *The Galileo Affair: A Documentary History* (Berkeley: University of California Press, 1989).

11. Quoted in De Santillana, *Crime of Galileo,* 312.

12. Ronald Numbers, ed., *Galileo Goes to Jail: And Other Myths About Science and Religion* (Cambridge, Mass.: Harvard University Press, 2009).

13. Additional scholarly works on Galileo, the trial, and his relationship with the church include: Rivka Feldhay, *Galileo and the Church* (New York: Cambridge University Press, 1995); Annibale Fantoli, *Galileo: For Copernicanism and for the Church* (Vatican City: Vatican Observatory Publications, 2003); William R. Shea and Mariano Artigas, *Galileo in Rome* (New York: Oxford University Press, 2003); Ernan McMullin, ed., *The Church and Galileo* (Notre Dame, Ind.: University of Notre Dame Press, 2005); Mario Biagioli, *Galileo's Instruments of Credit* (Chicago: University of Chicago Press, 2006); and Richard J. Blackwell, *Behind the Scenes at Galileo's Trial* (Notre Dame, Ind.: University of Notre Dame Press, 2006).

14. Pope John Paul II, "Fidei Depositum," *L'Osservatore Romano* 44, no. 1264 (November 4, 1992).

15. Quoted in Edwin Arthur Burtt, *The Metaphysical Foundations of Modern Science* (New York: Doubleday, 1954), 83.

16. Quoted in I. Bernard Cohen, *Revolution in Science* (Cambridge, Mass.: Harvard University Press, 1985).

17. Richard Feynman, quoted in "The Best Mind Since Einstein," *Nova*, WGBH Boston, 1993.

18. J. Stannard, "Natural History," in David Lindberg, ed., *Science in the Middle Ages* (Chicago: University of Chicago Press, 1978).

19. Allen Debus, *Man and Nature in the Renaissance* (New York: Cambridge University Press, 1978).

20. Francis Bacon, *Novum Organum* (1620), in E. A. Burtt, ed., *The English Philosophers from Bacon to Mill* (New York: Random House, 1939).

21. Ibid.

22. John F. W. Herschel, *Preliminary Discourse on the Study of Natural Philosophy* (London: Longmans, Rees, Orme, Brown and Green, 1830); William Whewell, *The Philosophy of the Inductive Sciences* (London: J. W. Parker, 1840); John Stuart Mill, *A System of Logic, Ratiocinative and Inductive, Being a Connected View of the Principles of Evidence, and the Methods of Scientific Investigation* (London: Longmans, Green, 1843).

23. Stephen Jay Gould, "The Sharp-Eyed Lynx, Outfoxed by Nature," *Natural History,* May 1998, 16–21, 70–72.

24. Quoted in Gould, "Sharp-Eyed Lynx," 19, translation by Gould.

25. Edward R. Tufte, *Beautiful Evidence* (Cheshire, Conn.: Graphics Press, 2006).

26. Edward R. Tufte, *Visual Explanations: Images and Quantities, Evidence and Narrative* (Cheshire, Conn.: Graphics Press, 1997), 106–8.

27. Gould, "Sharp-Eyed Lynx," 19.

Chapter 14: Cosmologies of Belief

1. Thomas Wright, *An Original Theory; or, New Hypothesis of the Universe* (London: H. Chapelle, 1750).

2. Immanuel Kant, *Universal Natural History and Theory of the Heavens*, trans. W. Hastie (Ann Arbor: University of Michigan Press, 1969), 61–64.

3. Marcia Bartusiak, *The Day We Found the Universe* (New York: Pantheon Books, 2009); Gale E. Christianson, *Edwin Hubble: Mariner of the Nebulae* (Chicago: University of Chicago Press, 1995); Timothy Ferris, *Coming of Age in the Milky Way* (New York: Harper Perennial, 1988).

4. Charles Messier, *Catalogue des Nébuleuses et Amas d'Étoiles Observées à Paris* (Paris: Imprimerie Royal, 1781).

5. William Herschel, "On the Construction of the Heavens," *Philosophical Transactions of the Royal Society of London* 75 (1785): 213–66.

6. William Herschel, "On Nebulous Stars, Properly So Called," *Philosophical Transactions of the Royal Society of London* 81 (1791): 71–78.

7. William Herschel, "Catalogue of a Second Thousand of New Nebulae and Clusters of Stars; with a Few Introductory Remarks on the Construction of the Heavens," *Philosophical Transactions of the Royal Society of London* 79 (1789): 212–55.

8. Earl of Rosse, "Observations on the Nebulae," *Philosophical Transactions of the Royal Society of London* 140 (1850): 499–514.

9. John P. Nichol, *The Stellar Universe* (Edinburgh: John Johnstone, 1848).

10. William Huggins and Lady Huggins, *The Scientific Papers of Sir William Huggins* (London: Wesley and Son, 1909), 106.

11. Agnes M. Clerke, *The System of the Stars* (London: Longmans, Green, and Co., 1890). A decade later Clerke further reinforced the nebular hypothesis in Agnes M. Clerke, *A Popular History of Astronomy During the Nineteenth Century* (London: Adam and Charles Black, 1902).

12. Arthur C. Clarke, "Hazards of Prophecy: The Failure of Imagination," in *Profiles of the Future: An Enquiry into the Limits of the Possible* (New York: Harper and Row, 1962), 14. But note as well, Isaac Asimov's corollary to Clarke's law: "When, however, the lay public rallies round an idea that is denounced by distinguished but elderly scientists and supports that idea with great fervor and emotion—the distinguished but elderly scientists are then, after all, probably right."

13. Edward A. Fath, "The Spectra of Some Spiral Nebulae and Globular Star Clusters," *Lick Observatory Bulletin* 149 (1908): 71–77.

14. I am grateful to the current director of the Lick Observatory, Michael Bolte, along with the astronomer Remington Stone, for a personally guided tour of the observatory and telescopes, along with a colorful narrative history of the construction, development, and history of this historic monument to science.

15. Quoted in Robert Smith, *The Expanding Universe* (New York: Cambridge University Press, 1982), 43.

16. A. C. D. Crommelin, "Are the Spiral Nebulae External Galaxies?" *Journal of the Royal Astronomical Society of Canada* 12 (1918): 46.

17. Vesto Slipher, "Spectrographic Observations of Nebulae," *Popular Astronomy* 23 (1915): 21–24.

18. From a letter dated June 8, 1921, Harvard University Archives, quoted in Bartusiak, *Day We Found the Universe*, 164.
19. Logbook, 100-inch Reflector, Box 29, 156. Quoted in Christianson, *Edwin Hubble*, 158.
20. Quoted in Christianson, *Edwin Hubble*, 159.
21. Quoted in Katherine Haramundanis, ed., *Cecilia Payne-Gaposhkin: An Autobiography and Other Recollections* (New York: Cambridge University Press, 1984), 209.
22. Quoted in Christianson, *Edwin Hubble*, 161.
23. Stenger has made this and similar arguments for the natural origin of the universe in several of his excellent books. See, for example, Victor Stenger, *The New Atheism* (Buffalo, N.Y.: Prometheus Books, 2009); Stenger, *God: The Failed Hypothesis* (Buffalo, N.Y.: Prometheus Books, 2008); and Stenger, *Quantum Gods: Creation, Chaos, and the Search for Cosmic Consciousness* (Buffalo, N.Y.: Prometheus Books, 2009).
24. Einstein solved this problem through his theory of relativity by demonstrating that space objects such as stars distort the space-time around them—planets are not "attracted" to the star because of a mysterious force called "gravity"; planets "fall" around the star by moving through the curved space-time around it.
25. Martin Rees, *Just Six Numbers: The Deep Forces That Shape the Universe* (New York: BasicBooks, 2000).
26. John D. Barrow and Frank Tipler, *The Anthropic Cosmological Principle* (New York: Oxford University Press, 1988), vii.
27. Philosopher Robert Lawrence Kuhn outlined the problem and at least twenty-seven different solutions to it in a brilliantly executed article: "Why This Universe? Toward a Taxonomy of Possible Explanations," *Skeptic* 13, no. 3 (2007): 28–39.
28. John Barrow and John Webb, "Inconstant Constants," *Scientific American*, June 2005, 57–63.
29. Sean Carroll, *From Eternity to Here: The Quest for the Ultimate Theory of Time* (New York: Dutton/Penguin, 2010), 50.
30. Martin J. Rees, *Before the Beginning: Our Universe and Others* (New York: Perseus Books, 1998); Rees, *Our Cosmic Habitat* (Princeton, N.J.: Princeton University Press, 2004); Rees, "Exploring Our Universe and Others," *Scientific American*, December 1999; John Leslie, *Universes* (London: Routledge, 1989).
31. Carroll, *From Eternity to Here*, 51, 64.
32. Paul J. Steinhardt and Neil Turok, "A Cyclic Model of the Universe," *Science* 296, no. 5572 (May 2002): 1436–39.
33. Alan Guth, "The Inflationary Universe: A Possible Solution to the Horizon and Flatness Problems," *Physical Review D* 23, no. 2 (1981): 347; Guth, *The Inflationary Universe: The Quest for a New Theory of Cosmic Origins* (Boston: Addison-Wesley, 1997); Andrei Linde, "The Self-Reproducing Inflationary Universe," *Scientific American*, November 1991, 48–55; Linde, "Current Understanding of Inflation," *New Astronomy Reviews* 49 (2005): 35–41; Alex Vilenkin, *Many Worlds in One: The Search for Other Universes* (New York: Hill and Wang, 2006).
34. Justin Khoury, Burt A. Ovrut, Paul J. Steinhardt, and Neil Turok, "Density Perturbations in the Ekpyrotic Scenario," *Physical Review D* 66, no. 4 (2002): 046005; Jeremiah P. Ostriker and Paul Steinhardt, "The Quintessential Universe," *Scientific American*, January 2001, 46–53.
35. Raphael Bousso and Joseph Polchinski, "The String Theory Landscape," *Scientific American*, September 2004.

36. Victor Stenger, *The Unconscious Quantum: Metaphysics in Modern Physics and Cosmology* (Buffalo, N.Y.: Prometheus, 1995); Stenger, "Is the Universe Fine-Tuned for Us?" in *Why Intelligent Design Fails: A Scientific Critique of the New Creationism*, ed. Matt Young and Taner Edis (New Brunswick, N.J.: Rutgers University Press, 2004).

37. Hugh Everett, "'Relative State' Formulation of Quantum Mechanics," *Reviews of Modern Physics* 29, no. 3 (1957): 454–62, reprinted in *The Many-Worlds Interpretation of Quantum Mechanics*, ed. B. S. DeWitt and N. Graham (Princeton, N.J.: Princeton University Press, 1973), 141–49; John Archibald Wheeler, *Geons, Black Holes & Quantum Foam* (New York: W. W. Norton, 1998), 268–70.

38. Stephen Hawking, "Quantum Cosmology," in Stephen Hawking and Roger Penrose, *The Nature of Space and Time* (Princeton, N.J.: Princeton University Press, 1996), 89–90.

39. Roger Penrose, *The Road to Reality: A Complete Guide to the Laws of the Universe* (New York: Knopf, 2005), 726–32, 762–65.

40. Stephen Hawking, "The Future of Theoretical Physics and Cosmology: Stephen Hawking 60th Birthday Symposium" (lecture at the Centre for Mathematical Sciences, Cambridge, U.K., January 11, 2002).

41. Lee Smolin, *The Life of the Cosmos* (New York: Oxford University Press, 1997). See also Quentin Smith, "A Natural Explanation of the Existence and Laws of Our Universe," *Australasian Journal of Philosophy* 68 (1990): 22–43. For an elegant summary see James Gardner, *Biocosm* (Maui, Hi.: Inner Ocean Publishing, 2003).

42. Stephen Hawking and Leonard Mlodinow, *The Grand Design* (New York: Bantam Books, 2010), 6–9, 46, 75, 83, 136, 179–80.

Epilogue: The Truth Is Out There

1. Thanks to Arthur Benjamin, professor of mathematics at Harvey Mudd College and the famous mathemagician, for this calculation. Art recommends this Web page for such calculations (where $N=52$ and $p=0.5$): http://www.stat.tamu.edu/~west/applets/binomialdemo.html.

2. "Lennart Green Does Close-up Card Magic," TED, February 2005, http://www.ted.com/talks/lang/eng/lennart_green_does_close_up_card_magic.html.

3. Frank J. Sulloway, *Born to Rebel: Birth Order, Family Dynamics, and Creative Lives* (New York: Pantheon Books, 1996), 336.

4. Jared Diamond, *Guns, Germs, and Steel: The Fates of Human Societies* (New York: W. W. Norton, 1997).

5. Jared Diamond, *Natural Experiments of History* (Cambridge, Mass.: Harvard University Press, 2010), 120–29.

6. *To the stars with difficulty.* Sometimes rendered *Per aspera ad astra.* The phrase originated with the Roman poet Seneca the Younger and was made famous on a plaque honoring the Apollo 1 astronauts who perished in a fire on the launchpad at Cape Canaveral.

Acknowledgments

The construction of a book is not unlike that of a building, in which the reading public sees only the finished edifice after the scaffolding is taken down and the construction crews have moved on to other projects. The foundation and assembly of this book—along with my work in general—were aided by a number of individuals, starting with my agents, Katinka Matson, John Brockman, and Max Brockman, who together continue to help shape the genre of science writing into what I call integrative science, which integrates data, theory, and narrative into a unified whole. And to my lecture agent, Scott Wolfman, and his ambitious team at Wolfman Productions, for having the foresight to market science and skepticism as a viable form of entertainment and education. Thanks as well to Stephen Rubin, Paul Golob, and Robin Dennis at Henry Holt/Times Books, who oversaw the project, and especially to my general editor, Serena Jones, who disciplined me into tightening the manuscript, and to the remarkable copy editor, Michelle Daniel, who went through the manuscript line by line and thereby saved me much literary embarrassment with her many excellent suggestions. And recognition goes to the designer of the book, Meryl Sussman Levavi, whose typography, layout, and design elevated the book to elegance, and to Maggie Richards in sales and marketing and Nicole Dewey in publicity for bringing the manuscript to market, the final and in many ways the most important step in the ever-changing world of book publishing.

I also wish to recognize the office staff of the Skeptics Society and *Skeptic* magazine, including Pat Linse, Nicole McCullough, Ann Edwards,

Daniel Loxton, William Bull, Jim Smith, Jerry Friedman, and Teresa LeVelle, as well as Senior Editor Frank Miele, Senior Scientists David Naiditch, Bernard Leikind, Liam McDaid, Claudio Maccone, and Thomas McDonough, contributing editors Tim Callahan, Harriet Hall, Phil Molé, and James Randi; editorial assistant Sara Meric; photographer David Patton and videographer Brad Davies for their visual record of the Skeptics' Caltech Science Lecture Series. I would also like to recognize *Skeptic* magazine's board members: Richard Abanes, David Alexander, the late Steve Allen, Arthur Benjamin, Roger Bingham, Napoleon Chagnon, K. C. Cole, Jared Diamond, Clayton J. Drees, Mark Edward, George Fischbeck, Greg Forbes, the late Stephen Jay Gould, John Gribbin, Steve Harris, William Jarvis, Lawrence Krauss, Gerald Larue, William McComas, John Mosley, Bill Nye, Richard Olson, Donald Prothero, James Randi, Vincent Sarich, Eugenie Scott, Nancy Segal, Elie Shneour, Jay Stuart Snelson, Julia Sweeney, Frank Sulloway, Carol Tavris, and Stuart Vyse. Special thanks to Pat Linse for rendering much of the artwork for the book.

Thanks as well for the institutional support for the Skeptics Society at the California Institute of Technology goes to Susan Davis, Eric Wood, Hall Daily, Laurel Auchampaugh, Christof Koch, Leonard Mlodinow, Sean Carroll, and Kip Thorne. Likewise, I appreciate the institutional support of the School of Politics and Economics at Claremont Graduate University, most notably Paul Zak, Wendy Martin, Mary Ellen Wanderlingh, Laura Beavin, Thomas Willett, Thomas Borcherding, and Arthur Denzau. As always, I acknowledge my friends at KPCC 89.3 FM radio in Pasadena, most notably Larry Mantle, Jackie Oclaray, Karen Fritsche, and Linda Othenin-Girard. I would like to acknowledge the generous support of the Skeptics Society from Jerome V. Broschart, Tom Glover, Tyson Jacobson, Matthew D. Madison and Sharon E. Madison, Ted A. Semon, Daniel Mendez, Robert and Mary Engman, and Whitney L. Ball. Finally, special thanks go to those who help at every level of our organization: Stephen Asma, Jaime Botero, Jason Bowes, Jean Paul Buquet, Adam Caldwell, Bonnie Callahan, Tim Callahan, Cliff Caplan, Randy Cassingham, Shoshana Cohen, John Coulter, Brad Davies, Janet Dreyer, Bob Friedhoffer, Michael Gilmore, Tyson Gilmore, Andrew Harter, Diane Knudtson, and Joe Lee.

Mariette DiChristina and John Rennie at *Scientific American* deserve special recognition for being such trusted friends and for shepherding the Skeptic column to fruition each month. My column in those august pages of what is now the longest published magazine in American history (165 years and counting) is the most fulfilling thing I do in my working days.

The dedicatee of this book, Devin Ziel Shermer, has now started her own life's journey, and I thank her for the opportunity to express unconditional love and for giving my life deep purpose and ultimate meaning as we contribute to the three-and-a-half-billion-year evolutionary imperative of life's continuity from one generation to the next, remembering always that there's no place like home . . .

Index

About the Author

MICHAEL SHERMER is the author of *Why People Believe Weird Things*, *The Science of Good and Evil*, and eight other books on the evolution of human beliefs and conduct. He is the founding publisher of *Skeptic* magazine, the editor of Skeptic.com, a monthly columnist for *Scientific American*, and an adjunct professor at Claremont Graduate University. He lives in Southern California.

TAKE A JOURNEY THROUGH THE SCIENCE OF SKEPTICISM WITH
MICHAEL SHERMER, "ONE OF OUR MOST COMMITTED CHAMPIONS
OF SCIENTIFIC THINKING IN THE FACE OF POPULAR DELUSION."
—SAM HARRIS, *NEW YORK TIMES* BESTSELLING AUTHOR

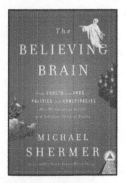

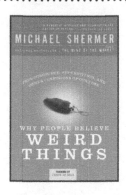

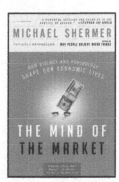

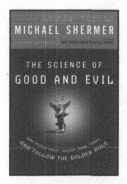

St. Martin's Griffin